Understanding Crop Production

Understanding Crop Production

NEAL C. STOSKOPF

A RESTON BOOK
PRENTICE-HALL, INC., Englewood Cliffs, New Jersey 07632

Library of Congress #81-4420
ISBN 0-8359-8028-6

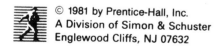
© 1981 by Prentice-Hall, Inc.
A Division of Simon & Schuster
Englewood Cliffs, NJ 07632

10 9 8 7 6 5 4

Printed in the United States of America

Contents

Preface

The *Oxford Dictionary* defines a principle as a general law; as a fundamental source; as a fundamental truth for a basis of reasoning.

This text is an attempt to develop principles relative to crop production, which may be considered a branch of the inexact science of biology. Because crop yields are the final summation of many interacting genetic physiological and environmental factors, most of which are beyond man's control, it is easy to see why crop production principles can never be stated with the authority or force that such things as the temperature—pressure laws of the exact physical sciences can be stated.

Despite the inexact nature of crop production, this book will develop basic principles predicated on the philosophy that when principles are understood, practices based on sound reasoning and understanding follow logically and with a degree of confidence. It is confidence that crop producers need to possess in order to be able to meet changing agricultural conditions; confidence to stay in business and to try to increase yields with fewer inputs under economic and energy-related pressures; confidence to discard outdated methods or unnecessary inputs and adopt new concepts; and confidence to cope with criticisms and some failures. Logical scientific techniques, interpreted correctly, can provide production techniques and principles that lead to success and confidence.

The approach in this book is to first of all understand the basis of all yield, i.e., photosynthesis—and then to build on this concept. The format of this book grew out of a course in crop production offered at the Ontario Agricultural College, University of Guelph, a course that has been well received by the students. However, the students in the course were frustrated by the lack of a text that built upon the basic photosynthetic equation. Existing texts on physiology were too technical and did not relate to farm practice; other texts adopted a recipe approach and systematically covered production practices crop after crop in terms of origin, distribution, economic importance, botanical description, cultural and rotational practices, harvesting, and storage. Such books were useful but were unsuited for an introductory principles course. Students also found conventional textbook approaches to crop production unexciting and unchallenging; instructors found the approach stifled effective pedagogy; and both students and instructors disparagingly referred to conventional textbooks as "a rate, date, and depth of seeding approach." Furthermore, experienced crop producers found such approaches of little or no value because they were never sure if the rate, date, and depth of seeding information was applicable to their particular situation.

Good students in the course accepted the principles approach because they became capable of coping with situations that deviated from the textbook situation. They felt they could modify practices to meet conditions of early or late seedings, wet or dry conditions, high- or low-fertility situations, or for a specific plant architecture and genotype. They felt confident they could cope with unexpected conditions at home, for an employer, or on an overseas assignment. A logical decision could be made based on a principle rather than on a memorized fact.

But a course based on principles and deductive reasoning did not appeal to academically weak students. To help these students develop a degree of confidence, labs and lectures in the course were fashioned after suggestions of educational psychologists.[1] College students are known to progress through at least four stages: namely, dualism, multiplicity, relativism, and commitment. This text attempts to help the student through these four stages (briefly described below):

Dualism.   Students at this early stage of intellectual development expect knowledge to be true or false, right or wrong, and a dualistic attitude about almost any topic is assumed. Relative to crop production, these students consider one cultivar to be best under all conditions; there is one ideal seeding rate, plant population, and row width, etc. They see no argument or no controversy with an established agronomic procedure. Information sources such as books, professors, and superiors are regarded as correct.

When dualistic students are confronted with the fact that in crop production there is no one best technique and no single superior procedure, they express fear and lack of confidence when they say: "I don't know what I am supposed to know in this subject," or "I don't know and I don't care." A typical reaction to an announced lecture topic is, "We already have covered that subject in the soils or pathology course," implying that they feel they know all there is to know about the subject. These students dread and do poorly on exams with questions that have no single correct answer or exams that require logical reasoning to reach a defensible conclusion.

Multiplicity.   At this stage of intellectual development, the students begin to realize that there is often more than one viewpoint, but as yet they have not established any criteria by which to evaluate the merits of one opinion as opposed to another. They use their own opinion as a key means of reaching what they consider to be the correct answer. Parker and Lawson quote two examples of student responses to illustrate this way of thinking:

Things can be a hundred different ways. Both sides can bring a ton of evidence to support their views. Both are equally right. Everybody is right. That's disillusioning.

Sometimes one professor will give an opinion and you have studied it differently. You see, he's a little bit wrong. But I just write down how he wants it remembered.[2]

Relativism.   The students now perceive that there are many points of view, many interpretations, and many frames of reference. They begin to take into consideration the various viewpoints and begin to analyse, compare, and evaluate. "It all depends" is a common expression of a relativist. Such individuals are free-thinkers, flexible in their approach, broad-minded and receptive to new ideas.

[1]See C. A. Parker and F. M. Lawson, "Individualized Approach to Improved Instruction," *NACTA Journal*, 22, no. 3 (1978); 14-28. (Includes discussion and review by J. J. Richter, N. C. Stoskopf, and J. F. Richards.)

[2]Ibid., p. 15.

Commitment. This fourth stage involves a step beyond relativism in which all the facts in a relative world are considered and applied to each individual case, and a personal responsible position or choice is adopted. It is recognized that everything is interdependent, and if one factor is changed, a new position or a modified course of action may be required.

Crop production is a complex science requiring reflective judgment, critical thinking, and ability to solve problems. By taking topics one at a time and giving reasons supported by scientific research, it is believed that the thinking reader can understand each step in crop production, can develop a degree of confidence, and can adopt the procedure, perhaps with modifications, to his own situation. Meticulously avoided here is a recipe approach without giving the reasons why.

To meet these stated pedagogic objectives, a set of challenging questions is included in each chapter. No answers are provided, but if the material in the texts is understood, if the principles are comprehended, a logical and correct answer can be reached. Careful consideration of each question will reinforce the material in the text, raise new issues, and build up confidence. In the classroom, answers to the questions can provide a useful discussion and hopefully will serve as a source of ideas to instructors developing laboratory assignments or examinations.

A stated objective of this text is to make available to crop producers (both farmers and students) a cross section of pertinent scientific research findings. The students themselves often fail to use the library to best advantage, but if they see a title that appeals, perhaps they will be encouraged to seek it out. Also volumes of research material have not been utilized by crop producers, often because such material is written in a complex scientific jargon, or it has no apparent immediate application, or perhaps it was conducted on an unfamiliar crop. Listing references and describing research findings, even from technical sources, are one way of unlocking this wealth of information, and if this information is unleashed in the hands of capable farmers it could be of enormous value.

References have been used throughout this text to illustrate and reinforce the principles. Crop production research is an ongoing field, and all the questions have not yet been resolved. However, the material in this book is presented in a forceful manner as if all the answers are known. Thus, the two-handed approach is generally avoided here—where an idea is presented with positive issues on one hand, and uncertainty on the other, leaving the reader without a clearly guided course of action.

Crop production is an exciting force in today's agriculture. This statement is based on the premise that it is by the action of individual farmers that food is produced. Agriculture has switched from being extensive to being intensive. The adage "Go west, young man" for purposes of farming has gone; there are no new inexpensive frontier lands left. But there is a recognized need today to understand yield-related inputs in order to increase crop production.

This book is intended as an introductory book to be accompanied by additional books or courses in forage and cereal crops. *Understanding Crop Production* is designed to provide the principles and understanding on which these other courses can be developed. Principles will not go out of date like current recommendations. For the skeptic who cannot envision just what a college course might contain and who is therefore critical of a son or daughter attending courses in agriculture, the content of this book should be closely examined. Crop production is a dynamic science, and without a basic understanding of related forces the very core of crop production increases can be overlooked.

Metric units followed by imperial values are presented throughout the text where applicable. Values converted from metric to imperial values have generally been rounded to the nearest whole value. Conversion tables are presented in Appendix B.

Finally, I wish to express my sincere appreciation to my wife Nora and my children Joanne and Michael for their understanding and patience without which this book could not have been completed.

N. C. STOSKOPF

chapter one

The Basis of All Crop Yield

In the book *Gulliver's Travels*, Jonathan Swift describes Gulliver's interest in the work of a member of the Academy of Laputa who "had been eight years upon a project for extracting sunbeams out of cucumbers, which were to be put into vials hermetically sealed, and let out to warm the air in raw inclement summers."[1] The thermodynamics are incorrect, but Swift truly got to the heart of **agriculture**[2]—to fix solar **energy** in such a way as to make it storable and usable elsewhere and at a later date.

Today we know that sunlight energy conversion takes place through a process that occurs in green plants called **photosynthesis;** it is the most important biological phenomenon on earth because it is the source of all primary energy for mankind. Photosynthesis is the very essence of agriculture, the basis of **crop production.**

All crop production practices and principles hinge around the fact that the yield of agricultural crops ultimately depends on the ability of plants to carry on photosynthesis. The aim of crop production therefore is to maximize photosynthesis. This point cannot be overemphasized and will be stressed again and again in this book. The science of crop production can be approached with the fundamental question—How will crop production practices affect the rate of photosynthesis? Decisions as to what crop to grow, what plant population, what row width, seeding depth, seeding rate, fertilizer practices, weed and pest control, etc.—indeed all considerations in crop production—can be approached in the context of how inputs will affect the rate of photosynthesis.

The fact that photosynthesis is the cornerstone of all crop production practices is intuitively obvious, yet seldom is crop production approached

[1]Jonathan Swift, *Gulliver's Travels*, 12th ed., M. K. Starkman ed. (New York: Bantam Books, Inc., 1978), p. 177.

[2]Words or terms that are defined or explained in the Glossary are set in boldface type the first time they appear in the text.

from this viewpoint. Crop producers must learn to view a field as the floor of a factory with the objective of utilizing the sunlight energy hitting the factory floor to best advantage. Plants growing in that field should be viewed in terms of how photosynthesis is proceeding, and how it can be maintained or increased through irrigation, fertilizer, or pest-control measures. At harvest the final yield should be considered as to how effectively the factory floor was utilized. Could a higher yield have been achieved if practices had been altered?

Animals grazing on a pasture field should also be viewed in terms of their impact on photosynthesis. Is the forage crop being overgrazed to the point where photosynthesis is being severely curtailed? What are the effects of trampling and droppings on photosynthesis? **Weeds** should be viewed in terms of how they are affecting photosynthesis in the crop plants, possibly by intercepting sunlight energy or utilizing soil moisture and nutrients that would otherwise be available to and needed by the crop. Adopting a photosynthetic perspective means that a truly scientific approach is being taken. Therefore, understanding the factors that affect photosynthesis is a big step toward understanding crop production and increasing crop yields.

PHOTOSYNTHESIS—THE SOURCE OF FOOD, CLOTHING, AND SHELTER

It is agriculture's responsibility to supply the food needs of mankind. All food comes from the products of photosynthesis through agriculture, and there are no shortcuts or alternatives to photosynthesis. Consider the following:

1. On land, crops such as wheat, rice, corn, **cassava,** and **canola**[3] (rapeseed) convert sunlight energy into products that serve directly as food sources for man. Feed crops like forages (grasses and **legumes**), silage corn, and coarse grains such as barley, oats, corn, and sorghum supply food for man indirectly by serving as energy sources for livestock to produce meat, milk, and eggs.

2. In the sea, microscopic organisms known as **phytoplankton** convert sunlight energy through photosynthesis. These organisms serve as the primary food source for single-celled zooplankton organisms, which in turn serve as food for larger organisms and so on up the food chain. At the top of the food chain is man. Oysters, scallops, lobsters, fish, both large and small, and whales are products in the food chain—all of which are dependent on photosynthesis for their food supply.

3. So-called synthetic foods are really substitute foods that are processed and texturized. Soybean and canola (rapeseed) seeds are a common source of synthetic foods and are a direct product of photosynthesis.

[3]The term *canola* was registered in 1978 as the name for rapeseed or any derivative of the crop.

Processing plant products into meat substitutes, meal extenders, margarine, vegetable oils, and others is an attempt to shorten the food chain. However, even with the most advanced technology available to man today, he cannot make something from nothing; there are no truly synthetic foods. All foods are the products of photosynthesis.

4. Fossil fuels are the products of photosynthesis formed millions of years ago into coal, gas, or petroleum by pressure and heat. Fossil fuels have replaced the slaves of early agriculture to fuel the machines that produce food for the exploding world population. It is as if modern man is relying on photosynthesis from another era for his food today. The diminishing and increasingly expensive petroleum fuels have caused many to view sunlight as a new energy source, but agriculture has always used solar energy. Agriculture also has the potential to produce energy for heat, electricity, or chemical fuels for industrial use; but agriculture's first responsibility is to meet the food, clothing, and shelter needs of man.

5. In addition to food needs, man's clothing and shelter needs are met through photosynthesis. Cotton fiber is produced by the cotton plant directly from the products of photosynthesis. Leather and wool are produced indirectly in the form of animal products supported by photosynthetic sources. Nylon, Dacron, Orlon, or polyester—"synthetic" fibers for clothing and shelter—are manufactured from plant oils or petroleum products originating from photosynthesis. Forestry, a branch of agriculture, produces building, heating, and food products. Diamonds, the symbol of love and said to be a "girl's best friend," are produced by heat and pressure on carbonaceous products of photosynthesis.

6. The air we breathe is comprised of 21% oxygen, an element essential for life on earth. Oxygen is a direct product of photosynthesis. A hectare of land producing a crop can add 5 to 7 tonnes[4] of oxygen to the atmosphere—enough oxygen to supply twelve people for a year—and at the same time it removes 7 to 8 tonnes of carbon dioxide from the atmosphere. Have city fathers overlooked the importance of photosynthesis as an oxygen source and as a user of carbon dioxide when they replace trees and green areas with asphalt and concrete in downtown areas? Is the invigorating feeling found in rural areas or in a large forested area related to the fresh, pure air produced from photosynthesis?

One of the largest oxygen sources on earth is the vast number of photosynthesizing **plankton** in the oceans. But pollution of the seas is reducing plant life in ocean waters and hence the amount of

---

[4]*Tonne* and the abbreviation *t* will be used throughout for metric ton, an international metric term. Refer to Appendices A and B for information on metric measurements.

oxygen released. On the land, forest removal, industrialization, and housing developments that reduce oxygen release and increase carbon dioxide (by burning fossil fuels) may influence the composition of the atmosphere. The long-term influence of man's continued industrial activities on the earth's atmosphere and on plant photosynthesis is uncertain.

7. Actively photosynthesizing plants transpire water (through the plants) into the atmosphere. A hectare of an actively growing crop releases enough water in a season to account for 20 to 40 cm of rainfall per hectare. To produce a tonne of dry matter of many common crops requires 700 to 1000 tonnes of water (635 to 907 tons of water/ton of D.M.). Much of this water is released from plants as water vapor (**transpiration**) into the atmosphere.

THE PROCESS OF PHOTOSYNTHESIS

Plants are composed of **cells,** the building blocks of all living things. Chloroplasts are important components of green plant cells as they are the sites where photosynthesis takes place. Chloroplasts contain **chlorophyll,** which is a substance that is capable of absorbing light energy and through a complex set of chemical reactions is able to convert this light energy into food materials.

For crop production purposes, it is necessary to understand general photosynthetic processes. The process of photosynthesis is shown schematically in Figure 1-1. The chemical reactions that occur in the chloroplast make up the process of photosynthesis and chemically stated are as follows:

Carbon dioxide (CO_2) plus water (H_2O) with the chloroplasts of green cells and in the presence of light energy produces **carbohydrate** or glucose sugar ($C_6H_{12}O_6$) and releases oxygen (O_2) and water (H_2O).

In chemical symbols, the balanced photosynthetic equation can be written:

$$6CO_2 + 12H_2O \xrightarrow[\text{energy}]{\text{sunlight}} C_6H_{12}O_6 + 6O_2 + 6H_2O$$

The plant uses the glucose molecules for growth and development through assimilation processes to form products we harvest for food or fiber.

Light energy is the energy source for photosynthesis. Light striking a plant is absorbed (produces heat), used in photosynthesis, transmitted through the plant tissues, or reflected. Only a narrow band of light—the visible spectrum —is used in photosynthesis. Visible light consists of a spectrum of colors— red, orange, yellow, green, and blue. The chlorophyll molecules normally absorb light from the red and blue portions and reflect or transmit the green portion of the spectrum. The unused green light strikes our eye and gives the impression that the plants are green. Chlorophyll is commonly recognized as giving plants their characteristic green color. High rates of photosynthesis are

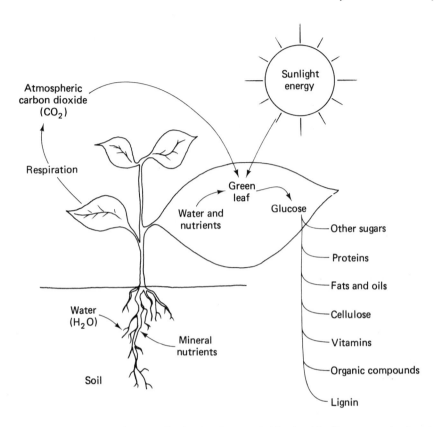

Figure 1-1. Photosynthesis is the basis of crop yields. As this diagrammatic sketch illustrates, the plant secures its energy from sunlight striking the leaf and activating chloroplasts. Carbon dioxide comes from the atmosphere, and water and minerals from the soil.

associated with plants that have a healthy, rich green color due to sound crop production practices such as fertilizer applications that encourage chlorophyll development. Fertilizer elements are essential to the development of green plants so that photosynthesis can proceed, but it is carbon dioxide that is the main ingredient in the photosynthetic process. Carbon dioxide products comprise 90% of the dry weight of a plant. Nutrients taken up from the soil account for the other 10%.

PHOTOSYNTHETIC EFFICIENCY

If crop production is to be improved by attention to photosynthesis, the first thing a crop producer needs to know is the level of photosynthetic efficiency at which his crop is operating. If a crop is at a low level of efficiency, presumably improvements can be made. If a crop is at a high level of efficiency, the crop producer then needs to know the upper limit of photosynthetic improvement and whether room for improvement exists. **Economic yield** such as grain, oil, or digestible dry matter is a common measure of the success of crop production rather than photosynthetic efficiency. There are two good reasons for this: first, photosynthetic efficiency is a complex entity that requires sophisticated scientific equipment if it is to be accurately determined; and second, determination of photosynthetic efficiency at any stage of crop development may be unrelated to final economic yield. If an integrated photosynthetic efficiency could be determined, it too may not be related to final economic yield because the division of food energy into economic yield, vegetative parts, and roots may vary among crops and species.

Conventional methods of measuring economic yield are an incomplete measure of photosynthetic production because the entire plant is seldom harvested. Photosynthetic efficiency calculations must be based on the entire dry matter produced, including root weights.

Corn is used here to calculate photosynthetic efficiency because it has a high yielding ability; it is also a full-season crop and responds to management and soil fertility. Presumably it has a photosynthetic efficiency greater than a lower yielding **cereal** crop. Calculations of photosynthetic efficiency are based on sunlight energy received for a geographic site and on dry matter produced. Calculations are based on average sunlight energy received at 43°41′ north latitude (Guelph, Canada) for a 6200 kg/ha of grain corn yield (100 bu/acre) at 15% moisture. The details of the calculation are shown to emphasize the factors considered, the magnitude of these factors, and to serve as a model in which values for other crops and regions can be inserted. More importantly, the calculation demonstrates photosynthetic efficiency on a field basis. The calculation is as follows:

Step I.   Determine total production per hectare (acre) on a dry-weight basis.

Water is picked up from the soil and is important to photosynthesis but is not a product of photosynthesis. Grain yields are normally reported on a 15% moisture basis, a level that allows for storage and maintenance of seed viability. A grain yield of 6200 kg/ha (100 bu/ac) (15% moisture) must be expressed on a dry weight basis:

$$6200 \text{ kg/ha (5600 lb/ac)} \times \frac{85}{100} = 5270 \text{ kg/ha (4705 lb/ac)}$$

Corn stover (leaves and stems) amounted to 6270 kg/ha, (5598 lb/ac) and root weights were estimated at 4480 kg/ha (4000 lb/ac). The total dry weight produced is as follows:

Grain	5270	(4705 lb/ac)
Stover	6270	(5598 lb/ac)
Roots	4480	(4000 lb/ac)
Total	16,020 kg/ha	(14,303 lb/ac)

Step II.   Determine dry weight produced by photosynthesis by deducting ash content.

Plant nutrients from the soil constitute about 10% of the dry weight per hectare (acre). If plant material is combusted, that which does not burn is known as the **ash** content, and it is removed from the weight because it is not a photosynthetic product. Weight produced from photosynthesis is:

$$16,020 \text{ kg/ha } (14,303 \text{ lb/ac}) \times \frac{90}{100} = 14,418 \text{ kg/ha } (12,873 \text{ lb/ac})$$

Step III.   Determine respiration losses.

In the production of 14,418 kg/ha (12,873 lb/ac), additional energy was used for respiration and may be regarded as a normal "cost of living" loss. A value of 25% was chosen, but respiration losses may vary according to environmental conditions. Under stress conditions, respiration losses may be higher than 25%. The total equivalent photosynthate produced is:

$$14,418 \text{ kg/ha } (12,873 \text{ lb/ac}) \times \frac{100}{75} = 19,224 \text{ kg/ha } (17,164 \text{ lb/ac})$$

Step IV.   Determine crop output by estimating energy contained in the crop.

The energy required for synthesis of one **kilogram** (2.20 lb) of glucose is 15,792 **kilojoules** (kJ) (3760 k calories). The energy contained in 19,224 kg/ha (17,164 lb/ac) crop is:

$$19,224 \ (17,164) \times 15,792 \ (3760) = 303,585,400 \text{ kJ } (72,282,240 \text{ k cal.})$$

Step V.   Determine energy input or how much sunlight energy was received per hectare (acre).

Estimated total solar energy striking a hectare (acre) of land during a 100-day growing season at Guelph is 20,593 million kJ. (4903 million k calories).

Step VI.   Determine photosynthetic efficiency by calculating energy output/energy input.

$$\frac{303,585,400 \text{ kJ } (72,282,240 \text{ k Cal})}{20,593,000,000 \text{ kJ } (4,903,000,000 \text{ k Cal})} \times 100 = 1.47 \text{ or } 1.5\%$$

This means that of the total radiation striking a hectare (acre), 1.5% is converted into chemical energy. Expressed in other terms, out of 100 units of sunlight energy falling on photosynthetically active green leaves, 1.5% is captured and carried through to dry matter. This low value may come as a surprise to many who think agriculture is highly efficient. The following points need to be emphasized:

1. The efficiency value would be considerably lower if calculated on a 365-day rather than on a 100-day growing season especially in extreme latitudes.
2. Corn is a high yielding crop with a higher photosynthetic efficiency than other crops adapted to the same region.
3. Grain yield accounted for 33% of the total dry weight harvested. If economic yield alone was taken as a measure of output, photosynthetic efficiency would be much lower. Crop residues are a source of energy or valuable organic matter and should not be omitted.
4. Ash content of nonphotosynthetic origin of 10% may be considered as a high value. To produce a 6200 kg/ha (100 bu/ac) grain yield, 139 kg (125 lb) of nitrogen, 57 kg (51 lb) of phosphorus, and 55 kg (49 lb) potassium were applied per hectare (acre). Uptake of all of these nutrients, which is highly unlikely, would account for 4% of the grain yield and 1.6% of the total yield. Additional elements such as sulfur, magnesium, or residue nitrogen, phosphorus, and potassium may have been absorbed. Trace elements would be absorbed also. Blackman (1962) suggested that as a broad generalization it can be stated that the mineral ash content of herbaceous species ranges from 4% to 15% of the total dry matter with an average of under 10%.

A photosynthetic efficiency of 1.5% can be regarded as a high estimate, and a value of about 1% is more widely accepted (Rabinowitch 1951).

Upper Limit of Photosynthetic Efficiency

Photosynthetic efficiencies of less than 1% were found to exist in many areas of the world (Army and Greer 1967), but the reasons are readily apparent. Deficiencies in mineral nutrition, water stress, extremes in temperature, poor cultural practices, inadequate pest control, and many other factors can be identified as reasons why. If these limitations are removed, the photosynthetic efficiency can be increased. But is it possible to have an efficiency of 2%, 3%, 5%, 10%, or higher?

On a purely theoretical basis, Bonner (1962) calculated a possible photosynthetic efficiency of 20% based on chlorophyll absorption of light energy within a narrowly defined spectral band. Such a high value cannot be expected under field conditions, but it does establish a goal.

Under actual field conditions, various photosynthetic efficiencies have been reported ranging from 2.0 to 2.5% for rice in Japan and wheat in Denmark (Bonner 1962), to 4.2% sugarcane in Hawaii (Blackman and Black 1959), and to 5.3% for a corn crop in California (Loomis and Williams 1963).

Values for selected periods during the growing period ranged from 2.9% for corn at Ithaca, New York (Lemon 1966) to values between 5.0 and 6.0% for several agricultural crops in Holland (Alberda 1962). Values higher than 6% were not expected under field conditions.

Photosynthetic efficiencies greater than 1.0 to 1.5% are possible. The challenge to crop producers is to duplicate the greater efficiencies achieved under carefully controlled field experiments on a commercial basis. This will not be accomplished with ease as a direct cause and effect relationship between photosynthesis and yield has yet to be established (Moss and Musgrave 1971). A doubling of photosynthetic efficiency cannot be expected to result in a doubling of economic yield. A more direct relationship between the amount of photosynthate and harvested yield may exist where a major portion of the plant is harvested rather than where grain, fruit, or fiber alone constitutes economic yield. Yield increases have occurred slowly and by small increments rather than by a doubling or tripling of yield, and this is not likely to change. If photosynthesis and grain yield could be equated, the need for expensive yield trials would be eliminated, and techniques to measure the highest photosynthetic rates would be adopted. No such shortcut for measuring yield has yet been devised. Without a high photosynthetic rate, however, an increase in crop yield is unlikely.

Photosynthetic Differences among Plants

Measurements of the rate of photosynthesis done on attached leaves on growing plants in the field show that plant species differ markedly in their capacity to utilize sunlight (Waggoner, Moss, and Hesketh 1963; Hesketh 1963) (Figures 1-2 and 1-3). Measurements are taken by enclosing leaves in transparent envelopes and measuring the amount of carbon dioxide that enters and exits from the envelope. Of seven species measured, corn had the highest photosynthetic rate. Sugar cane and sorghum, like corn, are highly productive crops in open, unshaded fields. These crops have a maximum photosynthetic rate approximately one-half to two-thirds higher than tobacco, a crop representative of the great majority of crop plants. This information is of value to a crop producer in selecting the most productive crop species for his area. Even under the most careful management a crop producer cannot induce orchard grass to achieve photosynthetic levels as high as corn. Dry matter yields of orchard grass cannot be expected therefore to equal dry-matter yields of corn under comparable conditions.

Maximum rates of photosynthesis in the group of plants represented by dogwood are achieved in one-fourth or less of full sunlight. This information is of value to the ornamental horticulturist to utilize the slow growth rates desirable for low-maintenance foundation plantings. The ability of plants like dogwood to survive in open areas with full sunlight as well as in shade makes these a good choice for ornamental use.

The reason(s) why photosynthetic rates differ so greatly under intense light is not fully understood. No differences in chlorophyll concentration were reported in the leaves among corn, sunflower, red clover and maple species studied. (Hesketh 1963). The similarities in the photosynthetic curves for corn and sunflower at 1000 parts per million (ppm) of carbon dioxide (Figure 1-4) suggest carbon dioxide movement into the leaf may be restricted at lower concentrations. Of interest at this point is the increase in photosynthesis as carbon dioxide levels are increased, a subject that will be dealt with in greater detail in Chapter Nine.

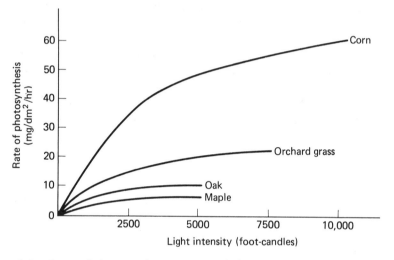

Figure 1-2.   Rates of photosynthesis at various light intensities of four plant species.

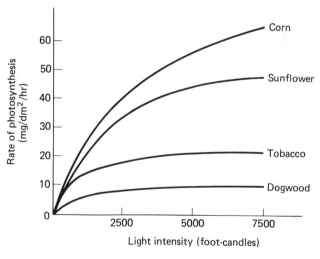

Figure 1-3.   Photosynthetic response of four plant species to increasing light.

QUESTIONS

1. Calculate the photosynthetic efficiency for a crop on your farm or region.
2. List the essential elements for photosynthesis.
3. Explain the statement, "Fossil fuels are the product of photosynthesis."
4. Why do plants appear green in color?
5. The task of agriculture is seemingly to produce food, clothing, and shelter to meet the needs of man. Discuss the possibility of agriculture as a source of energy beyond food, clothing, and shelter needs.
6. How commonly is a 6200 kg/ha (100 bu/ac) grain yield of corn produced in your region?
7. Explain how fish that live and grow by eating smaller fish can be said to be the product of photosynthesis.
8. Discuss how a field test aimed at maximizing photosynthetic efficiency would be conducted from the standpoint of crop production inputs.
9. How would you expect corn to produce under shaded conditions? Your answer should be based on Figures 1-2 and 1-3.
10. If photosynthetic rates of plants could be substantially increased as shown in Figure 1-4, would a danger develop that the CO_2 levels (normally slightly over 300 ppm) in the atmosphere would be reduced?

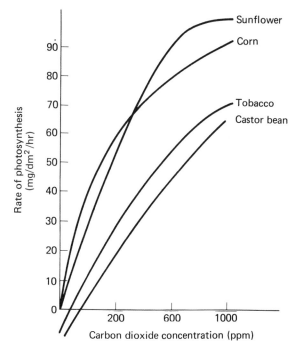

Figure 1-4.   Rates of photosynthesis at carbon dioxide levels of four plant species at a light intensity of 10,000 foot-candles.

11. Discuss the capacity of agriculture to meet the food, clothing, and shelter needs of the earth's exploding population.

12. If photosynthesis is regarded as the most important biological phenomenon on earth, what biological process is considered to be second (see Chapter Twelve)?

REFERENCES FOR CHAPTER ONE

Alberda, T. H. 1962. "Actual and Potential Production of Agricultural Crops," *Netherlands Journal Agricultural Science* 10:325-333.

Army, T. J., and F. A. Greer. 1967. "Photosynthesis and Crop Production Systems," in *Harvesting the Sun*, pp. 321-332, A. S. Pietro, F. A. Greer, and T. J. Army, eds. New York: Academic Press, Inc.

Blackman, G. E. 1962. "The Limit of Plant Productivity," *Annual Report East Malling Research Station 1961*, pp. 39-50.

Blackman, G. E., and J. N. Black. 1959. "Physiological and Ecological Studies in the Analysis of Plant Environment XII. The Role of the Light Factor in Limiting Growth," *Annals Botany* 23:131-145.

Bonner, J. 1962. "The Upper Limit of Crop Yield." *Science 137*, No. 3523:11-15.

Gaastra, P. 1962. "Photosynthesis of Leaves and Field Crops," *Netherlands Journal Agricultural Science* 10:311-323.

Hesketh, J. D. 1963. "Limitations to Photosynthesis Responsible for Differences among Species," *Crop Science* 3:493-496.

Lemon, E. R. 1966. "Energy Conversion and Water Use Efficiency," in *Plant Environment and Efficient Water Use*, pp. 28-48, W. H. Pierre et al., eds. Madison, Wis.: *American Society Agronomy Publication*.

Loomis, R. S., and W. A. Williams. 1963. "Maximum Crop Productivity: An Estimate," *Crop Science* 3:67-72.

Moss, D. N. 1967. "Solar Energy in Photosynthesis," *Solar Energy 11 (eleven)*, nos. 3 & 4: pp. 173-179.

Moss, D. N., and R. B. Musgrave. 1971. "Photosynthesis and Crop Production, *Advances in Agronomy* 23:317-336.

Rabinowitch, E. I. 1951. *Photosynthesis and Related Processes*, vol. 2, pp. 603-1208. New York: Interscience Publishing.

Waggoner, P. E., D. N. Moss, and J. D. Hesketh. 1963. "Radiation in the Plant Environment and Photosynthesis," *Agronomy Journal* 55:36-39.

chapter two

The Botany of Crop Production

Crop production encompasses many specialized aspects of botany including **plant physiology,** which deals with internal life-processes; **morphology** which deals with plant structure and shape of external parts; **anatomy** which emphasizes internal structure; **taxonomy** which classifies and names plants; **pathology,** the study of plant diseases; **cytology,** the detailed study of individual cells; and **plant breeding,** an applied aspect of plant genetics and inheritance. Each of these seven areas of specialization in botany represent major fields of study with each contributing to an understanding of crop production. The purpose of this chapter is to present selected items from each of these disciplines that are most applicable to field crops and to an understanding of crop production. Together these have been termed the botany of crop production.

PLANT CELLS

A description of a plant logically starts with the basic unit of plant structure—the cell. The complexity of a cell is emphasized by the fact that the unique ability to convert nonfood material into food, the process of photosynthesis, occurs in the cell.

Plant cells vary in size, shape and specialized function and hence in complexity and structure. Cell contents are contained within the cell wall. The cell may be variable in shape formed by pressures exerted upon it by adjacent cells. Cells are often portrayed in the form of a box having at least six faces (Figure 2–1). The cell wall contains the contents of cells which consists of living protoplasm and non-living compounds called inclusions.

Cell Wall

The cell wall is a relatively strong and semi-rigid structure but with a considerable degree of flexibility. Cell walls must be strong to give strength and support to the entire plant; however, their pliable nature allows plants to bend and whip in the wind and to return to their original upright position.

The wall separating two adjacent cells is laminated and has at least three distinct layers: The middle lamella is the center layer consisting of mucilagenous pectic substances that "glue" the cells together (Figure 2-2). It is these pectic substances in the middle lamella of cell walls in fruit that causes jelling or thickening when jams and jellies are produced. Protoplasmic deposits on the middle lamella form the primary wall (Figure 2-2). **Cellulose** is the main constitutent of cell walls. Fibers such as cotton, hemp and flax are largely comprised of cell wall material from cells with a high deposit of cellulose. As cells age, deposits may include **lignin** which makes the cell wall thick and hard, particularly those of wood. Forage digestibility and palatability may be reduced if cell walls have large deposits of cellulose and lignin. Mature cells found in hardwood trees or stone cells of plum and peach seeds have cell walls so thick that the cavity of the cell is almost completely replaced.

Waxes such as cutin and suberin are commonly formed on the outer surfaces of leaves and stems and help waterproof the cell. Limited amounts of

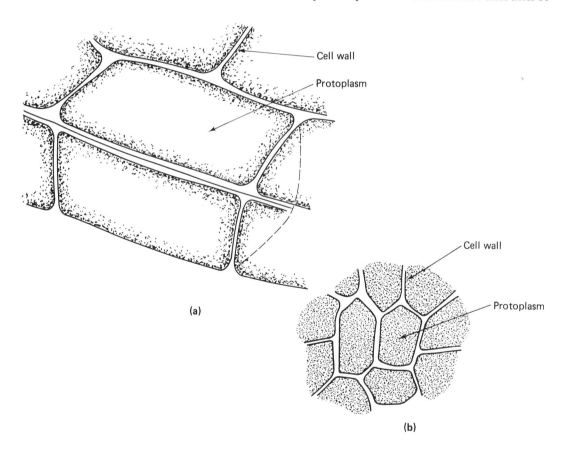

Cell wall

Protoplasm

(a)

Cell wall

Protoplasm

(b)

Figure 2-1. Cells are often portrayed as being six-sided or box-like as in (a). Cells may be variable in shape depending on pressures from adjacent cells as shown in (b).

foliar applied fertilizers can be expected therefore to be absorbed through leaf tissues. Some absorption is possible and plant deficiencies of micronutrients such as boron in apples and turnips, and manganese in soybeans and oats, can be corrected by applying these micronutrients on the leaves of the plants. Therefore, it is surprising that soybean yields were substantially increased by foliar fertilization with foliar applied major nutrients of nitrogen, phosphorus, potassium, and sulfur (Garcia and Hanway 1976). To determine the amounts absorbed, a subsequent study was conducted by Vasilas, Legg and Wolf (1980) with foliar applied radioactive nitrogen in the form of **urea**. When 84 kg N/ha (75 lb/ac) were applied in 21 kg N/ha (19 lb/ac) increments approximately 11 days apart, up to 67% or 56 kg N/ha (50 lb/ac) was taken up by the plant. Leaf burn from the fertilizer materials may reduce uptake. More research is needed before this can become a recommended practice.

Cell walls in the internal part of the plant may contain thin areas called pits that facilitate the movement of water and other food compounds from cell to cell. The same function is achieved in other cells by small canals filled with

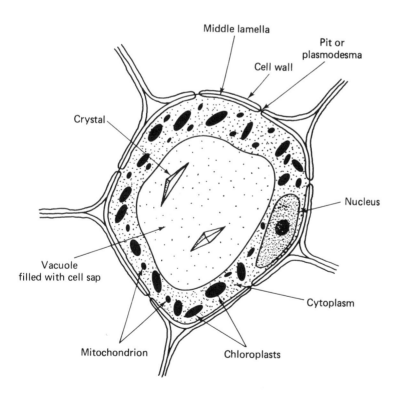

Figure 2-2. A green plant cell. The large central vacuole is filled with cell sap and contains stored food and other pigments. Two crystals are shown here. The protoplasm containing the nucleus is located in the outer edges of the cell.

protoplasmic strands called plasmodesmata (singular is plasmodesma) that connect one cell to that of another (Figure 2-2).

Cell Contents

The content of a cell is comprised of living substances called **protoplasm** where physiological processes occur. Protoplasmic material is comprised of **proteins** (a protein consists of amino acids formed from nitrogenous compounds) in a viscous gel-like form. Included in the protoplasm is the nucleus, a conspicuous entity in the protoplasm (Figure 2-2). The nucleus is the data control center of each cell as it contains the **chromosomes.** Body cells of plants contain a constant and specific number of chromosomes in each nuclei. Alfalfa, for example, has thirty-two, cotton fifty-two, corn twenty, barley fourteen, and wheat forty-two. Chromosomes are strand-like structures of combinations of four nitrogenous bases which collectively is called deoxyribonucleic acid or DNA. Molecular segments of DNA chromosomes are the units of inheritance (**genes**) that direct the functions of the cell through enzymes. An **enzyme** is a protein that functions as a catalyst and regulates cellular functions. Nearly all cellular functions depend on enzymes. The directive for the synthesis of a particular enzyme comes from genetic control. The nucleus governs the heritable characteristics through enzyme action in each cell and hence of the entire plant.

Students will commonly find the word cytoplasm in cell descriptions. All the protoplasm outside the nucleus is called cytoplasm (Figure 2-2).

Embedded in the protoplasm of green cells are chloroplasts (Figure 2-2) that contain chlorophyll where photosynthesis takes place. It is chlorophyll that gives plants their green color.

Mitochondria (plural for mitochondrion) are found in varying numbers and are the sites where products of photosynthesis can be broken down (Figure 2-2). Breakdown of photosynthetic products is called **respiration.** Photosynthetic products can be broken down through oxidative respiration to carbon dioxide (CO_2) and water (H_2O) with the release of energy for the normal functioning of the cell.

Under conditions of rapid cell enlargement, the cytoplasm fails to increase in quantity to fill the entire cell. The protoplasm moves to the outer edge of the cell, and the cavity created becomes known as a **vacuole** (Figure 2-2). The vacuole is surrounded by a membrane and is filled with cell sap containing dissolved substances that are stored for future use. Dissolved substances may include sugars, salts and pigments such as anthocyanins which give the red color to ripe, red tomatoes and red apples. Also included in the vacuole are food substances in solid form such as starch grains, protein bodies, oil drops, and crystals (Figure 2-2) formed from waste materials in the protoplasm.

Cells differentiate to perform specific functions which may vary among species. In sorghum, sugarcane, corn and many other grasses, bulliform or motor cells are commonly found (Figure 2-3). These large cells filled with cell sap shrink when the leaf wilts and cause the leaf to roll or fold across its width. Rolling helps reduce water loss through transpiration as a reduced leaf area is exposed to the heat of the sun. When adequate moisture is available, bulliform cells expand and return the leaf shape to its normal flat position. While rolling of the leaf is an important water conserving function, a more important function of bulliform cells is to keep the leaf flat when environmental conditions favor photosynthesis.

A group of cells performing essentially the same function and of similar structure is called a tissue. The most common and most abundant tissues found in virtually all organs of higher plants are parenchyma tissue. The pith of roots and stems consist of parenchyma cells that function chiefly in the storage of food and water. The storage areas of seeds (endosperm) are

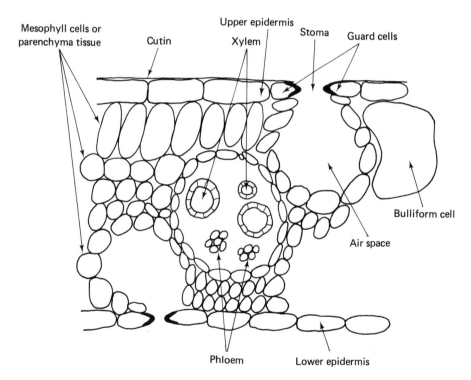

Figure 2-3. Cross section of a leaf showing the vascular bundle containing the xylem and phloem, guard cells which regulate the opening and closing of stomata, a bulliform cell which keeps the leaf flat or rolled up according to moisture supply, and mesophyll cells, all contained between two layers of cells called the upper and lower epidermis.

comprised mainly of parenchyma cells. The bulk of internal tissues in leaves are parenchyma cells that contain chloroplasts. Such leaf cells are commonly known as mesophyll cells (Figure 2-3). Photosynthesis is dependent on the free exchange of gases and transpiration of moisture in the leaf.

Since the outside epidermal layer of a leaf is covered by a wax-like material called cutin that restricts water loss, some controlled openings are essential for the functioning of plants. This is accomplished through small pores or openings in the epidermal layers called stomates (Figure 2-3). Surrounding the **stomata** (singular is stoma) are guard cells which open or close the stomata by expanding or contracting according to moisture supply. Stomata allow carbon dioxide to enter for use in photosynthesis. They are also the openings by which water taken in through the roots is transpired to the atmosphere. Stomata are generally present on both the upper and lower epidermis of leaves.

PLANT HORMONES

In addition to enzymes, plant **hormones** produced in plant cells also act as "chemical messengers" to direct the growth and development of plant cells in an orderly sequence and schedule throughout the life of the plant. Development of specialized tissue and reproductive organs, plant responses to environmental factors, or other stimuli are governed by growth regulators. Plant hormones can be described as naturally occurring substances produced in specific cells and transported to other cells to control plant development. The term "growth regulators" includes naturally occurring and man-made substances that affect growth and development. Herbicides are growth regulators that may act to either inhibit or promote plant growth.

The growing point of plants is the site of production of the well-known growth regulators called auxins. The most common natural auxin is indole acetic acid or IAA. The effect of auxins on plants can be demonstrated readily by exposing plant seedlings to unidirectional light. Within a few hours the seedlings will bend toward the light as a result of cell elongation on the unlighted side of the plant. Auxins on the lighted side of the seedling are destroyed by enzymes.

Similarly, growing plant parts subjected to low light intensity perhaps as a result of a dense stand, will result in the elongation of cells due to auxin concentrations. Plants with long, spindly stems will result, and this is known as cell **etiolation.** A tree seedling that has developed under a dense forest canopy will have elongated cells that will push it quickly through the canopy but the stem will be thin and spindly. Plants in a horizontal position as a result of **lodging** will curve upward as a result of auxin-induced cell elongation on the lower side of the stem.

Apical dominance appears to be a function of auxin distribution. Auxins produced in the apex of a plant inhibit the growth or formation of buds lower down on the stem. Removal of the lead shoot of garden flowers (such as geranium and snapdragon), and many crop plants will encourage lower buds to develop and produce lateral branching. In crop plants, the first ear to develop on a corn plant exerts a hormonal control over lower ears and prevents them from developing. If the first ear to develop is removed, hormonal inhibition is removed, and a lower ear (or ears) may develop.

Cereal plants capable of tillering or soybean plants that show a bushy growth habit do not possess a strongly dominant growing point, and as a result, lateral branch development is not suppressed.

A widely known man-made hormone is the auxin-type **herbicide** 2,4-D (2, 4-dichlorophenoxy acetic acid). Many broadleaved weeds are very susceptible to 2,4-D sprays while grass plants are unaffected. The selectivity of 2,4-D is dependent upon differences in the ability of plant species to absorb, translocate or breakdown the herbicide molecules. This selective action allows for broadleaf weeds to be removed from a cereal crop without damage to the crop.

Gibberellins are hormones that stimulate growth through cell division and elongation. Gibberellins influence seed dormancy, germination, and fruit development. Anti-gibberellins that act to prevent cell division and elongation are used commercially around industrial buildings and airports to inhibit grass growth.

Abscisic acid is an inhibitory compound that affects seed dormancy, and leaf drop as abscisic acid causes the vascular connections to the leaf to become plugged so that the leaf withers and eventually falls off.

Ethylene stimulates fruit ripening. The familiar pleasant odor of a ripe apple is from the ethylene produced. Fruit such as apples, citrus, and bananas that are picked in a green and immature state for storage and/or shipment can be induced to ripen by subjecting them to low levels of ethylene to replace the natural source of ethylene. Immature fruit such as avocados that are starting to ripen will mature best if sealed in a plastic bag to capture their natural production of ethylene.

Cytokinins in conjunction with auxins affect cell differentiation. Cells are induced to form root tissue, stem tissue, or leaf tissue depending on the level of hormones present.

Maleic hydrazide will prevent sucker development in tobacco after the apex has been removed to prevent seed development.

PLANT GROWTH AND DEVELOPMENT

Growth may be defined as cell division and cell expansion. Areas where cells are actively dividing are called **meristem regions.** The plant embryo contains only meristem tissue so that all plant parts—roots, stems, leaves and flowers—

are derived from the meristem. The unspecialized cells formed directly from the meristem become differentiated into specialized plant structures through the influences of genetic controls, hormones, mineral nutrition and other environmental factors. As the plant develops, cells may be differentiated into a stem to form a central axis to support the leaves, to connect leaves with the root, or to serve as a temporary storage organ.

Location of the meristem area or growing point is often useful in crop production. In the corn plant, the tip meristem of the shoot gives rise to leaves during the early stages of development. The tip meristem may be located below the soil surface and may be undamaged by early spring frosts, hail, or leaf removal. As the shoot develops, meristem areas become located at the base of each internode. High moisture levels, soil fertility and high rates of photosynthesis during internode elongation may promote tall plants. In the grasses, elongation of these internodes forces the apical meristem upward in preparation of flowering and seed development. If a newly emerged head of a grass plant is pulled vertically upward, the soft, succulent non-differentiated cells will readily tear at the meristem region.

When legume seeds germinate, the growing point quickly emerges from the soil and is subject to mechanical or frost damage.

As new leaves develop they increase in size because of marginal meristems.

TRANSLOCATION OF CARBOHYDRATES

When light energy is converted into food materials such as sugars in the chlorophyll of a cell, materials not needed in that cell must be moved or translocated to plant parts where they may be needed for cell division, growth of cells, structural development, or the normal functioning of the plant. If they are not immediately needed they may be stored.

Food energy may move between any two adjacent cells by interconnecting tube-like strands of plasmodesmata and pits, and eventually into conducting tissues called the **phloem** for transport to all parts of the plant.

The phloem is comprised of seive tubes appropriately named because sieve tubes have holes in their ends to efficiently act as part of the plant's plumbing system.

Whether photosynthetic products are retained in the leaf where they are produced or are transported elsewhere depends on leaf age. When meristem tissue is actively growing, food energy is needed for leaf growth. As the leaf expands and more food energy is produced than is required for growth of that leaf, outward transport begins.

Once sugars are in the phloem system, they move according to the concentrations of material in the sap. Movement is from a high to a low concentration, or in physiological terms, the flow is from **source** to **sink**. The leaf and other photosynthetic areas are the source, and the roots and developing seeds are the sinks. Export is generally to the nearest sink. In the soybean plant, upper leaves export to upper pods; lower leaves export to lower pods and to the roots. This is an important point to note because if lower pods are to be filled, adequate light penetration to lower leaves is important to stimulate sufficient food energy to fill the lower pods.

The relative position of a leaf changes as upper leaves develop. In the seedling stage, an upper leaf may supply the apex with food energy; as new leaves are added to the upper portion, the same leaf becomes a lower leaf that provides food energy to the roots.

It has been observed in corn that when upper leaves were frost damaged by an early fall frost, lower undamaged and photosynthetically active leaves may assume major proportions as a source to fill the ear.

In the cereal grains, up to 85% to 90% of the carbohydrate stored in the grain is produced by photosynthesis in green tissue in the head, **awns, glumes,** and the uppermost (flag) leaf and stem of the plant. This fact and supporting field observations led Briggs and Aytenfisu (1980) to suggest that in spring wheat, high yields may be associated with cultivars with long heads, a large flag-leaf lamina, and flag-leaf sheath.

Photosynthetic materials that are not needed for growth and that cannot be accommodated by a small or developing sink, may be stored and translocated at a later date. Surplus photosynthetic materials may be stored in the stalks of plants. Storage organs such as potatoes may continue to develop in crops even after the leaf area has turned old and brown from food energy stored in the stems.

As photosynthetic materials are moved out of stored areas into the sink, the stems or stalks become depleted of carbohydrate and a weakened stem or stalk results that breaks over readily and is susceptible to decay organisms.

Under high plant populations that restrict photosynthesis in lower leaves, insufficient food energy may be available for translocation to fill the sink, and abortion or sterility may occur. Mutual shading of the ear may cause poor seed development in rapeseed (canola), soybeans, or corn. Reduced rates of photosynthesis under stress conditions may prevent stalk storage from occurring and lodging may result.

Source to sink flow of photosynthetic materials suggests that the rate of flow increases when demand for food energy is high. In corn plants, a strong demand for food energy can be observed about a week after fertilization. This is the period when vegetative growth is completed and kernels are developing rapidly. Physiologists regard the developing ear of corn as a strong sink due to three possibilities: (1) because of its central location on the stalk relative to food energy produced in the leaves; (2) because it consists of a single concentrated unit, at least in the case of a single-eared plant; and (3) because there are no competing sources such as root nodulation or additional leaf growth. Up to 90% of the food energy translocated during the grain filling period may accumulate in the ear of a corn plant. Producers wishing to produce a multiple-eared corn cultivar must consider the adequacy of the photosynthetic area to produce sufficient food materials to fill two or more ears, as well as the effect of competing sinks on translocation.

In contrast to the strong sink of a single-eared corn plant, soybean pods located at numerous internodes on the stem are a weak sink. Not all pods may be at the same stage of development, and because the soybean is a nitrogen-fixing legume crop, a significant amount of food material is transported to the root to provide energy for nitrogen fixation.

In the rapeseed (canola) crop, the green seed capsule is the source of the majority of food energy to fill the seed.

TRANSLOCATION OF WATER AND MINERALS

Roots of plants are supplied with food materials conducted from the site of photosynthesis by means of the phloem. In exchange, water and minerals absorbed by the root from the soil are carried through the **xylem** tissues to the sites where they are required for photosynthesis or for normal cell functioning. The xylem and phloem tissues comprise the vascular system of plants that connects all plant parts. The vascular system of xylem and phloem of plants is similar to the veins and arteries in animals.

The xylem tissue is comprised of tracheid cells that are elongated, tapered cells with thick walls that collectively act as skeletal tissue to support the plant.

ROOT GROWTH AND DEVELOPMENT

Root systems differ among various crops. Generally the root system may be of two types—fibrous or tap. Roots may be deep or shallow, or sparse or dense, depending on the plant species and the soil environment.

The primary root is the first to emerge from a germinating seed. The primary root supports the seedling with water and nutrients picked up from the soil. As the primary root develops it may produce second, third, and fourth order root branches which may constitute all or part of the root system.

In corn, cereals, and other members of the grass family, the primary root system ceases its development with the development of secondary roots. Secondary or adventitious roots arise from stem nodes (often called nodal roots) generally below the soil surface, but in some cases they may arise above the soil surface. Prop roots in corn develop from nodes above the soil surface. Secondary roots frequently constitute the main root system of the plant.

Although the primary root is not impressive in terms of mass or length, it is essential to the establishment of the seedling. Krassovsky (1926) observed that the activity of primary roots in uptake of water and nutrients is greater than that of an equivalent length or mass of secondary root tissue. The essential nature of the primary root is clearly demonstrated under drought conditions when dryland grasses such as kleingrass (Tischler and Monk 1980) and blue grama (Hyder, Everson and Bemet 1971; Wilson, Hyder and Briske 1976) are seeded into moist soil before a prolonged dry period. Only those plants able to tap subsoil moisture with their primary root can survive. Secondary roots were found to form only when adequate moisture conditions prevailed near the surface. Without a deep and effective primary root, kleingrass and blue grama grass fail to establish under dry conditions.

In legumes, tap roots are formed from the primary root. In alfalfa and sugarbeets, the primary root is evident throughout the life of the plant. Secondary roots in legumes arise as branches from tissues within the root known as the pericycle.

Roots may also grow horizontally as is the case of Canada thistle. This weed, also called creeping thistle, spreads by horizontal roots which produce new shoots. **Rhizomes** of bromegrass, quackgrass, and bluegrass should not be confused with lateral roots. Rhizomes are underground stems which grow horizontally just beneath the soil surface. Lateral rhizomes send up new shoots which in turn establish new root systems.

Roots may form the harvested portion of field crops, as in the case of sugarbeets, turnips and horticultural crops such as garden beets and carrots.

ROOT FUNCTIONS

The three main functions of roots are absorption, anchorage, and storage. In legumes, a fourth very important function is to provide a site for nodulation and nitrogen fixation.

To absorb water, oxygen, and minerals from the soil, roots must have extensive and close contact with soil particles. Plant nutrients are essential to plant growth, but a few hundred kilograms (pounds) of nutrients may be distributed throughout millions of kilograms (pounds) of soil to a depth of several metres (feet). An extensive and effective root system is essential to pick up the necessary nutrients. Under field conditions, four cereal crops were shown by Pavlychenko (1937) to have extensive root systems (Table 2-1). As the plant grew to full size and as the demand for nutrients increased over the last forty days of plant development, the rate of root development ranged from a high of 576 cm (227 inches) per day in wild oats, to 380 cm (150 inches) per day in spring barley, to a low of 204 cm (80 inches) per day in spring wheat. Under most soil conditions, the movement of water through the soil to the roots is very slight; therefore, the importance of such a rapid extension of roots through the soil becomes evident. In addition to the availability of moisture, temperature and oxygen are also essential for nutrient uptake. Water absorption and nutrient uptake act independently of each other.

TABLE 2-1.   Lengths of entire root system in centimetres (inches) of cereal grains at four stages of development when grown under field conditions at Saskatoon Saskatchewan in 1934

Days after emergence	Wild Oats		Spring Wheat		Spring Rye		Spring Barley	
	(cm)	(in)	(cm)	(in)	(cm)	(in)	(cm)	(in)
5	89	(35)	170	(67)	366	(144)	541	(213)
22	4,636	(1,825)	6,208	(2,444)	10,457	(4,152)	14,351	(5,650)
40	15,842	(6,237)	17,752	(6,989)	20,630	(8,122)	25,337	(9,975)
80	38,885	(15,309)	25,908	(10,200)	33,749	(13,287)	40,549	(15,964)

Source: Data from Pavlychenko 1937.

Root Anchorage

Physical support of the plant by a well-developed root system is an important root function. A poorly developed root system may allow for lodging to occur in annual cereals. An inadequate tap root system may allow for frost heaving in perennials.

Root Storage

In crops that are grazed or clipped closely or are expected to recover after a dormant or overwintering period, an energy source is needed for re-growth when no green photosynthetic tissue exists to provide food energy. In such cases, previously stored food materials are relied upon to provide the necessary energy. Energy can be stored in the roots, especially tap roots, or the base of the stems (the crown region) of plants. Food storage in the roots of annual crops such as cereals and corn is not critical. In biennial crops such as sugar beets, the plant produces leaves (vegetative parts only with no reproductive organs) during the first year. The products of photosynthesis are stored in the fleshy tap root. Normally sugarbeets are harvested as an annual for the sugars they contain. If the sugarbeet is not harvested at the end of the first growing season, food energy stored in the root provides the energy source for plants to develop reproductive organs in the second season.

Root Structure

All roots arise from meristematic tissue located at the tip or apex of roots so that root anchorage is not upset by subsequent growth (Figure 2-4). The apical meristem, an area of rapidly dividing and hence undifferentiated and tender cells, is protected by a covering called the root cap (Figure 2-4). As the root cap penetrates hard soils, cells may slough off but are replaced by highly differentiated cells derived from the meristem.

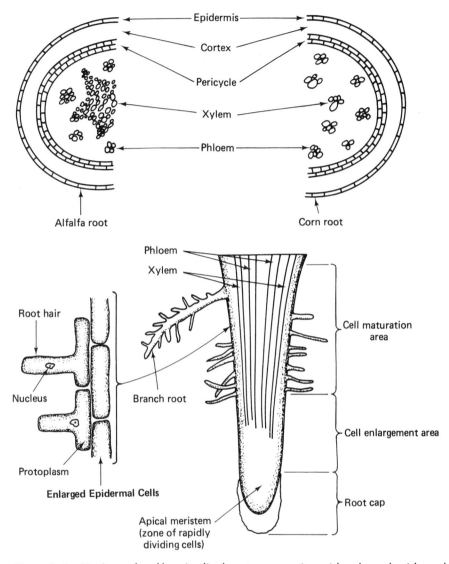

Figure 2–4.   Horizontal and longitudinal root cross section with enlarged epidermal cell showing root hair.

Behind the meristem area is an area of cell enlargement (Figure 2–4). As new cells grow and expand, they push the root-tip through the soil.

Behind the region of cell enlargement is a region of cell maturation where cells differentiate into highly specialized cells to perform specific functions (Figure 2–4). Specialized tissues include xylem and phloem, parenchyma or pith cells, and epidermal cells.

Root hairs are produced in the younger portion of the cell maturation zone (Figure 2–4). A root hair is an elongated, slender protuberance of the epidermal cell of roots. Root hairs are an extension of one cell and differ from branch roots which may be similar in size but arise from the pericycle of the root. Branch roots have their own root caps, meristem regions, xylem, and phloem.

Root hairs are capable of very rapid growth but may function for only a few days. Root hairs are usually only about 12 mm (1/2 inch) in length, but their total length is substantial. This was strikingly demonstrated by Dittmer (1937) who measured the total length of a 4-month-old rye plant grown in a container indoors. Rye has a fibrous root; after 4 months in a box entirely filled with fertile topsoil a large root system could be expected (Table 2–2). Although individual root hairs are short, in total they exceeded the length of the adventitious root system by seventeen times. Because of their abundance and speed of development, root hairs generally are effective in securing sufficient water and nutrients for plants. When seedlings with abundant root hairs, such as cucumber, are transplanted, and if the soil is allowed to fall away from the roots, the delicate root hairs may be broken off and transplants may die. Exposure to warm, dry air, or to bright sunlight can cause the very fragile root hairs to wither and die within a few minutes. Soil should be firmly packed around new transplants to assure that newly formed root hairs make contact with soil particles, moisture and nutrients.

TABLE 2–2. Total lengths of roots in metres and root hairs of a rye plant grown indoors at the University of Iowa, Iowa City in 1936

	(m)	*(yards)*
Total lengths of adventitious root system	623,089 m.	681,418
Total lengths of root hairs	10,627,552 m.	11,622,432

Source: Data from Dittmer 1937.

ROOT-TOP RELATIONSHIPS

The vascular system consisting of the xylem and phloem connects all parts of the plant forming one living organism with close relationships and interactions among the various components. As a plant develops, both top and root grow but seldom at the same rate. In general, the plant part nearest the source of an essential factor will have first call on that essential factor and suffer least by its deficiency. Conditions that reduce photosynthesis in green tissues by cutting or grazing, low light intensity, or carbon dioxide deficiency will retard root growth more than top growth. Under such circumstances the ratio of shoot to root increases.

When nutrients or water normally obtained by the roots become limiting, shoots farthest from the source will be affected first and to the greatest extent. Under these circumstances the ratio of shoot to root decreases.

Frequent defoliation that does not allow for the production of sufficient food energy to be used for root development will restrict root growth. Excessive defoliation can lead to the death of plants. This basic principle may be used in cultural weed control methods. Rhizomes of quackgrass can be cut up by a disk. When the rhizome sections produce a shoot, further cultivation will eventually deplete the rhizome of its food reserves, and it dies. Crop producers must manage crops to maintain a reasonable ratio of roots and shoots.

Other Root Functions

Roots are capable of synthesizing various compounds, including amino acids from nitrogen, hormones (which have been discussed), and alkaloids such as nicotine in tobacco. Nicotine can be moved from the root to the leaves; the largest quantities are transported to the closest leaves. Therefore, the lowest or sand leaves on tobacco plants can be expected to have the highest nicotine content.

As plant roots take up water and nutrients from the soil, they can exude a wide range of organic compounds. Some of these compounds cement the root hair to soil particles to facilities nutrient uptake. Other of the exuded compounds may stimulate certain soil micro-organisms and be toxic to others.

Nitrogen fixation in **nodules** formed on legume roots is a very important root function. In this **symbiotic** relationship, bacteria use energy produced by photosynthesis in the plant to convert inert atmospheric nitrogen into a form usable by plants. As the legume plants grow, the symbiotic bacteria responsible for the formation of the nodules on the root are able to use the inert nitrogen in the air and to multiply in the nodules. The nitrogen in turn becomes available to the legume plants and aids in its nourishment and growth.

Plant Classification

It is not the purpose of this book to confuse crop producers by scientific names or the taxonomic classifications of plants. Neither is it desirable to use the "brand name" of a group of plants, often called a "variety." To avoid confusion with the term variety, the term **cultivar,** a contraction of the words cultivated variety, is used. Some clarification on the meaning of crop cultivars as applied to different crops is useful. Cultivars are produced by one of the three following methods, according to the type of reproduction.

In self-pollinated crops, such as wheat and oats, the cultivar is a pure line of plants that breeds true to type, generation after generation. Because of the self-pollinated nature of sexual reproduction, these crops maintain their own characteristics.

In cross-pollinated crops, such as alfalfa, orchardgrass, and rye, the cultivar is a population or group of plants that can be distinguished on some morphological or physiological basis—such as flower color, seed size, leaf shape, disease resistance or seedling vigor. Seed is produced under controlled conditions of cross-pollination, or by isolation from other plants of the same species. This assures that plants in the next generation are more or less similar to previous generations, thereby maintaining the cultivar.

The term **hybrid** cultivar is applied to corn, a cross-pollinated crop, and refers to a population developed from a specific combination of inbred lines. A single-cross denotes a hybrid cultivar derived from two inbred lines; a double-cross is a hybrid cultivar synthesized from four specific inbred lines, i.e., two single crosses are combined to make a double-cross; a three-way cross involves three inbreds, i.e., a single-cross cultivar is crossed to an inbred.

MONOCOTS AND DICOTS

A distinction that has an important bearing on agricultural practices can be made between two classes of plants, namely **monocots** and **dicots.** These words are derived from the terms monocotyledon and dicotyledon: having one and two cotyledons, respectively. A cotyledon can be defined as the first leaf of the embryo that serves for food digestion and food storage.

A basic difference between monocots and dicots exists in the seed. In a monocot, such as corn, the seed contains an embryo that occupies about 10% of the seed; the other 90% is occupied by a starchy food reserve known as endosperm (Figure 2-5). When germination occurs, the cotyledon absorbs the food energy stored in the seed. Monocot seeds are surrounded by a layer of cells called the aleurone layer (Figure 2-5). Enzymes produced in the aleurone layer modify the starch of the endosperm into simple sugars that can be translocated to the embryo to meet its energy requirements.

In a dicot, such as a bean, the bulk of the seed is comprised of two cotyledons plus the embryionic root and shoot. When the bean germinates, the cotyledons are pulled through the soil, turn green after they emerge, and produce photosynthate until the first true leaves develop to carry on photosynthesis.

Monocots and dicots may be **annuals, biennials** or **perennials.** The most numerous member of the monocots are the grasses. Other monocots include the sedges, lilies, tulips, palms, and pineapples. Monocots are characterized by narrow leaves with parallel venation; they have fibrous roots, and the flower parts occur in groups of three—three stamens, three petals, and three sepals.

Legumes are dicots that typically exhibit broad leaves with the veins radiating out in a spreading or net pattern. The root system of dicots, especially the herbaceous legumes, is usually made up of a tap root and flower parts occurring in groups of two, four, or five.

The internal stem structure of dicots and monocots is quite different.

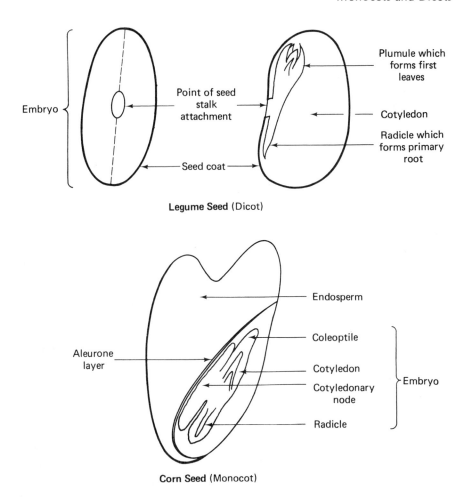

Figure 2-5. Seed structure of a dicot and a monocot.

Monocots have vascular bundles containing the xylem and phloem scattered throughout the parenchyma cells of the stem pith (Figure 2-6). Monocots lack the particular meristem known as **cambrium** which is responsible for secondary growth or thickening in dicot stems. Therefore, any thickening of the monocot stem occurs by cell expansion but not cell division.

Dicot stems are more complex than monocots and are almost always capable of secondary growth or thickening due to the presence of cambrium (Figure 2-6). Initially, the vascular bundles are arranged in a circle around a central core of pith. The cambrium which separates the xylem and phloem generates new cells and adds to the stem's girth. In perennial or woody dicots, annual ring growth results from the production of cells from the cambrium. Each ring represents one year of growth. Phloem tissue is regenerated, and old tissue forms the bark of trees. Removal of the bark to cambrium around the

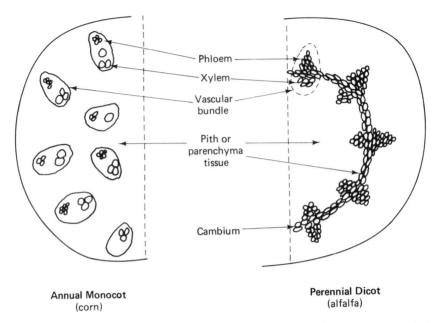

Figure 2-6. Stem cross sections showing xylem and phloem tissues and the cambrium from which new lateral growth is initiated in dicots.

girth of the tree will cut off the flow of food energy to the roots, and the tree will die. Trees must often be protected from the girdling action of mice under the snow in winter.

In monocots, the jointed stem of a grass is distinctly divided into **nodes** and **internodes.** The internode may be hollow, pithy or solid. The node or joint is always solid. Monocot leaves consist of a blade attached to a leaf sheath. The leaf sheath covers the stem to its origin at a node. Leaves have their vascular connections with the stem at the node (Figure 2-7). Prior to node development (jointing), meristem tissue is located at the base of the leaf, and if the grass blade is cut, the meristem at the base will regenerate blade tissue. When a lawn is cut, new growth at the base pushes the leaf blade upward.

In dicots, the leaves are attached to the stem by a **petiole** or stalk, and if removed, they will not be regenerated that season. When the stem of dicots is cut, new leaves and stems arise from the base of the stem or the top of the root known as the crown.

In addition to the vertical flowering stems of culms, many grasses have horizontal stems, usually below ground, called rhizomes. Rhizomes bear scales at the nodes and produce new shoots from their joints. Bromegrass, bluegrass, Johnsongrass, and quackgrass characteristically have sod-forming rhizomes, that often form the overwintering part and storage areas of perennial grasses.

Creeping stems above ground are called stolons. Bermudagrass, ladino clover, and buffalograss produce stolons.

Certain grasses have thickened lower internodes for storage called **corms.** Timothy has a modified corm at the base of the stem which serves as a means of identification (Figure 2-7).

Other structures are useful in plant identification. At the junction of the grass blade there are several structures which are useful in distinguishing among the grass species. The presence or absence, shape, and size of such structures as the ligule and the auricles can help identify a particular grass specimen (Figure 2-8).

The **ligule** is a thin filamentous tongue-like appendage or scale on the inside of the leaf of grasses at the junction of sheath and blade.

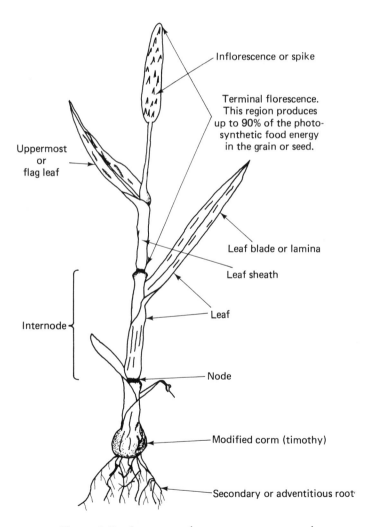

Inflorescence or spike

Terminal florescence. This region produces up to 90% of the photo-synthetic food energy in the grain or seed.

Uppermost or flag leaf

Leaf blade or lamina

Leaf sheath

Leaf

Internode

Node

Modified corm (timothy)

Secondary or adventitious root

Figure 2-7. Structure of a monocot or grass plant.

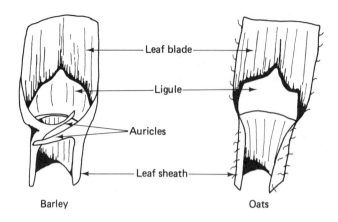

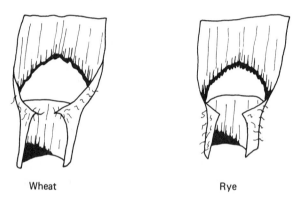

Figure 2–8. Cereal plants have distinguishing features of the leaves, ligules and auricle. Barley has long and clasping smooth auricles; the oat sheath has no auricles; wheat has short hairy auricles; rye has very short auricles. The sheaths of rye and oats and the leaf margins of oats have a fine pubescence or hairs. Differences in the shapes of the ligules are also apparent.

The **auricles** are ear-like or finger-like lobes at the base of the leaf blades and at the top of the sheath.

REPRODUCTIVE GROWTH AND DEVELOPMENT

Many food crops are grown for their seed or fruit to produce a form of stored food energy to meet man's food needs and to perpetuate the crop in subsequent years. All plants, even perennials, must reproduce. The most common form of reproduction is by means of seed. Seed production is achieved through sexual union of the male and female **gamete,** similar to reproduction in animals.

Plants can also reproduce asexually through rhizomes, stolons, tubers, grafts, or vegetative cuttings, but seed is also produced on plants that reproduce asexually. Potatoes, sugarcane, and cassava are large-scale field crops that are asexually reproduced on a commercial scale.

Apomixis is a form of asexual reproduction and occurs when seed and fruits develop without fertilization. Bluegrass seed is often produced apomitically.

Because progeny produced by apomixis resemble the maternal parent, it would be highly desirable to induce apomitic seed set to perpetuate a single desirable plant through the convenient mechanism of seed (Bashaw 1979). Sherwood, Young and Bashaw (1980) observed a culture of buffelgrass that produced 97% of its seed apomitically. Such a finding may be of value in studying this mechanism and learning how to stimulate apomitic seed set in other crops.

SEXUAL REPRODUCTION IN FLOWERING PLANTS

Sexual reproduction is dependent on the development of flowers that contain reproductive parts, specifically **stamens** and **pistil.** The stamen is the male reproductive organ and produces pollen. The pistil is the female reproductive organ and contains an ovary at its base which produces one or more egg cells (Figure 2-9). In corn, cereals, and the grasses, the ovary contains one egg cell, and fertilization of each ovary is required for seed set. The legume ovary contains more than one egg cell. Upon fertilization, each of these becomes a seed within the pod, the characteristic fruit of the legume family.

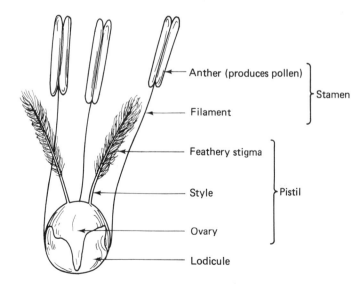

Figure 2-9.   Reproductive parts of a wheat plant normally contained between two glumes. The glumes are often referred to as chaff at threshing time.

In the corn plant, the pollen producing stamens are contained in the tassel at the top of the plant, and the ovaries in the developing ear which arises from a lateral bud in the axils of lower leaves. Each ovary bears a style or silk 10 to 30 cm (4 to 12 inches) in length. The silk has a sticky surface and is receptive to pollen for about two weeks. Silks have small hairs or stigmata (the singular form is stigma) along their entire length. The stigma is the part of the pistil that receives the pollen. Pollen can adhere to stigma along the entire length of the silk and send out a pollen tube and cause fertilization when male and female gametes unite. Unpollinated ovaries may produce silks 50 to 75 cm (20 to 30 inches) in length.

Considerable quantities of pollen are produced by a corn plant. An excessive amount of energy is required to produce such massive amounts of pollen. Under conditions of stress in which food energy is limiting (the source), the tassel may take priority as a sink because of its dominant position at the top of the plant. Such a situation could severely reduce grain yield; under such conditions, crop producers may find it economically advantageous to mechanically detassel corn plants. Modern corn hybrid cultivars have been selected with genetically controlled pollen production to help reduce the energy requirements of this non-economic component.

Corn is normally cross-pollinated. Pollen produced on the top of the plant is carried by wind currents to adjacent plants where it may settle on the female flowers midway up the stalk. Corn plants tend to shed pollen before the appearance of the silks. The emergence of silks may be delayed by stress from drought or inadequate nitrogen. This situation may be intensified by high plant populations, and when the shedding of pollen and the receptivity of the silks do not overlap a sterile ear, **nubbin** will develop.

Normally the pollen grain germinates and establishes a pollen tube within 5 to 10 minutes after it lands on the silk. Fertilization is accomplished within 15 to 36 hours after pollination.

Plants that have stamens and pistils in separate flowers on the same plant are said to be monoecious. Corn, squash, and cucumber plants have male and female flowers on the same plant. Only female flowers produce seed when pollen from a male flower unites with eggs of the female flower. Pollination is accomplished by insects or wind.

Plants having stamens and pistils in separate flowers of different plants are dioecious. Hemp, hops, and buffalograss are dioecious. Seed set is accomplished only if male and female plants are sufficiently close for wind or insects to effect pollination.

Types of Pollination

The following three types of pollination are commonly found in crop plants.

Naturally cross-pollinated crops. Corn, rye, clovers, alfalfa, vetches, trefoil, buckwheat, sunflowers, most perennial grasses, and some annual grasses are examples of normally cross-pollinated crops. Grasses are mostly wind-pollinated; alfalfa, trefoil, clovers, and buckwheat are insect-pollinated. In alfalfa, it is necessary to have an insect pollinator trigger the release of the pistil enclosed within the keel petals of the flower. As the pistil matures, some mechanical force or tension builds up. Honey bees searching for nectar release the pistil which strikes the insect, discouraging it from further interest in alfalfa flowers. Honey bees are therefore poor pollinators of alfalfa. The evolutionary significance of the forceful release of the pistil is to ensure cross-pollination by rubbing some foreign pollen off the bee.

Many normally cross-pollinated plants have a mechanism of self-sterility. Pollen tubes of pollen from the same plant are restricted in their growth, and fertilization may take place only if pollen from another plant is unavailable.

Naturally self-pollinated crops.   Wheat, oats, barley, tobacco, potatoes, flax, rice, field peas, soybeans, and cowpeas are examples of normally self-pollinated crops. In such plants, the pistil usually is pollinated by pollen from the same flower. Cross-pollination can be induced in naturally self-pollinated crops by male sterility or any other factor that interferes with normal self-pollination and fertilization. The development of hybrid cereal cultivars is dependent upon successful cross-pollination in these crops.

Often cross-pollinated crops.   In such crops, the pistil may be pollinated by pollen from the same flower or from another flower on the same plant or from another plant. Such crops include sorghum, cotton, several cultivated grasses, foxtail, and proso millets.

Reproductive Structures

A flower arises from a bud either at the apex of the stem, as in cereals, or in the leaf axils, as in soybeans. A flower of these crops is composed of the stamens and pistils. The stamens are supported on stalks called the filaments (Figure 2-9). The pistil consists of the feathery stigma, style, and ovary (Figure 2-9). Soybean and alfalfa flowers have showy flower petals consisting of calyx and corolla. The calyx consists of two sepals that enclose and protect the young flower. The corolla of alfalfa may be purple, blue, or white in color. Trefoil corollas are yellow.

In the non-showy inflorescence of the grasses, the calyx and corolla are replaced by the glumes, the protective covers of the reproductive organ, and eventually the seed.

Grasses and cereals have two small organs called lodicules that are located between the ovary and the surrounding glumes (Figure 2-9). The lodicules swell at the time of fertilization which forces open the glumes; the filaments or stalks of the pollen-producing stamens elongate and extrude out of the flower. Opening of the glumes encourages cross-pollination. Fertilization of an embryo can be accomplished only once, but if self-pollination has not been successful the open glumes may allow for cross-pollination to occur.

The basic unit of a grass inflorescence, a term used to describe the "flowering" parts of the grass plant, is the **spikelet.** If the spikelets are attached directly to the stem or rachis, the seed head is called a **spike** as is the case in wheat, rye, barley, quackgrass, and timothy (Figure 2–7). If the spikelets are attached to fine branches or subdivisions of the stem, then the inflorescence is called a **panicle.** Orchardgrass, bromegrass, oats, a corn tassel, barnyard grass, and fall panicum are examples of crops having a panicle.

The spikelet consists of two structures called glumes that totally or partially enclose one or more florets.

The floret is the seed-producing unit of the inflorescence and is enclosed by the lemma and palea. In wheat and rye, the **lemma** and **palea** are loose and are removed from the seed by threshing. In contrast, in oats, barley, and rye, the lemma and palea adhere to the seed to form a hull. The awn is a projection of the lemma and may be smooth or barbed. The awn is capable of photosynthesis and under stress conditions, awned cultivars may be of value over awnless cultivars.

Legume seeds are covered by a relatively firm and water resistant seed coat. Seeds that do not germinate because of a lack of water absorption are called "hard" seeds.

Seed Dormancy

Seeds may fail to grow immediately after maturity even under environmental conditions that favor germination. Seed dormancy may be attributed to a number of factors, including hard seed coats and germination inhibitors.

Hard seed coats.   Some legumes such as trefoil and alfalfa commonly exhibit hard seeds in which water absorption and germination are prevented. Soil acids and weathering action after planting in the field will usually soften hard seeds but germination is delayed. In cases of extremely hard seed, germination potential can be increased by mechanical scarification in which the seed coat is cracked or weakened by the abrasive action of a rough surface.

Germination inhibitors.   Chemical germination inhibitors may prevent germination immediately following harvest but normally undergo a natural chemical change under conditions of dry storage. Wheat cultivars with red pigment help prevent premature sprouting of the embryo under moist, warm conditions prior to harvesting. In white wheats, preharvest germination can be a serious problem as it reduces milling quality and harvesting ease. Some inhibition of germination may be therefore desirable.

Crops harvested before they are fully mature often have a high proportion of dormant seeds.

Once seeds start to germinate they can be induced into reproductive development by a process of **vernalization.** Winter annuals seeded in the fall are vernalized by low temperature conditions over the winter months. Vernalization is required by winter annuals to initiate reproductive development. Winter annuals seeded in the spring normally will remain vegetative and will not reach reproductive development without being subjected to cold or cool temperatures. Crops which are winter-hardy, that is, can withstand severe winter conditions, have a greater vernalization requirement (cold temperatures for an extended period) than do less winter-hardy crops, such as winter oats or winter barley.

CONCLUSIONS

The topics related to botany in this chapter represent only a portion of the extensive subject of plant botany. The reader will find the material selected for this chapter of considerable value in understanding crop production. Additional aspects of botany will be introduced in subsequent chapters.

For those with a sound background in botany this chapter will serve to emphasize the importance of plant structures in crop production.

QUESTIONS

1. Explain why bluegrass resumes growth after close clipping and is therefore a suitable lawn species. Would trefoil or ladino clover (white Dutch clover) make a satisfactory lawn species?
2. Explain the differences between a single-cross and a double-cross corn hybrid cultivar.
3. What is meant by an "open-pollinated" corn cultivar?
4. Explain source to sink flow of photosynthetic materials, and speculate what may happen to the flow of materials in a multiple-eared corn plant.
5. Discuss internal competition for flow of photosynthetic material from source to sink as it applies to single-eared corn and to soybeans.
6. Explain how unharvested potato tubers may increase in size after the leaves of the plant have turned brown and photosynthesis has ceased.
7. Review the functions of cotyledons.
8. When seed of a dicot is planted containing broken seeds (cotyledons separated) plant stands are usually thin. Why?
9. Why will the root of a seed placed upside down in the soil send a root downward and a shoot upward?
10. Explain the difference between a cultivar of hybrid corn and a cultivar of wheat.
11. How does a rhizome differ from a horizontal root?

12. Why is transpiration so important to plants? Why can transpiration be considered a compromise?

13. In corn production there is a compromise between stalk breakage and grain yield. Why?

14. Is stalk breakage a problem in silage corn that is harvested when the leaves and stalks are green? Why or why not?

REFERENCES FOR CHAPTER TWO

Bashaw, E. C. 1979. "Apomixis and Its Application in Crop Improvement." W. R. Fehr and H. H. Hadley (ed.) chapter 3, *Hybridization of Crop Plants*. Madison, Wisconsin: American Society of Agronomy Press.

Briggs, K. G. and Attinaw Aytenfisu. 1980. "Relationships Between Morphological Characters Above the Flag Leaf Node and Grain Yield in Spring Wheats," *Crop Science* 20:350-354.

Dittmer, H. J. 1937. "A Quantitative Study of the Roots and Root Hairs of a Winter Rye Plant (*Secale cereale*)," *American Journal of Botany* 24:417-420.

Garcia, L. R. and J. J. Hanway. 1976. "Foliar Fertilization of Soybeans During the Seed-Filling Period," *Agronomy Journal* 68:653-657.

Hyder, D. N., A. C. Everson and R. E. Bemet. 1971. "Seedling Morphology and Seedling Failures with Blue Grama," *Journal of Range Management* 24:287-292.

Krassovsky, Irene. 1926. "Physiological Activity of the Seminal and Nodal Roots of Crop Plants," *Soil Science* 21:307-325.

Pavlychenko, T. K. 1937. "Quantitative Studies of the Entire Root Systems of Weed and Crop Plants Under Field Conditions," *Ecology* 18:62-79.

Sherwood, R. T., B. A. Young and E. C. Bashaw. 1980. "Facultative Apomixis in Buffelgrass," *Crop Science* 20:375-379.

Tischler, C. R. and R. L. Monk. 1980. "Variability in Root System Characteristics of Kleingrass Seedlings," *Crop Science* 20:384-386.

Vasilas, B. L., J. O. Legg and D. C. Wolf. 1980. "Foliar Fertilization of Soybeans: Absorption and Translocation of the ^{15}N-Labelled Urea," *Agronomy Journal* 72:271-275.

Wilson, A. M., D. N. Hyder and D. D. Briske. 1976. "Drought Resistance Characteristics of Blue Grama Seedlings," *Agronomy Journal* 68:479-484.

chapter three

Yield—Progress, Problems, and Prospects

The modest photosynthetic efficiency of 1% found in most crop production is not an indication that yield advances have been unsuccessful in the past. After many centuries of yield stagnation, steady yield increases have taken place in the past half century. These increases are the result of a combination of scientific inputs in the hands of knowledgeable and capable crop producers. This chapter examines the magnitude of yield increases and considers the problems and prospects for further yield increases.

Almost any crop could be selected to demonstrate yield increases in the past 50 years. Soft white winter wheat in Ontario, for example, has shown a 70% yield increase in the past 38 years (Table 3-1). Yields are averaged over five-year periods to minimize yearly fluctuations caused by environmental factors. On the average, yields increased 35 kg/ha per year (½ bu/ac).

TABLE 3-1. Yield increases of winter wheat in Ontario, 1941–1979

Time	Yield		Yield Increase over 1941–1945 Period	
	(kg/ha)	(bu/ac)	(kg/ha)	(bu/ac)
1941–1945	1902	27.8	—	
1946–1950	1966	28.8	+64	+1.0
1951–1955	2204	32.3	+302	+4.5
1956–1960	2292	33.5	+390	+5.7
1961–1965	2520	36.9	+618	+9.1
1966–1970	2796	40.9	+894	+13.1
1971–1975	2950	43.7	+1048	+15.4
1976–1979 (4 years)	3230	47.3	+1328	+19.5

Source: Statistics Section; Economics Branch, Ontario Ministry of Agriculture & Food, Agricultural Statistics for Ontario. Metric Edition. 1941–1978. Dec. 1979.

Even more striking are yield increases that occurred in grain corn over the same period along with an increase in hectarage (hectares grown) (Table 3–2). The area of corn production increased over seven times, a situation often associated with minor increases in yield per hectare as inexperienced growers learn to produce a new crop and as the frontiers of production are pushed forward. However, yields increased nearly 100% along with this increase in hectarage. The increase for the 38-year period averaged 70 kg/ha (1.12 bu/ac) annually. Such increases are typical for North America.

TABLE 3–2. Grain yield and hectares grown of corn in Ontario, 1941–1979

Time Period	Yield (kg/ha)	Yield (bu/ac)	Yield Increase over 1941–1945 Period (kg/ha)	Yield Increase over 1941–1945 Period (bu/ac)	Hectares Produced	(acres)
1941–1945	2872	45.8	—	—	93,700	231,439
1946–1950	2959	47.2	+87	+1.4	98,400	243,048
1951–1955	3668	58.5	+796	+12.7	160,000	395,200
1956–1960	3712	59.2	+840	+13.4	190,500	470,535
1961–1965	4764	76.0	+1893	+30.2	218,000	538,460
1966–1970	5198	82.9	+2326	+37.1	360,000	889,200
1971–1975	5066	80.8	+2194	+35.0	480,000	1,185,600
1976–1979 (4 years)	5522	88.0	+2650	+42.0	682,000	1,684,540

Source: Statistics Section; Economics Branch, Ontario Ministry of Agriculture & Food, Agricultural Statistics for Ontario. Metric Edition. 1941–1978. Dec. 1979.

Such phenomenal yield increases raise several important questions. What caused these yield increases? Can yield increases of this magnitude be sustained? Is there an upper level at which yield increases are no longer profitable and hence not desirable? How much of the yield increase can be attributed to crop production inputs (fertilizer, weed control, cultural practices, mechanization), and how much resulted from genetic improvements (superior **genotype** and **hybrid vigor,** genetic resistance to disease, insects, and lodging, or a better ratio of grain to straw)?

In an attempt to identify genetic advances, eleven winter wheat cultivars developed over a 50-year period at Cornell University at Ithaca, New York were grown for a 10-year period in a "living museum" nursery. When all cultivars are grown under comparable fertility and cultural conditions, yield variations should be a reflection of genetic differences. The yield of Honor, a cultivar released in 1920, had an average yield of 2964 kg/ha (43.4 bu/ac) over the 10-year period 1966 to 1975 (Table 3–3). Ticonderoga, released in 1973, had an average yield of 4348 kg/ha, (63.6 bu/ac) 50% higher than Honor. For the period of 1925 to 1975, New York State farm yields increased from 1297 to

2641 kg/ha, (19.0 to 38.7 bu/ac) or about 100% (Jensen 1978). If geno-type contributed 50% to this increase, then cropping practices can be presumed to have contributed the other 50%. The lesson to be emphasized from this observation is that yield increases are a result of many factors. Crop production is most successful when all available factors are utilized advantageously.

TABLE 3–3. Mean yields (1966–1975) of winter wheat cultivars developed at Cornell University and grown in the living museum nursery, Ithaca, New York

Cultivar	Year of Release	Height (cm)	(in)	Yield (kg/ha)	(bu/ac)
Honor	1920	116	45.7	2964	43.4
Forward	1920	112	44.1	3212	47.0
Valprize	1930	116	45.7	2970	43.5
Yorkwin	1936	115	45.3	3252	47.6
Nured	1938	114	44.9	3091	45.2
Cornell 595	1942	112	44.1	3501	51.2
Genesee	1950	111	43.7	3528	51.6
Avon	1959	108	42.5	3615	52.9
Yorkstar	1968	95	37.4	3951	57.8
Arrow	1971	94	37.0	3851	56.4
Ticonderoga	1973	85	33.5	4348	63.6

Source: Data from Jensen 1978.

Producing old cultivars alongside the best current cultivars always makes for an interesting comparison. However, such comparisons can seldom be made because viable seed of old cultivars is hard to obtain. In one instance, seed of old oat and barley cultivars was discovered entombed in the cornerstone of the Nuremberg State Opera House in West Germany. When the building, erected in 1832, was demolished in 1955, the oat and barley seeds were viable and were successfully germinated to produce new seed that was multiplied and distributed (Aufkammer and Simon 1957). This seed could be similar to that which European immigrants brought with them as they departed for the New World. Such seed may have helped establish cereal production in North America. When grown in the field in North America, the old unnamed cultivars were taller than the present-day cultivars; however, they were very susceptible to fungus diseases such as rust and mildew and lodging occurred shortly after heading. The barley cultivar was of the two-row type, characteristic of most barley produced in that era. A visual comparison in the field clearly revealed the improvements that have taken place in the past 140 years in height, lodging, and disease resistance.

Further studies with these old cultivars with indoor pot experiments to observe fertility responses under disease-free conditions produced surprising results. Not only were the old cultivars capable of high yields under such conditions, but they also equalled the response of present-day cultivars. It might be assumed that low fertility would favor the old cereals and high fertility would favor the modern cultivars developed under such conditions. However, this assumption was not confirmed. Height and disease differences among the old and the new cereals were prominent, but other factors had changed little if at all.

Progress in advancing crop yields is slow and methodical, not rapid and dramatic. Further improvements and progress remain to be made. Perhaps responsiveness to fertility levels in the form of increased grain yield rather than vegetative growth may be possible. Genetically controlled dwarf plants with a favorable tillering capacity may allow crop producers greater scope for management practices to increase grain yield. Such an improvement may be possible if disease and lodging resistance and other genetic factors are favorable. Likewise, crop producers may be successful in increasing yield through greater fertility only if moisture is adequate and the crop is planted at a proper date, depth, and row width and if weeds are controlled and other production factors are favorable. An increase in one input without consideration of other inputs may not by itself be expected to increase yield. The complexity of the many interacting factors in crop production demands that a systems approach be adopted in which all factors are integrated.

ARE HIGHER YIELDS POSSIBLE?

The yields presented in Tables 3-1 and 3-2 are Ontario provincial averages. Capable individual growers may expect yields higher than the average. Outstanding yields have been reported for many crops but without a satisfactory explanation or understanding of why they were achieved and usually cannot be repeated. Record yields reported here are of academic interest, given to show what is possible but without a clear approach to repeating such yields.

A yield of 78,000 kg/ha (69,643 lb/ac) of sugar cane in Hawaii was reported as early as 1945 (Blackman and Black 1959), but since then, yields of this magnitude have not been reached on a consistent basis. A semidwarf winter wheat yielded 12,900 kg/ha (189 bu/ac) in Washington State (Anonymous 1966) and was attributed to the new cultivar Gaines. Several record yields of this magnitude were reported but were not produced regularly. In Michigan, a crop competition spurred one grower to produce 22,500 kg/ha (359 bu/ac) of grain corn (Anonymous 1978). A single crop of rice in the Philippines produced a yield of 10,573 kg/ha (155 bu/ac) and 26,000 kg/ha (380 bu/ac) under a four-crop sequence in one year. Cotton is reported to have produced 11,900 kg/ha (10,625 lb/ac) of lint and soybeans 7398 kg/ha (117 bu/ac).

Yield levels that are multiples of yields commonly obtained lend credibility to the suggestion in Chapter One that photosynthetic efficiencies up to 6% are possible. Record yield levels are encouraging because they indicate the potential magnitude of yield increases. However, record yields raise the question as to the upper limit to crop yields, a question deserving of some attention. Although there is no absolute upper yield limit, the nature of yield increases may be obtained by an examination of the biological or sigmoid growth curve, Liebig's law of the minimum, and Mitscherlick's law of diminishing returns.

The Biological or Sigmoid Growth Curve

Plotting the growth curve for divisions of a single-celled bacteria produces a typical S-shaped curve. An initial bacterial cell divides into 2, then 4, 8, 16, 32, 64, etc. Initial growth is slow because of the relatively low number of cells involved. As division continues, 8192 cells divide into 16,384, then into 32,768, etc., and a phase of rapid growth occurs—the grand growth period. In time, competition for space, nutrients, or other factors may limit division or growth of cells, and the curve levels off. The resulting S-shaped or sigmoid growth curve is typical of the growth pattern of individual cells, of the cells in a plant, or of the cells of the plants in an entire population. The typical growth curve of a cotton fiber is shown in Figure 3-1. In the cotton plant, age, senility, nutrient limitations, competition, or other factors may restrict growth of the fiber. If stress conditions occur during the grand period of growth (the sixth to the twentieth day), growth of the cotton fiber may be restricted.

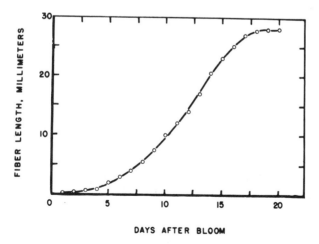

DAYS AFTER BLOOM

Figure 3-1. Growth curve of an Upland cotton fiber. Fiber elongation begins at about the time of flowering and continues slowly for about five days, more rapidly from about the sixth to fourteenth day, and again slower until about the eighteenth to twentieth day, when it ceases. (Chart from Tharp, 1960.)

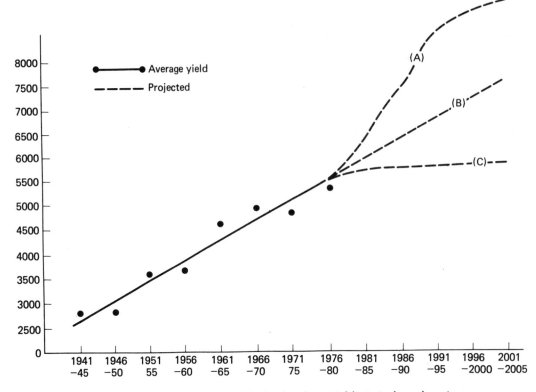

Figure 3-2. Corn yields (kg/ha) from Table 2-2 plotted against years.

If the corn yields from 1941 to 1979 (as shown in Table 3-2) are plotted (Figure 3-2), the response appears to be a straight-line increase. Projecting the line is done by guesswork because it is uncertain whether the curve is entering the grand period of growth (A), or will continue as a straight line (B), or whether yields are about to level off (C). It cannot be determined where on a typical sigmoid curve present yields are located. Jensen (1978) argued that wheat yields in New York State are levelling off and that further yield increases will be difficult to achieve. An already advanced yield base exists and further improvements will be difficult. When weed control is complete, little scope for further improvement remains. Man has attempted to remove the yield limitations within his control, but he cannot control adverse weather, which remains as a major bottleneck to yield increases.

Liebig's Law of the Minimum

In 1862, Justin Von Liebig, a German chemist, examined the sigmoid growth curve and developed the "law of the minimum" (Cannon 1963). This law states that yield depends basically on the growth factor that is available to the plant in the smallest relative amounts. Mitchell (1970) likened the law of the minimum to that of a barrel having staves of different lengths—the barrel cannot hold anything above the height of the shortest stave (the most deficient or limiting factor). In other words, if all factors are ideal for crop production, yields will be held down by the one factor in least supply. Liebig's law provides a simple guide through an otherwise puzzling maze of tests involving soil fertility, and Liebig's law still stands as a logical, simple, and successful predictor of crop responses. Modern mathematical techniques have verified the modern use of this nineteenth century law (Waggoner and Norvell 1979).

Liebig's law must be kept in mind in subsequent chapters when the application, modification, or addition of a single factor may appear to have a potential for yield increase. Yields are not likely to be increased by a single factor. Crop producers should not be discouraged by the failure of a single modification to increase yield but must find the factor limiting yield and modify it.

The Law of Diminishing Returns

A refinement of Liebig's law was made in 1909 by E. A. Mitscherlick, a German soil scientist. He observed that as fertilizer nutrients were applied, a typical response curve resulted. The greatest response occurred when fertilizer was applied at low fertility levels, whereas at high fertility levels crop response was less. This observation became known as the "law of diminishing returns" and was stated by Mitscherlick as follows: "The increase in any crop produced by a unit increment of a deficient factor is proportional to the decrement of that factor from the maximum" (Mitchell 1970, p. 97). In numerical terms it follows that if one unit of a nutrient or other factor produces half of the maximum yield, then two units will produce 75% of the maximum yield, three units will give 87.5% of the maximum, and so on. A typical response curve results, as shown in Figure 3-3, for the response of perennial ryegrass to levels of nitrogen fertility in England (Reid 1978).

YIELD—QUANTITY VERSUS QUALITY

High yields are desirable, but quality is also important in many crops. A generalization is that as dry-matter yields are increased either through breeding or crop management, quality aspects usually decline. Quality is a complex entity, however, and it is not easily described because it varies from crop to crop and species to species. Quality may be determined by color, texture, protein level, quality of protein in terms of its amino acid composition, gluten strength, digestibility, and nitrogen content, or by many other physical or chemical properties. Quality may be lowered by damage from disease or insects.

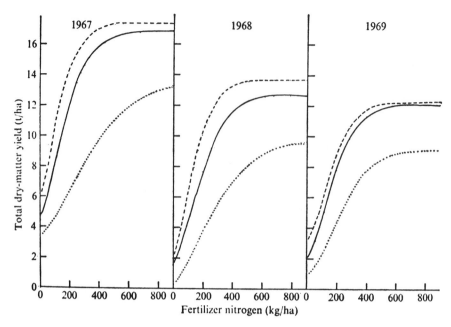

Figure 3-3. Curves relating total yields of herbage dry matter to nitrogen rates
within each frequency of cutting
.... Three cuts; _____ five cuts; ..... ten cuts per year.

Strong gluten is a desirable quality aspect for making bread because it can trap carbon dioxide released from yeast, which causes the dough to rise. High yields of wheat are desired also. Under conditions that restrict growth (drought, heat, or a combination of factors) grain yield may be restricted leaving a surplus of food energy to be transformed into gluten and protein products. Such a wheat is favored for bread-making and may command a premium price. Under conditions that favor high dry-matter yields, food energy is required for grain development with the result that gluten and protein levels are reduced. Hard red wheat produced under the dry-land conditions of a continental climate has consistently high bread-making properties from year to year, region to region. The relationship between grain yield and protein levels of four hard red winter wheats produced in Alberta is shown in Table 3-4.

Several lessons can be drawn from these results:

1. Sundance produced the highest yields and had the lowest protein levels.
2. Lower yielding cultivars had higher protein levels, but grain-protein relationships were not exact. This is the nature of **biological variation.**

TABLE 3–4. Mean yield (kg/ha) and grain protein (%) of four hard red winter wheat cultivars grown at Vulcan and Bow Island, Alberta, 1971–1974

Cultivar	Vulcan Yield (kg/ha)	bu/ac	Protein (%)	Bow Island Yield (kg/ha)	bu/ac	Protein (%)
Sundance	1810	26.5	12.3	2087	30.5	12.2
Kharkov 22 MC	1253	18.3	14.1	1705	25.0	13.8
Winalta	1234	18.1	13.7	1590	23.3	13.7
Jones Fife	1163	17.0	13.3	1396	20.4	13.2

Source: Data from Pittman and Tipples 1978.

In contrast to hard red wheat, soft white wheat is favored for the manufacture of pastry flour because of its low protein level and soft gluten. As yields are increased, protein levels generally decline, and the highly productive, adequate rainfall, cool temperature areas of eastern North America (Ontario, Quebec, Michigan, Washington and New York) are favored soft white winter wheat-producing areas.

The biological complexity between dry-matter yield and quality components such as digestible organic matter, protein, and nitrogen content is shown for 30 grass cultivars in Table 3-5. Yield ranged from 13,926 to 2250 kg/ha (12,434 to 2,009 lb/ac), percent digestible organic matter ranged from 50.6 to 60.0%, and percent nitrogen content ranged from 1.66 to 2.74%. Nitrogen is a requirement of all cells and has the closest relationship with yield. According to Lawrence (1978), as dry-matter yields increased, less food energy remained unused, protein content declined, and more cells had a nitrogen requirement so that nitrogen percent content declined.

Malting barley has precisely defined biochemical quality parameters. In the production of malting barley, attempts to increase grain yields with high levels of nitrogen fertility may promote high protein levels and reduce malting quality. Likewise, abundant vegetative growth may be promoted by high levels of nitrogen fertility on tobacco, although low quality may be the result.

A special-purpose milling oat, Hinoat, was licensed in Canada in 1973 (*Anonymous* 1973) because of a superior protein level. Hinoat, grown at 26 sites across Canada in 1970, averaged 3.4 percentage points higher in protein that the cultivar Victory.

At every location, Hinoat had a higher protein level than Victory, but the average yield potential of Hinoat is approximately 75% of standard cultivars. Therefore, Hinoat is not thought of as a cultivar that is competitive in the feed grain market because of its lower yielding ability (Table 3-6). On the basis of kilograms per hectare (pounds per acre) of protein produced, little difference was found to exist between Hinoat and the check cultivars (Table 3-7). If grain yield and protein are to be increased simultaneously, efforts should be directed

TABLE 3–5. Relationships among dry-matter yields (kg/ha) and three quality criteria of 30 grass cultivars grown at Swift Current, Saskatchewan Results are for first harvest year, 1971

Entry Number	Total Dry-Matter Yield (kg/ha)*	(lb per acre)	Percent Digestible Organic Matter	Percent Protein	Percent Nitrogen Content
No. 1	13,926	12,434	55.4	8.0	1.89
No. 2	12,688	11,329	57.4	12.0	1.98
No. 3	11,740	10,482	53.8	10.9	1.85
No. 4	10,670	9,527	53.4	10.9	1.76
No. 5	10,404	9,289	53.0	13.1	2.03
No. 6	9,770	8,723	50.6	11.8	1.66
No. 7	9,643	8,610	54.8	10.9	1.82
No. 8	8,518	7,605	58.8	15.9	2.51
No. 9	8,410	7,509	53.8	14.3	2.16
No. 10	8,164	7,289	55.0	10.5	1.79
No. 11	7,796	6,961	55.6	12.9	2.12
No. 12	7,635	6,817	55.0	14.6	2.32
No. 13	7,612	6,796	55.0	14.4	2.43
No. 14	7,277	6,487	54.8	13.8	2.05
No. 15	6,895	6,156	58.4	19.1	2.74
No. 16	6,818	6,088	57.8	12.2	2.22
No. 17	6,800	6,071	55.2	11.2	1.84
No. 18	6,497	5,801	51.8	13.2	2.06
No. 19	6,225	5,558	58.0	12.1	2.17
No. 20	5,792	5,171	53.0	9.5	1.62
No. 21	5,541	4,947	59.2	11.8	2.29
No. 22	5,448	4,864	54.8	14.1	2.23
No. 23	5,129	4,579	53.0	10.7	1.71
No. 24	5,080	4,536	58.8	12.8	2.22
No. 25	5,013	4,476	56.8	12.0	2.08
No. 26	4,516	4,032	55.6	14.0	2.30
No. 27	4,493	4,012	54.7	13.1	2.27
No. 28	3,578	3,195	52.0	11.2	1.82
No. 29	3,380	3,018	57.8	9.7	1.93
No. 30	2,250	2,009	60.0	12.1	2.21

Source: Data based on Lawrence 1978.
*Mean yield of 23 cuttings.

toward increasing the total amount of food energy in the plant. This may require an increase in photosynthesis.

In horticulture, quality is often a prime consideration. The apple may epitomize quality aspects because it has features that appeal to all of man's senses—sight, sound, smell, touch, and taste. Consider:

TABLE 3–6. Average yield of Hinoat and six check cultivars in 1970 at all eastern and western Canadian sites and the calculated loss to producers by growing Hinoat rather than control cultivars

Cultivar	Yield (kg/ha)	bu/ac	Loss (%) of Control
Eastern Canadian sites			
Hinoat	2363	62.1	—
Six check cultivars	3102	81.5	23.8
Western Canadian sites			
Hinoat	2666	70.0	—
Six check cultivars	3534	92.8	24.6

Source: Data from *Agriculture Canada* 1973.

TABLE 3–7. Average protein yields of Hinoat and check cultivars

Region of Canada	Protein produced			
	Check Cultivars		Hinoat	
	kg/ha	lb/ac	kg/ha	lb/ac
Eastern	470*	428	448	400
Western	596**	532	512	457

Source: Data from *Agriculture Canada* 1973.
* Based on 90 entries.
**Based on 66 entries.

1. An apple is selected that fits the palm of the hand and is firm. The appropriate size is determined by thrusting the apple into the palm of the hand and checking and squeezing it gently.
2. To satisfy the sense of sight, the apple is polished to remove the "bloom" so its full color can be seen.
3. Before biting into the polished cheek of the apple, the aroma is savored. A mature apple releases ethylene, which satisfies the senses of smell and gives an assurance that the apple is ripe.
4. A resounding crack when the apple is bitten into gives an assurance through the sense of sound that the apple is firm and of the highest quality.
5. Finally, the sense of taste is satisfied as the apple is chewed.

Apple producers, attempting to produce such an apple, may limit the number of apples produced per tree so that a few large apples rather than a lot of small ones are produced. This is accomplished by rigorous pruning to allow air currents to penetrate so as to discourage disease, to allow for more effective penetration of sprays, and to facilitate sunlight penetration. Irrigation water will be controlled and timed to promote growth and to encourage development of color (differentiation).

If an apple producer is interested in apples for processing (juice), he may try to maximize yields to the full capacity of the tree and yet avoid biennial bearing, a situation caused when the crop is so extensive and demands for food energy allow little if any food energy to be directed to bud development for the next crop year. Quality demands may also dictate cropping practices.

A fundamental difference exists between crop producers and horticultural producers. Field crops are used to meet the energy and protein needs of man and animals, and they are needed in quantity. Quality is important but surely secondary to quantity. Horticultural crops are produced to aid nutrition and to meet aesthetic needs; for example, a green salad in midwinter with a red tomato or radish for garnish may have greater appeal for aesthetic rather than nutritional needs.

This book is devoted to maximizing yield in the recognition that food needs come first, quality second. It is up to the producer to recognize the economic limitations of production inputs, as well as the quality standard required for a satisfactory product, and to govern his production program accordingly. Dryland wheat producers must choose between high yields with reduced quality under irrigation, and low yields with high quality and premium price under nonirrigated conditions.

GROWTH AND DIFFERENTIATION

In plant development, **growth** results from cell division and cell enlargement, and if quantity is desired, then advantage should be taken of such factors as **fertilization**, irrigation water, and warm temperatures that favor growth processes. **Differentiation** includes the development of small, thick-walled cells and the development of lignin, protein products, oils, pigments, and other differentiation products, some desirable, some not. Latex in lettuce, for example, is an undesirable differentiation product as it is bitter and makes lettuce tough and leathery. Latex is formed by conditions that discourage growth and favor differentiation, such as hot and/or dry weather conditions.

In hot, dry years, forage producers may be alarmed by the low number of hay bales produced compared to the larger number produced in wetter, cooler seasons. Because the low yields are associated with a highly differentiated product, the hay produced is a quality product, and limiting livestock intake may result in satisfactory livestock production. Quality in this case may be associated with protein levels and the percentage of total digestible nutrients.

An understanding of growth and differentiation is essential for fall management of forages to allow for build-up of root food reserves to ensure winter-hardiness and will be covered in Chapter Ten in more detail.

YIELD AND ADAPTABILITY

In terms of crop production, **adaptability** refers to good crop performance over a wide geographic region under conditions of variable climatic and environmental conditions. Adaptability is a measure of crop reliability. Modern agricultural technology demands more than a cultivar capable of high yields only under the best of conditions. A widely adapted cultivar will produce a satisfactory yield when conditions are less than ideal. Specifically desired is a cultivar possessing the ability to yield well under both good and adverse conditions, and this may not always be possible.

Yield is an important criterion in evaluating adaptability. Cultivar recommendations are commonly based on average yield performance calculated from field trials in a number of regional (environmental) sites over a number of years. Good yield response to environmental variation is a measure of wide adaptability. Wide adaptability lowers risks for growers and allows the transfer of related technology to other environments without extensive experimentation at specific sites.

High yield and wide adaptability do not always go together. Because farmers aim for maximizing yield, specific cultivars—especially those of full-season crops—must be selected. Wide adaptability may be equated to a quantity factor: with wide adaptability, yields may decline; with narrow adaptability, high yields will be obtained under favored conditions.

First-time producers may choose wide adaptability, whereas more experienced producers wishing to maximize yield may choose a specific, narrowly adapted cultivar but one known to perform well on their farms. Testing agencies provide farmers with recommendations based on regional tests, and these are adequate for most growers. Some producers, however, will test a number of selected cultivars on their farms to allow for an evaluation under their own conditions and to evaluate factors of threshability, standability, and drying and storage difficulties or other peculiarities that can be evaluated best under farm field-scale conditions. To specifically define yield stability or to identify the environmental variables that favor a given cultivar is a complex issue.

POSITIVE APPROACHES TO CROP PRODUCTION

The history of crop production is filled with crop failures caused by disease and insect attacks such as grasshoppers, wheat rust, potato blight, army worms, and many others. Wheat production in the Midwest of North America is a classical example of such problems. Losses from pests provide vivid reminders that encourage efforts to eliminate such problems. However, there is a risk that too much effort may be directed toward developing factors influencing yield stability and not enough toward developing yield ability.

Often the result is a loss of positive approaches to crop production at the expense of inputs that eliminate negative factors. Recognition and emphasis of positive approaches to crop production are likely to be most successful in increasing yields. Crop producers must learn to recognize and emphasize the positive factors.

Consideration of factors that increase photosynthesis must be considered a positive approach. This seems obvious, but many approaches have considered the amount of grain produced per head, the size of the seed, and the number of heads per hectare (acre) as the "components of yield." Numerous attempts to increase one or more of these factors have led to disappointment because often as one factor is increased, another one will decline. This is to be expected, for the amount of grain produced per head, the size of the seed, and the number of heads per hectare (acre) only describe yield. All yield factors depend on available food energy to produce yield. Such components draw on a fixed pool of food energy, and if yield is to be increased, photosynthesis must be increased. This is a positive approach to crop production. Yield may be increased most effectively if the reasons for each input or each crop production practice are considered in terms of how they will affect photosynthesis.

QUESTIONS

1. Trace the yield of a common crop in your region over a selected number of years. Have yield trends shown an increase or a decrease? What magnitude of change has occurred?

2. Can you find any evidence that yields are leveling off in your region after a period of steady increases?

3. Can you suggest any evidence that higher yields can be achieved?

4. List factors that can be said to have a positive impact on crop production.

5. Give examples of crops or production inputs where quality considerations take priority over high yield.

6. Single-cross hybrid corn cultivars that offer yield ability are often preferred over four-way cross hybrid corn cultivars that offer yield stability. Discuss the potential consequences and merits of a trend toward single-cross hybrid corn cultivar production.

7. Crop production has been referred to as being more of an art than a science. Discuss this concept in the light of the material presented in this chapter.

8. Do crops produced in your area have wide adaptation or yield stability? A cultivar grown extensively over a wide geographic region or one that has been grown for many years may be regarded as having yield stability. Explain why this may or may not be true.

9. Identify why higher yields are not produced in your area. Can the limiting factors be eliminated?

10. Suggest various ways in which the record yields reported in this chapter may have been achieved.

REFERENCES FOR CHAPTER THREE

Anonymous. April 1978. "A 352 Bushel Corn Crop Made Him World Champion," *Cash Crop Farming* pp. 36-37.

Anonymous. 1973. Production and Marketing Branch Description of Variety, Hinoat. Agriculture Canada License No. 1458.

Anonymous. 1966. "New Record Wheat Yields 216 Bushels per Acre," *Farm Journal* 90:42.

Aufkammer, G., and U. Simon. 1957. "Die Samen landwirtschaftlicker Kulturpflanzen im Grundstein des ehemalizer Nurnberger Stadttheaters und ikre Keimfahigkeit," *Zeitschrift fur Acker-u Pflanzenbau* 103:454-472.

Blackman, G. E., and J. N. Black. 1959. "Physiological and Ecological Studies in the Analysis of Plant Environment XII. The Role of the Light Factor in Limiting Growth," *Annals Botany* 23:131-145.

Cannon, G. G. 1963. "Baron Justus Von Liebig," *Great Men of Modern Agriculture*, pp. 148-159. New York: Macmillan Publishing Co., Inc.

Jensen, N. F. July 1978. "Limits to Growth in World Food Production," *Science* 201:317-320.

Lawrence, T. 1978. "An Evaluation of Thirty Grass Populations as Forage Crops for Southwestern Saskatchewan," *Canadian Journal Plant Sci.* 58:107-115.

Mitchell, R. L. 1970. Crop Growth and Culture. Ames, Iowa: Iowa State University Press.

Pittman, U. J., and K. H. Tipples. 1978. "Survival, Yield, Protein Content and Baking Quality of Hard Red Winter Wheats Grown under Various Fertilizer Practices in Southern Alberta," *Canadian Journal Plant Sci.* 58:1049-1060.

Reid, D. 1978. "The Effects of Frequency of Defoliation on the Yield Response of a Perennial Ryegrass Sward to a Wide Range of Nitrogen Application Rates," *Journal of Agricultural Science. (Camb.)* 90:447-457.

Tharp, W. H. 1960. *The Cotton Plant. How It Grows and Why Its Growth Varies.* Agriculture Handbook No. 178. Washington, D.C.: USDA.

Waggoner, P. E., and W. A. Norwell. 1979. "Fitting the Law of the Minimum to Fertilizer Applications and Crop Yields," *Agronomy Journal.* 71:352-354.

chapter four

Yield and Duration of Photosynthesis

Photosynthesis is the basis of all crop yield, and since a full-season crop has a longer duration of photosynthesis than a short-season crop, the former can be expected to outyield the latter. In areas where spring and fall temperatures delineate the growing season, crops that utilize the full photosynthetic period are full-season crops in contrast to short-season crops such as cereals that mature partway through the growing season. This chapter considers aspects of photosynthetic duration from the standpoint of understanding yield, production practices that maximize photosynthetic duration, and the basis of choosing crop species and cultivars for photosynthetic advantage.

LENGTH OF THE GROWING PERIOD

The corn crop may be regarded as a full-season crop. In northern areas it may be seeded on May 15, after the last spring frost, and it is capable of reaching physiological maturity by September 10, hopefully prior to the first fall frost. **Physiological maturity** is the point in the development of a crop at which no further dry matter is added. The duration of photosynthesis for this corn crop may be considered as follows:

Example A—Duration of photosynthesis of full-season corn crop

  8 days in May—seeded May 15 with emergence on May 22
      when photosynthesis begins
 30 days in June
 31 days in July
 31 days in August
 <u>10</u> days in September to physiological maturity
110 days total

55

In a corn crop that does not utilize the full season because of delayed seeding or early maturity, the duration of photosynthesis may be as follows:

Example B—Duration of photosynthesis of short-season corn crop

Normal Seeding

  8 days in May as in Example A
30 days in June
31 days in July
<u>23</u> days in August to physiological maturity
92 days total

Delayed Seeding

30 days in June—seeded May 24 followed by emergence
        about June 1 when photosynthesis begins
31 days in July
<u>31</u> days in August to physiological maturity
92 days total

The difference between a full-season and a short-season corn hybrid cultivar according to these examples is 18 days or a 20% difference in the duration of photosynthesis. If the full-season hybrid yielded 6200 kg/ha (100 bu/ac), the short-season hybrid cultivar may yield about 20% less or 4960 kg/ha (80 bu/ac).

Under actual practice this situation may be realistic. When the yield results of the Hybrid Corn Performance Trials[1] were compared, hybrid cultivars tested in areas of Ontario with a high **corn heat unit** rating and hence long growing seasons generally outyielded hybrid cultivars with a lower corn heat unit rating (Table 4–1). The results in this table represent the mean yields of an average of 238 hybrid cultivars tested each year for three years and support the general conclusion that a full-season crop is likely to outyield a short-season crop. A corn producer in the 2800–3000 corn heat unit zone could select a hybrid cultivar with a lower than recommended corn heat unit rating. This would provide some assurance to the producer that the crop will mature, that it can be harvested early in the fall, and that drying is less of a problem than with full-season crop. But because grain yield can be expected to be less than the full-season crop, the full-season hybrid cultivar is invariably preferred. If a corn hybrid cultivar is produced that is rated as requiring more corn heat units than is normally available, the producer is gambling that the crop will reach physiological maturity before the fall frosts. Immature corn may be unsatisfactory for grain production and may require considerable energy and expense to reduce moisture to a level at which it can be safely stored. Producers of full-

[1]Standardized corn tests conducted by trained personnel at selected sites in each of the heat unit zones.

season crops must choose cultivars that will utilize the full growing season but will reach physiological maturity with some degree of certainty.

TABLE 4-1. Corn grain yield comparisons among seven maturity zones according to the corn heat units available in Ontario for three years

Maturity Group According to Corn Heat Unit Rating	Corn Yield (t/ha) or (tons/ac)							
	1977		1978		1979		Mean	
	t/ha	t/ac	t/ha	t/ac	t/ha	t/ac	t/ha	t/ac
2400–2500	6.27	2.82	5.40	2.43	5.15	2.32	5.67	2.55
2500–2700	8.97	4.04	7.34	3.30	6.03	2.17	7.44	3.35
2700–2900	7.78	3.50	7.78	3.50	7.85	3.53	7.80	3.51
2800–3000	7.84	3.53	7.34	3.30	7.34	3.30	7.31	3.29
2900–3100	8.03	3.61	8.16	3.67	8.41	3.78	8.20	3.69
3100–3400	9.95	4.48	9.29	4.18	9.43	4.24	9.56	4.30
3300–3500	8.90	4.01	5.71	2.57	10.02	4.51	8.21	3.69

Source: Reports of the 1977, 1978 and 1979 Ontario Hybrid Corn Performance Trials conducted by the Ontario Corn Committee.

Canola (rapeseed) is a crop capable of effectively utilizing the full, although short, growing season in the northern fringe area of crop production on the Canadian prairies. Canola (rapeseed) can mature in as little as 90 days and can tolerate cool temperatures; it has become Canada's major oilseed crop. Canola (rapeseed) refers to two species: Polish or turnip rape, and Argentine rape, with winter cultivars available from both. However, these winter cultivars are not hardy enough to survive Canadian winters. Spring Argentine types are produced in the northern prairie region. The Polish or turnip rape tolerates frosts in the seedling stage better but is lower yielding. Turnip or Polish cultivars mature earliest and are produced in regions with a marginal season suitable for the crop.

DURATION OF VEGETATIVE AND REPRODUCTIVE PERIODS

In addition to the total period or duration of photosynthesis, the contribution of photosynthesis to yield must be considered in terms of the duration of vegetative and reproductive periods. In some crops a definite division between the vegetative and reproductive periods can be found. These are called **determinate** species, and in such crops the vegetative growth period begins when the seed germinates and continues until **fertilization** occurs, the point at which a **pollen** grain unites with the **egg**. It is during the vegetative period that roots, leaves, and stems are produced and reproductive organs developed. After fertilization, vegetative growth ceases, and the reproductive period begins (Figure 4-1). Photosynthate produced during the reproductive period is no longer needed for vegetative growth and can be used to produce economic or grain yield. In many crops it is maximum reproductive development or economic yield that is desired.

Morphological stages in the development of a determinate crop and the relative length of the stages are shown in Figure 4-1. During the seedling or juvenile stage, the plant is small, sunlight interception is incomplete, and the amount of photosynthate produced is low. The vegetative growth period is essential to develop a plant capable of utilizing sunlight energy to best advantage and to provide time for reproductive development. A vegetative growth period of excessive duration may lead to excessive leaf growth and premature shading and yellowing of lower leaves (senescence) and allows only a short period of time for reproductive development. Although the vegetative period is essential in plants, it is during the reproductive period that economic yield is achieved. In a rice experiment noted by Yoshida (1972), 68% of the grain yield was from photosynthate produced during the reproductive period, 20% was respired during the reproductive period, and 12% was used to maintain the plant. About 20% of the carbohydrate in the rice grain was translocated from photosynthate produced and stored in vegetative parts prior to the onset of reproductive development. The suggestion is that yield can be increased if food energy availability during the reproductive period is increased.

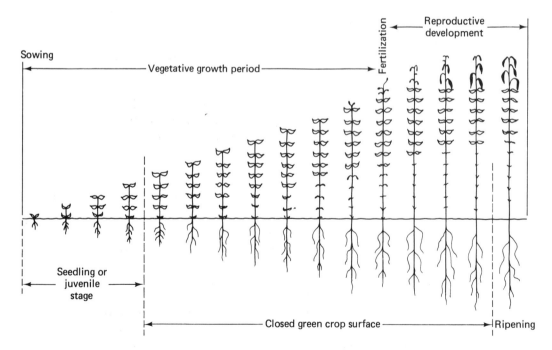

Figure 4-1. Schematic presentation of the developmental stages of a determinate crop plant.

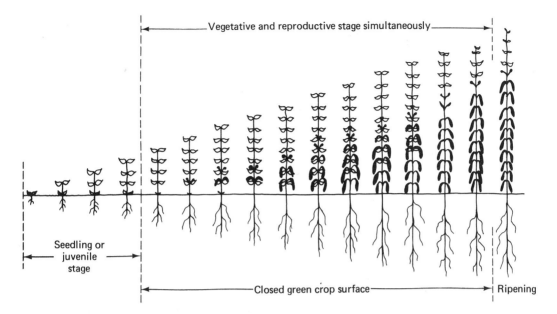

Figure 4–2. Schematic presentation of the developmental stages of an indeterminate crop plant.

In contrast to crops that have distinct vegetative and reproductive periods (determinate), some crops have overlapping vegetative and reproductive periods and are termed **indeterminate** plants. Flowers and reproductive development may occur in the lower leaf axils even while new leaves and stems are being formed (Figure 4–2). Northern strains of soybean, canola (rapeseed), faba beans, field peas, and forage legumes show indeterminate growth, and the flowering period may extend over several weeks. The length of the seed development or reproductive period in soybeans (estimated as the period when the pods are 2 cm (¾ in) in length at the top four nodes to the time when the pods start to yellow) was most closely related to seed yield when variations among 119 soybean cultivars were evaluated over a three-year period (Dunphy, Hanway, and Green 1979). Because seed pods may be at varying stages, precise stages for harvesting indeterminate plants are less apparent than with determinate crops. In birdsfoot trefoil, for example, reproductive development may vary from overripe seed pods to flower buds, which complicates harvesting for seed production.

Considerable attention is often paid to early vegetative plant development. Although this period is important to later plant development, the importance of the reproductive period should not be overlooked.

Consider the total 110-day duration of photosynthesis in Example A in terms of the vegetative and reproductive periods. Corn is a determinate crop, and fertilization can be expected to occur in early August. The vegetative growth period and the reproductive growth period for corn may be as follows:

Example C—Duration of vegetative growth period for corn

8 days in May—seeded May 15 followed by emergence
    about May 22 when photosynthesis begins
30 days in June
31 days in July
69 days total

Example D—Duration of reproductive growth period for corn

31 days in August
10 days in September to reach physiological maturity
41 days total

The importance of the duration of reproductive development to grain yield was determined by Daynard, Tanner, and Duncan (1971). The "effective filling period duration" was estimated among corn hybrid cultivars differing in grain yield over a two-year period (Table 4-2). The effective filling period duration varied from 28 to 36 days rather than 41 days as in Example D. In 1966, Pride 4 had the lowest yield at 4790 kg/ha (76.4 bu/ac) and had the shortest effective filling period duration of 28 days. An average yield increase of 171 kg/ha (2.7 bu/ac) was produced each day. United 108 had the highest yield of 6360 kg/ha (101.4 bu/ac) an effective filling period duration of 36 days, and an average yield increase of 176 kg/ha (2.8 bu/ac) each day. In 1977, Pride 4 was again the lowest yielding, producing an average yield increase of

TABLE 4–2.   Grain yield effective filling period duration for selected hybrid cultivars in 1966 and 1967 at Guelph, Ontario

Hybrid Corn Cultivar	Effective Filling Period Duration (days)	Grain Yield (kg/ha)	(bu/ac)
1966 Results			
Pride 4	28	4790	76.4
Pride 5	31	6090	97.1
United 108	36	6360	101.4
1977 Results			
Pride 4	28	5560	88.6
Pride 5	28	6200	100.0
Ox 306	32	6710	107.0
United 108	32	6810	108.6
United 10	33	6740	107.5
Funks G43	34	6810	108.6

Source: Data from Daynard, Tanner, and Duncan 1971.

199 kg/ha (3.17 bu/ac) each day. Funks G43 was among the highest yielding hybrid cultivars and at an average yield increase of 200 kg/ha (3.18 bu/ac) per day was similar to Pride 4. Clearly, duration of the filling period is a major factor for yield differences among these corn cultivars.

In wheat, Spiertz, ten Hag, and Kupers (1971) found that 61 to 83% of the variation in grain yield could be predicted from the duration of photosynthesis of the flag leaf and sheath. The flag leaf is the final uppermost leaf to develop on cereal plants and is formed just prior to fertilization. Photosynthetic food energy produced in this leaf has generally been observed to be of major importance as a source of food energy to fill the grain.

In the peanut crop, new cultivars developed in the past 40 years have more than doubled the yield potential of that crop. An analysis by Duncan et al. (1978) revealed that the division of food energy between vegetative and reproductive parts and the length of the filling or reproductive period accounted for most of the yield differences found among the five cultivars examined.

Grain filling rate has been suggested as a further refinement to grain filling duration (Daynard, Tanner, and Duncan 1971; Johnson and Tanner 1972; Carter and Poneleit 1973; Cross 1975; Thomas and Raper 1976). This factor may account for variations in yield in Table 4-2 not accounted for by duration alone. Grain filling rate may be regarded as a measure of efficiency within the plant; it helps explain yield differences, but in the studies reported, duration of the grain filling period was related more closely to yield than filling rate. In biology, exceptions are likely to exist.

A striking demonstration of grain filling rate was presented by Jones, Peterson, and Geng (1979) with rice. In fifteen rice cultivars, grain filling rate was found to be of greater consequence than grain filling duration (Table 4-3). Grain filling duration varied from 31.0 to 36.9 days and was a factor in yield. Mean filling rate varied from 54.0 to 90.9 mg per head per day and because of this large variation was also a factor in yield differences.

TABLE 4-3. Grain filling parameters for rice in California in 1975

Rice Identification Number	Grain Filling Duration (days)	Mean Filling Rate (mg/day) per Head	Yield— Weight per Head in Grams
No. 7	34.4	54.0	1.86
No. 3	31.0	62.9	1.95
No. 11	36.9	69.7	2.57
No. 9	34.5	68.1	2.35
No. 13	35.0	75.4	2.64
No. 5	36.7	78.6	2.88
No. 2	35.0	90.9	3.18
Mean of 15 rice cultivars	34.9	68.5	2.40

Source: Data from Jones, Peterson, and Geng 1979.

In oats, the time between **anthesis** and physiological maturity was considered by McKee et al. (1979) to be the "fill period" or the reproductive development stage. The duration of this period and the rate of fill for nine oat cultivars in Pennsylvania were evaluated (Table 4-4).

Grain filling rate may provide an additional measure of physiological efficiency, but it is a process that cannot be visually evaluated or influenced by a crop producer. Its use to a crop producer is to provide an insight, an understanding of why yield variations among cultivars are found.

Both duration and rate of grain filling are genetically controlled factors; the expression of both, however, may be influenced by the environment. Direct modification of these two yield-influencing factors is beyond the control of the producer. Appreciation of how and when yield is produced is of value to critically evaluate cultivars and to select a cultivar most suited to a particular set of farm conditions. Duration of the reproductive period is a discrete entity that can be evaluated by a producer. The time of fertilization is readily apparent by the extrusion of anthers immediately following fertilization (Figure 4-3) and in determinate plants marks the beginning of the reproductive period. The end of the reproductive period coincides with physiological maturity. Crop producers should have an understanding of when completion of grain filling occurs.

PHYSIOLOGICAL MATURITY

Physiological maturity is defined as the completion of the reproductive period, the point at which no further increase in dry weight takes place. Following physiological maturity, the economic yield dries down as moisture

TABLE 4-4. Length of the fill period, rate of fill, and grain yield of nine oat cultivars in Pennsylvania

Cultivar	Fill Period (days)		Daily Rate of Fill 1975		Grain Yield 1975	
	1975	1976	(kg/ha)	(bu/ac)	(t/ha)	(bu/ac)
Astro	23	25	99	2.6	2.29	60.6
Clintford	21	24	98	2.6	2.05	54.2
Dal	23	—	92	2.4	2.12	56.1
Mariner	24	24	88	2.3	2.11	55.8
Noble	23	24	90	2.4	2.08	55.0
Orbit	24	25	100	2.6	2.40	63.5
Otee	23	23	93	2.4	2.14	56.6
Pennfield	22	26	97	2.5	2.14	56.6
Stout	20	23	98	2.6	1.96	51.9
Mean	23	24	95	2.5	2.14	56.6

Source: Data from McKee et al. 1979.

Figure 4–3. A head of bromegrass at anthesis. The anthers are extruded from the florets and fertilization has just occurred. In determinate plants, fertilization marks the end of vegetative growth and the beginning of reproductive development. (Courtesy of Ontario Ministry of Agriculture and Food.)

is lost as part of the ripening process. At the time of physiological maturity the grain may appear soft and immature, and the plant remains green and succulent. At this point the plant can be harvested as the flow of moisture is severed.

Crop producers may wish to identify physiological maturity not only to compare the duration of reproductive periods among cultivars but also to establish a time to harvest. Although further drying will occur after physiological maturity, the crop can be harvested as wet grain to be stored as high moisture silage, or be left to dry in the swath, or be harvested immediately to avoid excessive shattering with direct combining, a problem commom to canola (rapeseed).

Physiological maturity of any crop can be determined by a systematic sampling and dry-weight determination. When no further increase in grain dry weight is observed, the plant is said to have reached physiological maturity. This method is laborious and time-consuming. A more simple and accurate technique is found in corn. Physiological maturity is reached when the "black layer" develops.

Black layer development in corn can be likened to a suberized barrier around the base of the seed (Figure 4-4), a layer that can be observed by the naked eye to develop in three days or less once physiological maturity is reached. The formation of a black layer disrupts further flow of assimilate into the kernel and represents a precise, accurate, and simple indicator that maximum kernel dry weight has been achieved (Daynard and Duncan 1969; Carter and Poneleit 1973).

No such recognizable indicator exists in cereals to help identify physiological maturity. One method discussed by Lee, McKee, and Knievel (1979) was to use a water-soluble dye sprayed on oat leaves, which indicated physiological maturity when dye uptake into kernels ceased. Thus, crop producers can, through experience, become adept at identifying physiological maturity in various crops. The soybeans in Figure 4-5 are approaching maturity.

The following example is included to help illustrate how the knowledge and understanding of a physiological factor(s) can help explain unexpected yield variations. In 1977, corn producers in many areas of North America experienced record high yields, a situation shown for Ontario in Table 4-5. In

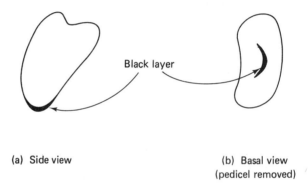

Black layer

(a) Side view

(b) Basal view
(pedicel removed)

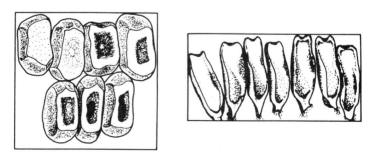

Figure 4-4.   Sketches of side (a) and basal (b) views of corn kernels showing black layer, an indicator of physiological maturity.

Figure 4-5. The soybeans in this photo are approaching maturity as indicated by leaf drop and a browning and drying of stems and pods. Harvesting could take place at physiological maturity, the point at which no further increase in dry weight occurs without a yield loss. Harvesting may be delayed to allow moisture levels to be reduced to minimize drying costs. (Courtesy Ontario Ministry of Agriculture and Food.)

response to the question of why the 1977 corn yields were the highest on record, most answers credited "favorable environmental conditions" as the reason. More specifically, however, favorable weather encouraged rapid vegetative growth during May and June. Rapid vegetative growth and a temporary drought in July produced stress conditions that stimulated the onset of reproductive development. The effective filling period duration was lengthened by about two weeks. This more than any single factor appeared to contribute to the record yield in 1977. Daynard, Tanner, and Duncan (1971) had suggested that for every effective day that the reproductive period can be extended, a yield increase of approximately 3% could result. A fourteen-day advantage could possibly have increased grain yield 42% (3% per day × 14 days). Indeed, many producers who normally anticipated a 6200 kg/ha (100 bu/ac) yield did produce yields 42% higher or 8800 kg/ha (140 bu/ac).

Crop production practices such as plant population, fertilizer availability, seeding date, and disease and insect control must be aimed at maximizing photosynthesis not just during vegetative growth but during reproductive growth, a period often neglected and ignored. Premature loss of leaves from

TABLE 4-5.   Corn grain yield comparisons in
Ontario for a four-year period

Year	Corn Grain Yield	
	(kg/ha)	(bu/ac)
1976	5712	91.1
1977	6384	101.8
1978	5100	81.3
1979	5700	90.9

Source: 1980 Report of the Ontario Corn Committee.

age or yellowing (senescence) from low soil fertility during reproductive growth should be avoided as food energy may be reduced. Concern about premature senescence in wheat led Patterson and Moss (1979) to postulate that higher yields might be achieved under conditions that reduced the rate of leaf senescence, thereby extending photosynthetic duration during reproductive development. In wheat, photosynthetic activity of a leaf canopy was found to reach a maximum after anthesis and declined rapidly thereafter.

DURATION OF PHOTOSYNTHESIS AS INFLUENCED BY DAY LENGTH

The amount of light received by plants for photosynthesis is determined by the intensity of light, the length of the growing season, and the day length. Day length has a profound effect not only on photosynthesis but on plant development. Considerable variation in day length can be found. At 60° north latitude for example, the day length is of 5 hours and 52 minutes duration on the shortest day of the year (December 21), although this date is not considered as part of the growing season for outdoor plants. On the longest day of the year (June 21) at 60° north latitude, day length is of 18 hours and 54 minutes duration. At the equator, day length remains a constant 12 hours and 7 minutes (Table 4-6).

Light intensity at various latitudes may vary according to the angle of incidence of the sun's rays. Long days may compensate, however, for a short growing season and low light intensity. At Beaverlodge, Alberta, for example, located at 55° north latitude, the total amount of sunlight received during a relatively short growing season is about the same as at Madison, Wisconsin, located at 43° north latitude and with a longer growing season (Carder 1957). The effective heat, however, at Beaverlodge was about 60% of that at Madison, Wisconsin.

In tests to compare plant growth and development at the two locations, nine cultivars of wheat, oats, barley, and peas were grown satisfactorily at the two sites. Cool temperatures at Beaverlodge resulted in slow and limited vegetative growth of the millet plants. Millet, however, proliferated vegetatively and did

TABLE 4–6.   Approximate length of day on various dates at different latitudes north of the equator

Latitude (North)	December 21 (winter solstice)		March 21 (spring equinox)		April 21		May 21		June 21 (summer solstice)		July 21		August 21		September 21 (autumn equinox)	
	(hr)	(min)	(hr)	(min)	(hr)	(min)	(hr)	(min)	(hr)	(min)	(hr)	(min)	(hr)	(min)	(hr)	(min)
0°	12	7	12	7	12	7	12	7	12	7	12	7	12	7	12	7
10°	11	33	12	7	12	24	12	37	12	43	12	38	12	24	12	8
20°	10	56	12	9	12	42	13	9	13	19	13	11	12	43	12	10
25°	10	35	12	10	12	53	13	28	13	42	13	29	12	54	12	10
30°	10	11	12	10	13	4	13	47	14	4	13	48	13	6	12	12
35°	9	48	12	12	13	17	14	9	14	32	14	12	13	18	12	12
40°	9	20	12	12	13	31	14	35	15	2	14	36	13	32	12	13
45°	8	46	12	14	13	48	15	4	15	38	15	6	13	49	12	15
50°	8	4	12	15	14	9	15	41	16	24	15	44	14	9	12	17
55°	7	10	12	17	14	34	16	29	17	24	16	33	14	35	12	19
60°	5	52	12	18	15	8	17	38	18	54	17	41	15	9	12	21

not reach reproductive development. Temperature differences at the two sites were less pronounced during reproductive development and hence were favorable for photosynthesis and grain yield development in all crops but the millet. Higher wheat yields were produced at Beaverlodge than at Madison, but in every instance a more favorable grain-to-straw ratio was found at Beaverlodge (Table 4–7). Generally, grain yields at Beaverlodge were associated with much less vegetative growth.

It is interesting to speculate on the influence of light duration and intensity on crop production at the equator, at Madison, and at Beaverlodge; but direct comparisons cannot be made because of differences in crop cultivars, soil fertility, effective rainfall, temperature, and other variables. It is tempting to conclude that because of the long growing season and warm temperature conditions at the equator, crop production would be more favorable there. On the contrary, in terms of sunlight energy for photosynthesis, a 100-day crop at Beaverlodge, growing during May to late August, is subjected to nearly 18 hours of sunlight per day; at Madison approximately 14.5 hours; and at the equator, 12 hours. The hours of photosynthesis for a 100-day crop would be 1800, 1450, and 1200 at Beaverlodge, Madison, and the equator, respectively. Thus, 150 days would be required at the equator to equal the period of light duration at Beaverlodge. Best (1962) reported that total radiation received by annual crops in the temperate zone may be 1.5 times as high as in the tropics and suggested that yields correspond.

TABLE 4–7.   Yield of grain and ratio of grain to straw of nine crop cultivars grown at Madison, Wisconsin and Beaverlodge, Alberta in 1952

Crop and Cultivar	Yield of Grain per 3m/row (oz/10 ft of Row)				Ratio of grain to straw	
	Beaverlodge		Madison		Beaverlodge	Madison
	g	oz	g	oz		
Wheat—Gaines	241	25.9	173	18.6	103.9	40.0
Red Bobs	227	24.4	107	11.5	86.3	27.3
Marquis	207	22.3	120	12.9	73.9	27.0
Oats—Clinton	202	21.7	340	36.6	69.9	62.6
Beaver	260	28.0	343	36.9	92.5	62.3
Victory	325	35.0	303	32.6	105.9	48.2
Barley—Olli	224	24.1	355	38.2	147.4	96.7
Montcalm	262	28.2	365	39.3	104.0	77.8
Peas—Chancellor	208	22.4	506	54.4	85.6	45.0
Mean	240	24.1	290	31.2	96.6	55.2

Source: Data from Carder 1957.

The importance of day length to photosynthesis means that crop producers should seed crops early to be able to benefit from the long days of June. Perhaps even more importantly, if reproductive development can be made to coincide with the June period of high sunlight, energy yields may be increased. Such a situation does exist with winter annuals (winter cereals), perennial forages, and early seeded annuals. In plants such as cereals and faba beans, that can tolerate spring frosts, early seeding is imperative to benefit from the light duration in June. Frost-sensitive crops such as soybeans and corn cannot be seeded, however, until the danger of spring frost has passed.

DAY LENGTH AND REPRODUCTIVE DEVELOPMENT

The reproductive development of plants is influenced by the relative lengths of day and night (Garner and Allard 1929). Plant reactions to day length are called *photoperiodism* or *photoperiodic responses*. For some plants to reach reproductive development a relatively long day (14 hours) is required, and such plants are called *long-day plants*. Wheat, oats, barley, rye (not rice), canola (rapeseed), and some forages (red clover) respond to the long days of June in the northern and southern hemispheres. Under short-day conditions, such plants remain vegetative. After harvest in late August in areas where days are getting shorter, volunteer cereal plants—those that grow from the seed lost that season—may remain vegetative until destroyed by fall frosts.

In contrast, short-day plants require a short day for reproductive development to be induced. Short-day crops include corn, soybeans, rice, millet, and sorghum. Short-day crops remain vegetative when grown under conditions of long days.

A third category of crops is termed *day-neutral,* and these crops will reach reproductive development under long- or short-day conditions. The widespread and free exchange of seed around the world has resulted in strong selection pressure toward cultivars rather indifferent to day length. Corn, for example, is regarded as a short-day crop but is adapted to areas with long days because of selection against photoperiodic sensitivity.

Photoperiodic sensitivity provides the explanation why crop response is considered unusual when photoperiodically sensitive seeds are exchanged among diverse latitudes. Short-day corn cultivars from the equator will grow vegetatively under a long-day photoperiod to a height of 3 to 4 metres before day length becomes sufficiently short late in the season for reproductive development to be initiated (Figure 4-6). Troyer and Brown (1972, 1976) observed that most corn cultivars from the tropics and subtropics developed excessive vegetative growth and exhibited delayed floral initiation when grown in a long-day photoperiod.

When corn plants adapted to long-day conditions of a high latitude are grown under short-day conditions near the equator, they are stimulated to reproductive development early in the seedling stage. The plants are of small stature and produce a small ear.

A practical example of the problems encountered by day-length sensitivity was demonstrated in an oat cultivar developed at Ottawa that was sensitive to long-day conditions. In an attempt to multiply seed under the relatively short-day conditions in northwestern Mexico, the oat cultivar from Ottawa was

Figure 4-6. Under long-day conditions in the northern hemisphere, the photo-periodically sensitive short-day corn in the background remained vegetative until late in the season. The corn in the foreground is photoperiodically insensitive and developed normally even under long-day conditions. (Courtesy Ontario Ministry of Agriculture and Food.)

seeded in October and remained vegetative until the lengthening days of March induced reproductive growth. Under such conditions an excessive vegetative growth of over 2 metres in height, was produced and reproductive growth and grain yield were limited in volume and quality.

The discovery that a critical determinate in the photoperiodic reaction is the length of the dark period (Borthwick and Hendricks 1960) led to a solution on a small-scale basis for the handling of long-day crops in tropical and semitropical regions. By interrupting the dark period for a period of about one hour in the middle of the night with outdoor lights, reproductive development can be triggered. The lighting system used may be similar to that used to light an outdoor sports park.

Although the length of the dark period is more critical to plants than the length of the light period in terms of reproductive development, the terms *short-* and *long-day plants* continue to be used.

UTILIZING PHOTOPERIODIC SENSITIVITY

Photoperiodic sensitivity is a critical determinate for crop producers not only for the exchange of seed but also for crop production in their region because it can influence the photosynthetic duration of the vegetative and reproductive periods.

In the production of long-day cereal crops under long-day conditions, early seeding is very important. Early seeded cereals (April 15) will almost invariably outyield a late (May 25) seeded crop. Despite a 40-day difference in seeding dates, maturity will only show a difference of from 5 to 7 days. Because of photoperiodic sensitivity, reproductive growth in the early seeded crop will begin about June 25 and in the late seeded crop about July 10. The developmental regime could be illustrated as follows:

Example E—Duration of vegetative growth
April 15 Seeding

April— 8 days (emergence April 22)
May —30 days
June —25 days (onset of reproductive development)
     63 days total

May 25 Seeding

Emergence June 1
June—30 days
July —10 days (onset of reproductive development)
     40 days total

Example F—Duration of reproductive development
April 15 Seeding

June— 5 days
July —31 days
Aug.— 1 day (physiological maturity)
     37 days total

May 25 Seeding

July —21 days
Aug.— 5 days (physiological maturity)
     25 days total

The early seeded crop has an 11-day advantage of reproductive development to provide food energy to produce grain. If the 63-day period of vegetative development is considered excessive, the 40-day period in the late seeded crop may be inadequate. Under cool early spring conditions, growth is slow and root development is encouraged, tillers are initiated, and well-differentiated cells result in a short, lodging-resistant plant.

Additional reasons for yield differences may exist. An early seeded crop may escape disease, drought, and hot weather, common in mid- and late summer.

Fall-seeded crops such as winter wheat or barley may normally be expected to yield more than spring-seeded cereals because of an extended reproductive period. Winter wheat in Ontario may reach reproductive development in early June and mature about July 25. Not only is the reproductive period of 45 to 50 days duration longer than the 25 to 40 days in Examples E or F, but reproductive development coincides precisely with the longest days of the season. Photosynthetic potential is high, and grain yield is also potentially high.

To maximize food energy produced during the relatively short growing season of the northern prairie region in Canada, canola (rapeseed) may be pelleted or encapsulated in material that prevents the uptake of moisture in the fall. Pelleted seed can be fall planted but germination in the fall is prevented. Frost and weathering action during the winter cause the pelleting material to decompose to allow for spring germination. Plants may be actively photosynthesizing by the time normal seeding could take place. Fall seeding with pelleted seed not only allows for maximum use of a short growing season but also allows for utilization of sunlight energy during the long days of June.

SINK-SOURCE RELATIONSHIPS

No discussion of the importance of photosynthetic duration of vegetative and reproductive development would be complete without reference to the **sink-source** relationship. Controversy exists whether yield is limited by the size of the photosynthetic factory, referred to as the *source,* or by the size of reproductive organs, referred to as the *sink.* Consider Example C. Did a 69-day duration of vegetative growth produce a larger set of ears (sink) than could be filled in a 41-day period of reproductive development (the source)? In contrast was the size of the sink a limiting factor for cereals when seeded on May 25 in Example E?

In theory, in the corn crop seeded May 15, a switch from vegetative to reproductive development of Example C by July 10 could result in a favorable yield situation as follows:

Example G—Duration of vegetative growth
May— 8 days for photosynthesis
June—30 days
July —10 days to plant fertilization
    48 days total (versus 69 in Example C)

Example H—Duration of reproductive growth
July   21 days
Aug. 31 days
Sept. 10 days
      62 days total (compared to 41 in Example D)

Perhaps such a drastic change, if it is even possible, could result in small plants, and small ears and with a photosynthetic area incapable of filling a normal sized ear. However, ear size per plant is not the most important consideration, but rather yield per hectare (acre). Small plants may tolerate high population densities.

At a plant population of 38,400 pph (15,546 ppa) (plants per hectare or acre) with large plants, each plant may produce 228 grams (½ lb) of grain. Grain yield per hectare would be:

$$\frac{38,400 \text{ pph (15,500 ppa)}}{} \times \frac{228 \text{ grams (½ lb) grain per plant}}{1000} = 8755 \text{ kg/ha}$$

$$(141 \text{ bu/ac}$$
$$\text{or}$$
$$7817 \text{ lb/ac})$$

If ears were half as large, the same yield might be achieved with twice the plant population of 76,800 pph (31,000 ppa), a situation considered realistic.

On the theory that many small corn plants per hectare may be more productive than fewer large plants, Major et al. (1973) tested the performance of 44 single-cross hybrids and their component inbred lines over a wide range of plant populations. The small inbred lines yielded only 60% of the large-sized hybrid cultivars, suggesting that the source may be a limiting factor.

On a purely theoretical basis and on the premise that sink (reproductive period) was the limiting factor, Army and Greer (1967) suggested that if the grain filling period was increased from the present 41 days (Example D) to possibly 75 days, along with satisfactory plant features and cultural practices, a yield potential of over 40,000 kg/ha (600 bu/ac) might exist. They also theorized that in the southern United States where a longer growing season exists, a yield of over 67,000 kg/ha (over 1000 bu/a) might be possible.

Evidence exists that in cereal and corn crops, if all the kernels initiated acted as a sink to store food energy, under most environmental conditions the photosynthetic source may be inadequate to fill all the kernels. The result could be incomplete development of all seeds. Black layer development may cut off the flow of assimilate into some of the newly developed kernels, possibly as a survival mechanism (Daynard and Duncan 1969). Some kernels thus abort, whereas others develop fully according to the food energy available. Blast in oats is considered a form of abortion, and occurs in response to the development of a larger sink than can be filled. In wheat, the number of kernels per spikelet may vary from two to six depending on the food energy available.

In a study on barley, neither the source nor the sink presented an overriding limitation to grain yield under four plant populations and other treatments. Moreover, photosynthetic rate may adjust to the requirements of the grain (King, Wardlaw, and Evans 1967; Evans and Rawson 1970). Perhaps both sink and source may be the limitation to yield, but under some circumstances either may predominate (Bingham 1969; Gifford, Bremner, and Jones 1973).

QUESTIONS

1. Explain why growers generally try to select the latest maturing cultivar of a full-season crop that they feel will mature in their region. Can such action be justified?

2. Describe the growth and development of a long-day plant grown at the equator.

3. Describe the growth and development of a short-day plant grown at the equator.

4. Describe the growth and development of a short-day plant at a 40° north latitude.

5. Describe the growth and development of a long-day plant at a 40° north latitude.

6. Speculate why millet did not reach reproductive development when grown at Beaverlodge, Alberta.

7. On the basis of the sink-source relationship, consider the merits of prolific corn, i.e., a corn capable of producing more than one ear per stalk.

8. Discuss how a corn grain yield of 40,000 to 67,000 kg/ha (600 to 1,000 bu/ac) might be produced. Consider the level of nutrient fertilization, dry matter produced per day, photosynthetic efficiency, and related production problems and approaches.

9. Record the dates of seeding, emergence, vegetative growth period, reproductive growth duration, and date of physiological maturity for a crop in your region and relate events to final economic yield produced. Compare crops in your area with the examples in this chapter.

10. In crops where the economic yield is periodically picked, such as in cucumbers, tomatoes, and eggplants, speculate what effect the picking action has on sink-source relationships.

REFERENCES FOR CHAPTER FOUR

Army, T. J., and F. A. Greer. 1967. "Photosynthesis and Crop Production Systems," in *Harvesting the Sun*, pp. 321–332, A. S. Pietro, F. A. Greer, and T. J. Army, eds. New York: Academic Press, Inc.

Best, R. 1962. "Production Factors in the Tropics," *Netherlands Journal of Agricultural Science* 10:347-353. (special issue)

Bingham, J. 1969. "The Physiological Determinants of Grain Yield in Cereals," *Agricultural Progress* 44:30-42.

Borthwick, H. A., and S. B. Hendricks. 1960. "Photoperiodism in Plants," *Science* 132(3435):1223-1228.

Carder, A. C. 1957. "Growth and Development of Some Field Crops as Influenced by Climatic Phenomena at Two Diverse Latitudes, " *Canadian Journal Plant Science* 37:392-406.

Carter, M. W., and C. B. Poneleit. 1973. "Black Layer Maturity and Filling Period Variation among Inbred Lines of Corn (*Zea mays* L.)," *Crop Science* 13:436-439.

Cross, H. Z. 1975. "Diallel Analysis of Duration and Rate of Grain Filling of Seven Inbred Lines of Corn," *Crop Science* 15:532-535.

Daynard, T. B., and W. G. Duncan. 1969. "The Black Layer and Grain Maturity in Corn," *Crop Science* 9:473-476.

Daynard, T. B., J. W. Tanner, and W. G. Duncan. 1971. "Duration of the Grain Filling Period and Its Relation to Grain Yield in Corn, *Zea mays* L.," *Crop Science* 11:45-48.

Duncan, W. G., et al. 1978. "Physiological Aspects of Peanut Yield Improvement," *Crop Science* 18:1015-1020.

Dunphy, E. J., J. J. Hanway, and D. E. Green. 1979. "Soybean Yields in Relation to Days between Specific Developmental Stages," *Agronomy Journal* 71:917-920.

Evans, L. T., and H. M. Rawson. 1970. "Photosynthesis and Respiration by the Flag Leaf and Components of the Ear during Grain Development in Wheat," *Australian Journal Biological Science* 23:245-254.

Garner, W. W., and H. A. Allard. 1929. "Effects of the Relative Length of Day and Night and Other Factors of the Environment on Growth and Reproductive Development in Plants," *Journal Agricultural Research* 18:553-606.

Gifford, R. M., P. M. Bremner, and D. B. Jones. 1973. "Assessing Photosynthetic Limitation to Grain Yield in a Field Crop," *Australian Journal Agricultural Research* 24:297-307.

Johnson, D. R., and J. W. Tanner. 1972. "Comparison of Corn (*Zea mays* L.) Inbreds and Hybrids Grown at Equal Leaf Area Index, Light Penetration, and Population," *Crop Science* 12:482-485.

Jones, D. B., M. L. Peterson, and S. Geng. 1979. "Association between Grain Filling Rate and Duration and Yield Components in Rice," *Crop Science* 19:641-644.

King, R. W., D. F. Wardlaw, and L. T. Evans. 1967. "Effects of Assimilate Utilization on Photosynthetic Rate in Wheat," *Plants* 77:261-277.

Lee, H. J., G. W. McKee, and D. P. Knievel. 1979. "Determination of Physiological Maturity in Oats, *Agronomy Journal* 71:931-935.

Major, D. J., et al. 1973. "Comparison of Inbred and Hybrid Corn Grain Yield Measured at Equal Leaf Area Index," *Canadian Journal Plant Science.* 52:315-319.

McKee, G. W., et al. 1979. "Rate of Fill Period for Nine Cultivars of Spring Oats," *Agronomy Journal* 71:1029-1034.

Patterson, T. G., and D. N. Moss. 1979. "Senescence in Full Grown Wheat," *Crop Science* 19:635–640.

Spietz, J. H. J., B. A. ten Hag, and L. J. P. Kupers. 1971. "Relationship between Green Leaf Duration and Grain Yield in Some Varieties of Spring Wheat," *Netherlands Journal Agricultural Science* 19:211–222.

Thomas, J. F., and C. D. Raper, Jr. 1976. "Photoperiodic Control of Seed Filling for Soybeans," *Crop Science* 16:667–672.

Troyer, A. F., and W. L. Brown. 1972. "Selection for Early Maturing in Corn," *Crop Science* 12:301–304.

———. 1976. "Selection for Early Flowering in Corn: Seven Late Synthetics," *Crop Science* 16:767–772.

Yoshida, S. 1972. "Physiological Aspects of Grain Yield," *Annual Review of Plant Physiology* 23:437–464.

chapter five

Net Assimilation Rate

When calculating the photosynthetic efficiency of sunlight conversion in Chapter One, respiration was considered to consume 25% of the food energy produced. Mitchell (1970) estimated a 25 to 50% loss due to respiration for plants growing in the field and suggested an average value of 33%. Some loss due to respiration cannot be avoided because respiration is similar to breathing in animals, a necessary life function. Respiration, the breakdown and release of energy, is essentially the opposite of photosynthesis, the build-up and storage of energy. Respiration converts glucose sugar ($C_6H_{12}O_6$), or in general terms, carbohydrates (CH_2O) produced by photosynthesis, into available energy forms and releases carbon dioxide and water. The simple respiration equation is:

$$CH_2O + O_2 \longrightarrow CO_2 + H_2O + energy$$

with the balanced equation as:

$$C_6H_{12}O_6 + 6O_2 \longrightarrow 6CO_2 + 6H_2O + energy$$

Consider an apple left for several days at room temperature. The apple continues to respire even after being picked from the tree, and the loss of carbon dioxide, water, and energy will mean that the apple will soon become soft and wrinkled. If placed in refrigerated storage, respiration will be slowed up, and if placed in a controlled atmosphere unit containing carbon dioxide and reduced oxygen level, respiration losses can be reduced further and the apple can be stored successfully for an extended period of time. Fruit and vegetable storage units should be regarded as a system for controlling respiration and are named **controlled storage** systems.

A number of factors can influence the rate of photosynthesis and respiration in plants, and the term *net assimilation rate* or *NAR* commonly is used as a measure of the rate of photosynthesis minus respiration losses.

Factors that influence net assimilation rate are temperature, light, carbon dioxide, water, leaf age, mineral nutrients, chlorophyll content, and genotype. NAR should be recognized as a factor influencing yield, and crop producers should incorporate NAR and its implications into their cropping program, vocabulary, and thinking. The fact that it sounds like a complex physiological term need not deter the crop producer.

An example of how photosynthesis and NAR may be increased in a farm field is illustrated by the study of tassel removal to reduce shading of leaves in corn. In a field, massive amounts of pollen are produced on the tassel for fertilization. Tassel size could either be reduced, or the tassels could be removed following fertilization. Under stress conditions of high plant populations, low soil fertility, or drought, yield increases have resulted when the tassel was removed (Chinwuba, Grogan, and Zuber 1961; Grogan 1956; Hunter et al. 1969). Results from an Ontario study (Hunter et al. 1969) show that the increase was larger and more consistent at high plant populations (Table 5-1). A large portion of the increase in yield resulting from **detasseling** was due to the elimination of light interception caused by the tassels.

TABLE 5–1.   Effect of various detasseling treatments on corn grain yield (t/ha) using United 108 hybrid conducted in 1967 at various locations in Ontario

Plant Population		Row Width		Yield of Check Plots		Yield of Detasseled Plots	
(pph)	(ppa)	(cm)	(in)	t/ha	bu/ac	t/ha	bu/ac
40,000*	16,194	50	20	4.7	75.5	5.6	90.0
45,000*	18,220	43	17	6.9	110.8	7.0	112.5
80,000*	32,388	34	13	8.6	138.2	9.1	146.2
124,000*	50,202	27	10	8.8	141.4	9.9	159.0
64,000**	25,910	68	27	6.8	109.2	6.7	107.6

Source: Data from Hunter et al. 1969.
*Treatment applied at time of tassel emergence.
**Treatment applied one week after anthesis.

DARK AND LIGHT RESPIRATION

During dark periods, respiration continues although photosynthesis is no longer active. In addition to dark respiration, there is also a form of light respiration or *photorespiration*. There are two distinct types of photorespiration, and it is essential that crop producers understand the two so that yield differences among crops or treatments can be understood. However, it is not necessary to comprehend the advanced biochemistry involved.

In the late fifties it was understood that at least part of the photorespiration process was initiated via biochemical pathways activated by light. These same biochemical pathways do not function during respiration in the day. Two types of photorespiration, C_3 and C_4 types, were identified according to their different pathways of carbon dioxide fixation (Goldsworthy 1970; Jackson and Volk 1970). Plants whose first carbon compound in photosynthesis consists of a three-carbon atom chain are called C_3 plants. C_3 plants such as soybeans, cereals, and most forage crops have a low NAR due to a high photorespiration rate. Plants whose first carbon compound in photosynthesis consists of a four-carbon atom chain are called C_4 plants and include warm-season crops such as corn, sorghum, certain tropical grasses, and pigweed. Because C_4 plants have a very low photorespiration rate, they generally are extremely efficient and have a high NAR. C_4 crops have a remarkable ability to efficiently utilize the sun's energy, especially at high light intensity compared to the less efficient C_3 species. C_3 and C_4 plants may differ greatly in NAR and yield, yet the difference between the two groups is governed by a single enzyme—ribulose diphosphate carboxylase.

APPLICATION TO CROP PRODUCTION

Crop producers should recognize that warm temperatures for C_3 crops will increase both dark and light respiration, and therefore they should avoid exposing such crops to high temperatures by seeding them early. They should understand why cool summers favor C_3 cereals but offer little advantage to warm-loving C_4 plants. At northern locations such as Beaverlodge, Alberta or Fairbanks, Alaska, long days and cool temperatures can result in an excellent cereal crop, but corn production is totally unsatisfactory.

In areas where corn can be grown, higher yields invariably will be associated with corn, a C_4 crop, than with C_3 cereals. No amount of management will ever produce higher cereal yields than corn in warm summer areas.

Comparisons between corn yields at the equator under a 12-hour day with corn yields at a 40° north or south latitude will show that lower yields under otherwise equal conditions will occur at the equator because of a longer dark respiration period.

In C_3 crops grown and compared at the equator and at a more distant latitude, high dark respiration losses at the equator coupled with high photorespiration losses can account substantially for the poor performance of C_3 crops at the equator. Dark respiration losses were reported to vary from 29 to 71% (Zelitch 1974).

Dark respiration losses cannot be eliminated because they are a part of normal plant functioning, but reduction in respiration can give a productivity gain. A comparison between two corn hybrids showed that one had 50% faster growth than the other (Heichel 1971). The slower dark respiration in the leaves of the more efficient inbred was related to the greater growth rate.

Comparison among crops shows the advantage of corn, due in part to its C_4 type or low photorespiration rate (Table 5-2). Photorespiration in the less efficient C_3 species occurs at rates up to 50% of net carbon dioxide uptake, whereas photorespiration is barely detectable in the C_4 species, and this is the basis for at least part of the yield differences noted in Table 5-2.

Yield differences can be explained further by considering light duration as discussed in Chapter Four. At Guelph, corn reached reproductive development on August 1 and had 41 days for reproductive development (see Example D in Chapter Four). Winter wheat had 45 to 50 days for reproductive development with the further advantage of coinciding with the advantageously long days of June. Despite these two advantages, the C_4 advantage of corn is one reason why corn outyielded the C_3 wheat crop. In simple terms, high C_3 photorespiration rates are a waste of energy compared to the C_4 types. To substantiate this waste, Brown (1978) estimated that the forage specie, tall fescue, lost about 40% of its photosynthetic products due to photorespiration whereas a tropical C_4 grass, *Panecium maximum*, lost only 3%.

TABLE 5-2. Yield comparisons (kg/ha) among crops in Ontario for the four-year period 1976 to 1979

		(k/ha)	(bu/ac)
Corn (grain)—a C_4 crop		5,522	88.0
Winter wheat		3230	47.3
Spring wheat		1735	27.7
Spring oats	C_3 crops	1847	48.5
Spring barley		2685	49.9
Soybeans		2108	33.6
White beans		1280	20.4

Source: Ontario Ministry of Agriculture & Food Statistics Section.

If the desirable enzyme system of C_4 crops could be incorporated by C_3 crops, Zelitch (1975) estimated that elimination of wasteful photorespiration in such crops as wheat and soybeans could boost yields by as much as 50%. Since it is debatable whether photorespiration performs a useful function within the plant, researchers could aid crop producers by controlling photorespiration either by biochemical or genetic means.

A further advantage of C_4 crops was presented by Begg and Turner (1970) as higher water-use efficiency. They suggested that C_4 species are generally twice as efficient as C_3 species (Table 5-3), and that the difference increases with temperature over the range of 20° to 35°C (68° to 95°F). Greater water-use efficiency is achieved under high light and temperature conditions as a result of higher rates of photosynthesis. Under low light conditions greater efficiencies also are associated with lower rates of transpiration.

TABLE 5–3.   Water-use efficiency averaged over a number of C_3 and C_4 species and expressed as grams of dry matter produced for each 1000 grams of water used

	Water-Use Efficiency (g/kg)	
	C_4 *Species*	C_3 *Species*
Dicotolydons	3.44	1.59
Grasses (Monocotolydons)	3.14	1.49

Source: Begg and Turner 1970.

COOL-LOVING CROPS

Every plant process operates within reasonably well-defined environmental limits. Temperature is one environmental factor that exerts an influence on photosynthesis and hence on growth. After a basic minimum temperature starting point, an increase in the activity of a physiological process (respiration, growth, photosynthesis, etc.) occurs, and the rate of increase is known as the Q_{10} value for plant processes; i.e., with each 10°C rise in temperature, a Q_{10} value can be determined. A Q_{10} value of 2 means that as temperature increases by 10°C, the rate of physiological processes double. As well as a Q_{10} value, three **cardinal temperature** points can be defined: (1) the minimum required to initiate a process; (2) the optimum at which the activity proceeds at the highest rate; and (3) the maximum at which activity ceases. Cardinal temperature points vary widely among species and throughout the life cycle of a plant. Mitchell (1970) gave generalized cardinal temperatures for corn NAR (Table 5–4). Cardinal temperatures for germination of corn were given by Wilsie (1962) (Table 5–5). In other words, NAR for corn will be optimum at

TABLE 5–4.   Cardinal temperatures for corn NAR

	°C	°F
Maximum	45	113
Minimum	10	50
Optimum	30–35	89–95

Source: Data from Mitchell 1970.

TABLE 5–5.   Cardinal temperatures for corn germination

	°C	°F
Maximum	40–44	104–111
Minimum	8–10	46–50
Optimum	32–35	89–95

Source: Data from Wilsie 1962.

temperatures between 30° and 35°C. Germination will be optimum between 32° and 35°C. Above 35°C, NAR will decline, and a positive NAR value will cease to exist at 45°C (the cardinal maximum point) due to excessive respiration or photorespiration, perhaps compounded by other stress factors such as moisture.

Photorespiration increases with temperature and shows a greater relative stimulation to increasing temperatures than is characteristic of dark respiration (Zelitch 1971). Goldsworthy (1970) and Daubenmire (1959) reported that at high temperatures, photorespiration increases faster than photosynthesis; i.e., the optimum temperature for photosynthesis is distinctly lower than the optimum for respiration. To illustrate the point, Daubenmire reported that in the white potato, the rate of photosynthesis rises to a maximum at 20°C (68°F), but respiration at this temperature is only 12% of its maximum rate. With an increase in temperature to 48°C (118°F), respiration reached its highest rate, but the photosynthetic rate had declined to zero at that temperature. Plants are clearly at a disadvantage whenever the temperature rises above the optimum for photosynthesis because NAR will decline. Daubenmire did not, however, distinguish between dark respiration and photorespiration.

In alfalfa, Stanhill (1962) found that production was far more closely related with solar radiation than with temperature, although temperature is a function of radiation. Respiration rate had a Q_{10} of 1.46; and NAR had a Q_{10} of 1.18 over the temperature range studied, indicating that respiration is more temperature-dependent than photosynthesis. Over the temperature range of 9° to 26°C (48 to 79°F), the positive effect of temperature in increasing the rate of photosynthesis was approximately one-fifth of the negative effect caused by increasing the respiration rate.

C_3 type crops are more adversely affected by high temperatures than C_4 type crops. Temperature responses in temperate and tropical grass species (*Gramineae*) differ strikingly (Moss and Musgrave 1971). Temperate species show optimal growth and NAR values at 20° to 25°C (68 to 77°F), whereas tropical grasses, including corn, have optimal temperatures of 30° to 35°C (86 to 95°F). A 1°C rise in temperature of the temperate C_3 spring wheat about the time of the onset of reproductive development was associated with a 4% reduction in grain yield (Fischer and Maurer 1976). The reduction was related largely to fewer grains per spike. During grain filling, temperature increases and related respiration reduced kernel size in some varieties.

Temperature considerations of NAR may be clarified by cereal yield comparisons between equatorial and temperate zones. Not only do equatorial crops have the disadvantage of short day length and warmer night temperatures during the 12-hour night with increased dark respiration, but they also experience high photorespiration rates under high day temperatures. Because of the superior performance of C_3 cereals under cool temperate zone conditions, cereals have become widely known as "cool-loving" cereals. Early seeding to benefit from cool spring temperatures is therefore desirable.

AGE OF LEAVES, PLANT POPULATION, AND NAR

Shading and aging of leaves may lower NAR values because of reduced rates of photosynthesis, but respiration continues as long as the leaves remain alive. Donald (1961) suggested photosynthetic and respiration rates for plants with related NAR values per leaf and for all leaves (Table 5-6) for a young plant with 4 leaves and an older plant with seven leaves.

Total photosynthesis in the older plant is higher (32) than in the young plant (31), but total respiration in the older plant is also higher (13), so NAR values are 21 and 24 for the old and young plants, respectively. This idealized situation should be considered in respect to a shortened vegetative period in determinate crops to reduce negative NAR values at the beginning of the reproductive period, but age of leaves as well as mutual shading may contribute to negative NAR values.

Crop production efforts are aimed at maximizing NAR:

1. By establishing a plant population in a spatial arrangement that reduces mutual shading yet effectively intercepts sunlight to maximize photosynthesis. This is possible to a degree in annual plants with precision planters, although leaf size or tiller number and lateral branching may vary according to environmental conditions.
2. By managing, clipping and grazing in perennial forage crops, so as

TABLE 5-6. Idealized plant in a community of plants showing relationship of increasing leafage to net assimilation of the foliage

	Leaf Number	Rate of Photosynthesis	Respiration Rate	NAR
Young plant (4 leaves)				
	1	12	2	10
	2	10	2	8
	3	7	2	5
	4	3	2	1
		31	8	24
Older plant (7 leaves)				
	1	12	2	10
	2	10	2	8
	3	7	2	5
	4	3	2	1
	5	2	2	0
	6	0	2	−2
	7	0	1	−1
		32	13	21

Source: Donald, 1961.

to favorably influence NAR. The situation where the three lower leaves reduce NAR in Table 5-6 in the older plant could be avoided by removing the crop before lower leaves reach senescence due to age or mutual shading. In forage legumes, lower leaves may fall off or be lost at harvest if allowed to senesce before harvesting begins. Watson (1952) also noted that the decline in NAR of the lower leaves was due to aging of the leaf as the season progresses and to mutual shading by upper leaves. Timeliness of clipping is representative of sound management.

Negative NAR values indicated in the older plant in Table 5-6 suggest that plants may lose weight or have negative growth as they grow older. If growth is defined as both cell division and cell enlargement, which means an increase in plant size and fresh weight, then a loss of weight may occur through shrinkage in cell size, perhaps as a result of water loss, which may be considered as negative growth. This is a debatable point however.

In Table 5-6, leaf 5 has a zero NAR value and is at the **compensation point.** Leaves 6 and 7 have NAR values each less than zero and are considered **parasitic** on the plant. Parasitism may be defined as occurring when lower leaves use assimilates at a greater rate than they photosynthesize. Such a situation may not be avoided and may in fact be beneficial. If leaves 5, 6, and 7 have a zero or negative NAR due to mutual shading alone, then a reduced population may be desirable. If on the other hand, leaves 5, 6, and 7 have low NAR values due to old age, the addition of fresh new leaves may be advantageous. The decline in NAR may not be directly proportional to the increase in leaf number. Leaves 5, 6, and 7 possibly were essential for effective light interception in the juvenile stage of plant development, and perhaps food energy was moved out of leaves 6 and 7 for use elsewhere before senescence began. In such a case the overall contribution of leaves 6 and 7 was greater than losses during senescence. Work by Shibles and Weber (1965) found that lower, shaded leaves of the soybean canopy were not parasitic and did not detract from the yield of the plant. The debate continues as to whether or not leaves are ever truly parasitic under field conditions.

A high yield of any crop can be achieved only when a proper combination of cultivar, environment, and agronomic practices are obtained. Understanding NAR and associated C_3 and C_4 types and dark respiration helps determine the best combination of the above three factors, helps reveal what improvements can be made to achieve a further increase in yield, and helps explain yield differences once they occur.

QUESTIONS

1. Winter barley is seeded in the fall, resumes growth in the spring, and vegetative growth ceases about June 10 when reproductive growth begins. Grain filling is complete and maturity is reached about July 10 at

Guelph. Spring barley seeded in early May starts to develop about June 25 and completes grain filling about July 25.

(a) Explain why winter barley will usually outyield spring barley by 25 to 35%.

(b) Occasionally winter and spring barley yields are about the same. Explain how this might occur.

(c) Under what conditions might spring barley exceed winter barley yields?

2. A technique to screen for efficient (C_4) plants was proposed by Menz et al. (1969). They reasoned that C_4 plants such as corn and sorghum (both efficient species), had "compensation concentrations" of 0 ppm, whereas inefficient plants such as cereals and soybeans had "compensation concentrations" of 60 ppm. Explained otherwise, in a sealed illuminated chamber, the process of photosynthesis would consume CO_2, and photorespiration would give off so little CO_2 that eventually 0 ppm in the chamber would be reached. In C_3 plants, photosynthesis would consume CO_2, but a CO_2 concentration of 60 ppm would be reached from respiration. These are the compensation concentrations.

(a) Explain what would happen if C_4 plants were placed in a sealed illuminated chamber with C_3 plants, and give reasons (see Table 5-7) (Menz et al. 1969).

(b) Which crop has the greater efficiency—wheat or soybean?

(c) Why would the two plants on which the cotyledons were not removed remain alive?

(d) What would happen if only wheat or soybeans were placed in the chamber? (See Table 5-8) (Menz et al. 1969).

(e) Explain the results in Table 5-8.

(f) Speculate what the CO_2 concentrations might be after day 6 in Table 5-8.

(g) What would you expect to happen if corn or sorghum alone was placed in the chamber?

TABLE 5-7. CO_2 concentrations (ppm) in chambers containing mixtures of C_3 and C_4 plants

Days in Chamber	Sorghum and Wheat	Status of Wheat	Sorghum and Soybeans	Status of Soybeans
1	38	Healthy	30	Healthy
2	30	Healthy	20	Yellowing
3	20	Yellowing	20	10% dead
4	10	90% dead	10	50% dead
5	2	95% dead	2	90% dead
6	2	All dead	3	All dead*

Source: Data from Menz et al. 1969.

*Except for two plants where cotyledons were not removed.

TABLE 5–8. CO_2 concentrations (ppm) in chambers containing C_3 plants alone

Days in Chamber	All Wheat	All Soybeans	Status of Pure Stand
1	153	155	Healthy
2	123	100	Healthy
3	98	92	Healthy
4	85	65	Healthy
5	75	45	Healthy
6	70	45	Healthy

Source: Data from Menz et al. 1969.

   (h) What would you expect the sorghum plants to look like on day 6 in
   Table 5-7?

3. In the Great Lakes region of North America, alfalfa can be cut either two
   or three times a season.

   (a) Which would you expect to yield more—two cuts or three cuts prior
   to September? Explain your answer. Use Table 5-9 to provide you
   with a clue.

   (b) How many days would elapse between cuts when harvested two times
   and when harvested three times?

   (c) Based on the observations in Table 5-10, would you expect four cuts
   to outyield three cuts? Explain.

TABLE 5–9. NAR values of alfalfa leaves of different ages

Days Since Unfolding	NAR (microlitres of $CO_2/hr/cm^2$)
5	119
12	121
17	112
22	80
28	17

Source: Data from Feuss and Tesar 1968.

4. The effect of age and temperature on NAR of orchard grass is shown in
   Table 5-10.

TABLE 5–10.   NAR as influenced by leaf age and temperature regimes, mean of three orchard grass cultivars (microlitres of $CO_2/hr/cm^2$)

Leaf Age Days	21°C (70°F)	29°C (84°F)
1	59.6	70.2
12–16	72.5	83.6
20–26	57.7	70.2
28–36	22.7	24.9

Source: Data from Treharne, Cooper, and Taylor 1968.

(a) Explain why orchard grass, a C₃ species, showed a positive temperature response in Table 5–11.

(b) Compare NAR values of orchard grass with alfalfa and determine which of two or three seasonal cuttings would produce the highest yield.

(c) Would orchard grass produce the highest yields at 21° (70°F) or 29°C (84°F) according to Table 5-10?

5. Temperature effects on NAR in 22 races of corn are shown in Table 5-11.

TABLE 5–11. Average values of NAR (mg $CO_2/dm^2/hr$)

Temperature		
(°C)	(°F)	NAR
15	59	25
18	64	39
21	70	46
24	75	48
27	81	51
30	86	52
33	91	55
36	97	53

Source: Data from Duncan and Hesketh 1968.

(a) Is an optimum temperature suggested for corn in Table 5–11, and how does this compare with that suggested in Table 5–4?

(b) At 36°C (97°F), the NAR value was 53. Is this likely to be the maximum cardinal temperature for NAR in corn?

6. What effect might prolonged, dull, cloudy weather have on crop yields? Explain your answer.

7. What effect might high night temperatures have on crop yields?

8. It is a well-known fact that the temperature near the soil in summer is higher than at a height of 1 or 2 metres (3¼ to 6½ feet). As plant height is reduced, consider the possible effect of plants growing in a higher temperature regime (Fischer and Maurer 1976).

(a) What effect might higher temperatures in the crop microclimate have on vegetative development of wheat?

(b) What effect might higher temperatures in the crop microclimate have on reproductive development of wheat?

(c) What effect might this have on yields of wheat 30 cm (12 in.) vs. 60 cm (24 in.) vs. 90 cm (36 in.) in height?

(d) How might temperature effects be expressed in the plant?

(e) Would the same impact of temperature be felt in dwarf sorghum or dwarf corn as that found in wheat? Explain.

9. (a) Explain why tassel removal just before anthesis produces a greater

yield advantage than when removed after anthesis (give two reasons). (Review Chapter Two.)

(b) Why is the tassel not removed several weeks before anthesis? (Review Chapter Two.)

REFERENCES FOR CHAPTER FIVE

Begg, J. E., and N. C. Turner. 1976. "Crop Water Deficits," *Advances in Agronomy* 28:161–217.

Brown, R. H. April–May 1978. "Newly Discovered Plant A Photosynthesis Key," *Crops and Soils*, pp. 5–6.

Chinwuba, P. M., C. O. Grogan, and M. S. Zuber. 1961. "Interaction of Detasseling, Sterility, and Spacing on Yield of Maize Hybrids," *Crop Science* 11:279–280.

Daubenmire, R. F. 1959. *Plants and Environment*, 2nd ed., pp. 176–177. New York: John Wiley & Sons, Inc.

Donald, C. M. 1961. "Competition for Light in Crops and Pastures," *Symposia Society Experimental Biology* 15:282–313.

Duncan, W. G., and J. D. Hesketh. 1968. "Net Photosynthetic Rates, Relative Leaf Growth Rates, and Leaf Numbers of 22 Races of Maize Grown at Eight Temperatures," *Crop Science* 8:670–674.

Fischer, R. A., and R. Maurer. 1976. "Crop Temperature Modification and Yield Potential in a Dwarf Spring Wheat," *Crop Science* 16:855–859.

Fuess, F. W., and M. B. Tesar. 1968. "Photosynthetic Efficiency, Yields and Leaf Loss in Alfalfa," *Crop Science* 8:159–163.

Goldsworthy, A. 1970. "Photorespiration," *Botanical Review* 36:321–340.

Grogan, C. O. 1956. "Detasseling Responses in Corn," *Agronomy Journal* 48:247–249.

Heichel, G. 1971. "Confirming Measurements of Respiration and Photosynthesis with Dry Matter Accumulation," *Photosynthetica* 5:93–98.

Hunter, R. B., et al. 1969. "Effect of Tassel Removal on Grain Yield of Corn (*Zea mays* L.)," *Crop Science* 9:405–406.

Jackson, W. A., and R. J. Volk. 1970. "Photorespiration," *Annual Review of Plant Physiology*. 21:385–432.

Menz, K. M., et al. 1969. "Screening for Photosynthetic Efficiency," *Crop Science* 9:692–694.

Mitchell, R. L. 1970. *Crop Growth and Culture*. Ames, Iowa: Iowa State University Press.

Moss, D. N., and R. B. Musgrave. 1971. "Photosynthesis and Crop Production," *Advances in Agronomy* 23:325.

Shibles, R. M., and C. R. Weber. 1965. "Leaf Area, Solar Radiation Interception and Dry Matter Productivity by Soybeans," *Crop Science* 5:575–577.

Stanhill, G. 1962. "The Effect of Environmental Factors on the Growth of Alfalfa in the Field," *Netherlands Journal Agricultural Science*. 10:247–253.

Treharne, K. J., J. P. Cooper, and T. H. Taylor. 1968. "Growth Response of Orchard Grass (*Dactylis glomerata* L.) to Different Light and Temperature Environments. II. Leaf Age and Photosynthetic Activity," *Crop Science* 8:441–445.

Watson, D. J. 1952. "The Physiological Basis of Variation in Yield," *Advances in Agronomy* 4:101–145.

Wilsie, C. 1962. *Crop Adaptation and Distribution,* pp. 196–201. San Francisco: W. H. Freeman & Company Publishers.

Zelitch, I. 1971. "Photosynthesis, Photorespiration and Plant Productivity," New York: Academic Press. 347 p.

———. 1974. "Improving the Energy Conversion in Agriculture," *American Society of Agronomy Special Publication 22:* A New Look at Energy Sources, pp. 37–50.

———. 1975. "Improving the Efficiency of Photosynthesis," in *Food: Politics and Economics, Nutrition and Research,* pp. 171–178.

chapter six

Leaf Area and Plant Architecture

THE PROBLEM

Milthorpe (1956) stated:

All aspects of agricultural production are intimately associated with the growth of leaves. Whilst this is immediately obvious with such crops as grass (forage), cabbage or kale, it is perhaps less so with those crops, such as sugar beet or corn, of which the harvested product forms but part of the plant. But even with these crops, the yield ultimately depends on the rate of addition of dry matter per unit area of land, i.e., on the efficiency of the photosynthetic process and on the extent of the photosynthetic surface. Even a casual examination of agricultural writings reveals much confusion as to the relative importance of these two components of energy production with consequently much misdirected effort in plant breeding, fertilizer and other agronomic programmes. Thought still seems to be dominated largely by the rate and mechanism of the photosynthetic reactions. Yet these are relatively unimportant in determining yield compared with the effect of variations in the extent of the surface in which these reactions proceed. Further, over a fairly wide range of the environments normally experienced in agriculture, leaf growth is itself largely independent of the rate of assimilation.

Despite the apparent simplicity of the relationship between yield and photosynthetic area, the problem of controlling the photosynthetic area for maximum advantage is the central theme of this chapter. Efforts to increase yield by the obvious means of greater plant density have not always been successful because of the number of factors involved.

A suggestion of just how complex the problem is was indicated when no less than nine factors related to light utilization alone were included in a computer model under the assumption that minerals, water, and temperature were not limiting (Duncan et al. 1965). The nine factors concerned the following properties of plant leaves:

1. Leaf area.
2. Leaf angle.

91

3. Vertical position.
4. Light reflected from leaves.
5. Light transmitted through leaves.
6. The light response curve, i.e., the physiological relationship between illumination and photosynthesis.
7. Elevation of the sun above the horizon.
8. Solar intensity.
9. Skylight brightness.

Because of the complexity of considering the mutual effects of so many interrelating factors, individual leaves have often been studied under controlled environments, and a great deal is known about photosynthetic capabilities in single leaves. Much less is known about the rate and efficiency of the photosynthetic process as it applies to plant communities and to economic yield. The dry matter that enters the grain is produced by photosynthesis largely after reproductive development has begun. Watson (1956) reported that in cereals, the leaf area present before reproductive development makes no direct contribution to grain yield. But the indirect effects cannot be ignored, and considerable attention must be directed toward the low efficiency of the overall utilization of incoming solar radiation caused by the inability of small developing seedlings to intercept the light and by the fact that so much of the light penetrates to the soil surface and is lost to photosynthesis. If light interception is important during the vegetative stages, then attention must be directed toward factors such as seed size and depth of planting—to make a plant capable of photosynthesizing as soon as possible—plus seed number per unit area (seeding rate), and time of sowing. There is little question that attention to these factors may considerably alter the course of dry-matter production and hence final yield. If seeding rates are directed toward maximizing the capture of sunlight in the seedling stage, a point would be reached at a later stage of crop development at which the net assimilation rate (NAR) would decline as a result of mutual shading of the leaves or some other interrelating competitive factor. Thus, a compromise must be reached between management considerations of seedlings versus adult plants.

To determine some of the physical properties of light interception plastic plants have been used in the field. By varying the leaf size and angle, light interception in an artificial crop canopy can be observed (Figure 6-1).

INCREASING PLANT POPULATION

It has been said that every generation of crop producers gets excited about increasing plant number per hectare (acre) to promote higher yields, especially of row crops. Recommended seeding rates, row widths, and spatial arrangements of plants in a population have been developed over many years

Figure 6-1. Plastic plants have been used to experiment with leaf size and leaf angle in relation to sunlight interception. (Courtesy Ontario Ministry of Agriculture and Food.)

of often **empirical** testing. Recommended seeding rates may represent the upper limit of population density stresses that the crop can tolerate under most conditions. Increasing seeding rates in conjunction with narrow rows, or just reducing plant spacings and increasing population pressure without regard to other factors, will probably not result in a yield increase and may result in a reduction of both yield and quality. In one study conducted to maximize yield of dry matter (**biomass**) per unit of land area without regard to quality or grain yield, Crookston et al. (1978) indicated that increasing plant population did not increase dry-matter production. The study suggested that field crop production practices may be utilizing a plant population near the upper limit of the existing genotypes. It is highly likely that plant populations commonly used today are frequently in excess of the optimum. Such being the case, if higher plant populations are to be successfully used, additional modifications or inputs will be needed.

MODIFICATIONS FOR HIGH PLANT DENSITIES

A weed-free environment, made possible by the application of modern herbicides, has set the stage for changes in cropping practices. No longer is it necessary to plant row crops such as corn, sorghum, and soybeans in wide rows to allow for interrow cultivation for weed control. But the big, tall, leafy cultivars selected for 70- to 100-cm (28 to 40 in) wide rows to allow for cultivation, which could assist in weed control by shading weeds that escaped the cultivator, and which are needed to "catch" more sunlight in wide row spacings, will probably not be the most efficient *cultivars* when grown in narrow rows or under increased plant population. Plant types adapted to high plant populations and narrow rows will have to be developed; and if obtained, consideration must be directed toward other factors that may in turn become limiting, such as fertility level, moisture availability, duration of effective light utilization, duration of vegetative and reproductive periods, NAR values, and parasitic leaves, and whether machines capable of seeding and harvesting such populations are available. It must be stressed again and again that before a farmer modifies his cropping practices to include narrow rows and high plant populations, he must be sure he is getting all the yield possible from his existing cropping program. If he is determined to proceed, he must be sure that he has complete weed control, genotypes capable of responding to new practices, and machines available to handle the field operations and whatever will be the next limiting factors. Changes should not be abrupt and drastic. Interest in higher plant densities is reported by Troyer and Brown (1976) to have increased over the past four decades in Iowa, accompanied by higher fertilizer utilization.

It is in the short-season zones where early maturing full-season crops are grown that increased plant populations may make the greatest impact, for it is there that small-statured crops due to a comparatively short season, and hence with a shortened vegetative period, are most likely to be found.

Leaf Area

Any consideration of plant populations must recognize that it is the photosynthetic area that is being modified, and changes in plant size or shape may be required if high populations are to result in an increase in NAR. Williams, Loomis, and Lepley (1965) suggested that the complex problem of accounting for variations in the growth of crops may be simplified by the consideration of two components of growth in a plant community—these are NAR and **leaf area index (LAI)**.

Leaf area generally has been adopted as the measure of the size of the photosynthetic system, but in adopting leaf area, it is recognized that other plant parts are capable of photosynthesis and may account for a sizable portion of total dry-matter production. In addition to the leaf blade, photosynthesis occurs in all green plant parts, including the stem or culm, the leaf sheath, **awns, glumes**, husks, pods, etc. Some of these areas are at the upper part of the plant and are not subject to shading. Mitchell (1970) believed that leaf sheaths and the inflorescence in cereals had a photosynthetic efficiency of 50 to 100% of the leaves. Thorne (1959) found that the leaf sheath of barley plants contributed from 15 to 40% of grain yield, and the head contributed from 9 to 40% for awnless and awned types, respectively. Photosynthetic areas of the plant other than the leaves are hard to measure, and so the sum of all leaf laminae, upper surface only, has become the accepted method to express photosynthetic area.

EXPRESSION OF LEAF AREA MEASUREMENTS

Various elaborate techniques have been developed to measure leaf area. It is of little interest to compare the leaf area per plant because such measurements are highly dependent on plant spacings. A more useful measure is to express leaf area as a measure per unit area of land, both areas being expressed in the same units. The resulting value gives no indication of the physical area of leaves but rather a ratio or index—hence leaf area index (LAI). Thus, if the total leaf area of a crop grown on a hectare (acre) of land adds up to 10,000 square metres (1 hectare) (or 43,560 sq. feet—one acre), then the LAI is 1. A LAI of 4 means that four square hectares of leaf laminae are present on each one hectare of land.

Comparisons of LAI's are more meaningful if related to specific growth stages. With determinate crops, LAI is often expressed at the beginning of reproductive development. With indeterminate crops, the upper limit of LAI may be used, or comparisons can be made among cultivars if all are measured at a common calendar date. LAI may also be expressed in terms of light intercepted per treatment.

A Russian worker, Nichiporovic, suggested that every crop has an **optimum leaf area** index, which he considered to be between 2.5 and 5.0 (Black and Watson 1960). The leaf area at which dry-matter production is maximized is regarded as the optimum leaf area. Below the optimum LAI, dry-matter production is less because not all the sunlight is intercepted and photosynthesis is not at a maximum rate; above the optimum LAI, dry-matter production is less because lower leaves are shaded and NAR is reduced. The concept of an optimum leaf area may be meaningless if used indiscriminately to explain yield differences, but may be useful in the following contexts:

1. For crops in which the objective is grain yield (cereals), the optimum LAI must be associated with a stage of plant development; in other crops (forages) the attainment of an optimum LAI may signal the ideal harvesting (cutting) time.
2. The greatest dry-matter production could be expected to occur when the optimum leaf area coincided with conditions most favorable for photosynthesis, such as long days or when moisture stress is removed.
3. It may signal the point at which NAR is highest.
4. It may signal the leaf area that is deemed to capture as much of the incoming radiation as possible. Brougham (1956) suggested that the

"critical" LAI was the point at which 95% of the light energy hitting the land area at noon is intercepted.

5. In addition to LAI leaf area duration (LAD) has been proposed as a measure to describe the length of time the leaf area is functional (Watson 1947).

LAI UNDER FIELD CONDITIONS

Based on field observations, but using a computer to simulate the effects of increasing corn populations, Duncan (1971) showed that theoretical increases in yield were achieved as LAI increased up to 4.0 with no further yield increase above a LAI of 4.7. The LAI at the time when reproductive growth was initiated was found to be related to grain corn yield (Eik and Hanway 1966). A LAI of 3.3 in conjunction with duration of the grain filling period up to 155 days was effective.

At recommended corn populations of 43,000 to 53,000 pph (17,400 to 21,500 ppa), a LAI of 2.8 to 3.4, respectively, was found at Guelph. At higher populations a larger LAI was found, but grain yields declined (Daynard, Tanner, and Duncan 1971)(Table 6-1). This finding was based on three hybrids—Pride 5, Pride 4, and United 108. In 1967, the test included three additional hybrids, and highest grain yields increased as plant population and LAI increased (Table 6-2).

The contrasting results between 1966 and 1967 at Guelph indicate the complexity of trying to establish a plant population to reach a desired LAI. The variation between the two years may be attributed to differences among hybrid cultivars and biological variability between years, which resulted in a different leaf shape (length, width, curvature, angle of display, plastic responses to wind and light, and possibly leaf number per plant). Relationships between grain yield and LAI of corn were found to vary appreciably among different genotypes (Rutgers, Francis, and Grogan 1971). When yield was divided by LAI to calculate a ratio, wide differences among genotypes were found. Leaf area indices in this test averaged 2.5 and 3.9 at population densities of 34,600 and 65,200 pph, respectively (14,000 and 26,400 ppa).

TABLE 6–1. Relationship of plant populations of corn to LAI and grain yield at Guelph, 1966

| Plant Population | | LAI | Grain Yield— | |
(pph)	(ppa)		(kg per ha)	(bu/ac)
69,900	28,300	3.2	6290	100.3
79,500	32,186	3.4	6020	96.0
91,300	36,964	3.9	5570	88.8
105,900	42,875	4.2	5100	81.3

Source: Daynard, Tanner and Duncan 1971.

TABLE 6–2.   Relationship of plant populations of corn to LAI and grain yield at
Guelph, 1967

| Plant Population | | LAI | Grain Yield— | |
(pph)	(ppa)		(kg per ha)	(bu/ac)
44,700	18,100	2.2	5430	86.6
79,500	32,186	4.2	6940	110.7
124,300	50,324	5.1	7110	113.4

Source: Daynard, Tanner and Duncan 1971.

In broadcast forage stands, it was not until a LAI of 5.0 was reached that growth rate reached a plateau (Brougham 1956). Whyte (1960) concluded that LAI's of pastures may exceed considerably those of grain crops, which have a range of LAI of 2.4 to 5.0, whereas pasture mixtures or pure stands of subterranean clover had LAI's of 6.2 to 8.9.

At population densities of 34,600 and 65,200 pph (14,000 and 26,400 ppa), Mason and Zuber (1976) measured LAI's of 2.47 and 4.00, respectively for corn. Variation at the low population density was minimal and varied from a low of 2.20 to 2.63 among the fifteen cultivars. At high population densities, LAI's varied among the fifteen cultivars from a low of 3.45 to a high of 4.61. The greater variation of LAI's at the high population density among cultivars indicates the problem of managing leaf area through plant populations.

CONTROL OF LEAF AREA

The work of Mason and Zuber (1976) shows the difficulty of controlling leaf area by population density. Other factors influence leaf growth and development and make accurate control of leaf area impossible; but because leaf area and photosynthesis are synonomous, the possibility of managing crops to optimize leaf area offers an enormous potential. Attempts to favorably influence leaf area may be made through the following approaches:

1. *Seeding rate.* A limited degree of control of leaf area is provided by seeding rate. In species that tiller such as cereals, control is more limited than in corn, and by empirical trials the recommended seeding rate may produce the highest yield because it approximates the optimum leaf area.

2. *Planting pattern.* At a fixed population, wide versus narrow rows will influence leaf area.

3. *Differences in plant morphology.* Corn hybrid cultivars differ in leaf number, leaf length, and width and internode length, and individually or in combination can influence leaf area. Corn selected and adapted to short-season areas will normally produce 9 to 18 leaves and may grow to about 60 to 70 cm (24 to 28 in.) in height. In the U.S.

cornbelt, 18 to 21 leaves may be found on adapted hybrids, and these may range in height from 100 to 270 cm (39 to 106 in) (Kuleshov 1933).

Comparisons between tall and short lines of a crop, sometimes differing by only one gene for height, have often produced results favoring the tall line even though similar leaf areas are found to exist in both lines. Usually dwarf plants have the same number of leaves as their taller counterparts, the only difference being shorter internodes. It is commonly found that under similar conditions the dwarf line yields less than the tall line, and this has been attributed to poorer sunlight interception resulting from the shorter internodes and increased mutual shading of leaves because dwarf plants may be less plastic in their responses to wind and air movement. A crop producer, knowing the morphological features of a cultivar may modify his seeding rate or planting pattern accordingly.

4. *Date of seeding.* Early seeding usually results in a smaller plant superstructure than an average seeding date in temperate regions because cool spring temperatures favor differentiation rather than growth (cell division and elongation). Late seeding may result in small plants as a result of a shortened vegetative growth period due to a photoperiodic response. Under conditions of early seeding, a corn grower will select a higher seeding rate than with subsequent later seedings. A cereal grower will recognize that early seeding will produce shorter plants but may also encourage tillering.

5. *Thinning and transplanting.* In labor-intensive regions of the world, sunlight interception is maximized in rice paddy fields by overseeding and transplanting as the plants develop. Intensive crop management of high-value market garden crops may include overseeding of onions or radishes to allow for periodic thinning and consumption of the removed plants.

   Transplanting of full-season crops in short-season areas (tobacco, tomatoes, cabbages) is an effective way of avoiding spring frosts yet allowing for more effective sunlight utilization once the danger of frosts has gone.

6. *Controlled cutting and grazing.* In forage crops, leaf area may be maintained near the optimum by controlled cutting and grazing. Lawns can be maintained as a green and healthy dense mat by frequent cutting at a controlled height.

7. *Natural control.* Competition for light and other variables may be so severe in densely seeded populations that some plants are crowded out, resulting in some control of leaf area. In cereals, high plant densities will reduce tillering and restrict leaf area development. In a dense stand, lower leaves may turn yellow and die. Natural reductions in leaf area may prevent a producer from achieving very high LAI's and will probably not reduce leaf area to the optimum level.

8. *Soil fertility.* Leaf growth can be stimulated by nitrogen fertilizer applications. Winter cereals that have been damaged by a severe winter should be given a top-dressing of nitrogen fertilizer to encourage leaf growth and photosynthesis in order to produce a satisfactory crop. Under circumstances of good winter survival, sufficient nitrogen should be applied to encourage photosynthesis, but high levels of nitrogen may result in excessive leaf growth and shading of lower leaves and lower internodes, which may result in lodging and reduced yields due to the reduced photosynthetic rates.

9. *Judicial timing of irrigation water.* Under dry-land conditions, water may be applied to favor or restrict leaf growth during vegetative periods. Water applied at the beginning and during reproductive development may be more beneficial than abundant water during vegetative growth, which promotes excessive leaf area and lodging, and reduces NAR and yield.

PLANT ARCHITECTURE

To the layman all plants within a species appear the same, but to the agriculturist even a cursory examination of plants within a species reveals substantial variation in morphological features. The manner in which leaves are displayed, differences in tillering habit, plant height, etc., all describe plant architecture. Quantitative variations in the vegetative characters of corn were reported by Kuleshov (1933) as follows:

1. Height of the plants—variations from 60-70 to 650-700 cm (2-22 ft) were found.
2. Number of leaves on the principal stalk varied from 8 to 48.
3. Leaf length ranged from 30 to 152 cm (12 to 60 in).
4. Leaf width variations from 4.0 to 15.0 cm (1½ to 6 in) were found.
5. Number of stalks (suckers, tillers, or stools) varied from 1 to 12.

Evolutionary forces and the law of survival of the fittest resulted in plants suited to the environment in which they existed. As plants were domesticated, selection for or against certain features took place, and plant architecture was modified. Either by conscious or unconscious selection pressure, the seed shattering habit at maturity was eliminated, large-seeded lines may have been favored, and prostrate types may have been eliminated by cutting methods and upright types favored. Perhaps high tillering was favored because certain plants produced higher yields or were more competitive in a solid stand.

Plant selections differing in morphology from conventional types must be carefully evaluated so as not to overlook a potential advantage. For example, dwarf lodging-resistant lines may be overlooked if not subjected to high fertility levels. Also, only those cultivars with broad, long lax leaves, i.e., a large-plant architecture, may perform satisfactorily under conditions of wide-row spacings commonly used for evaluating corn hybrid cultivars (Tanner and Stoskopf 1967). A potential cultivar with a small-plant architecture or with upright leaves may go unnoticed because of low yield when tested in conventional row widths. Observations of upright and floppy leaf types in wheat and oat trials in Ontario vividly demonstrated differences in competitive ability. At one location where weeds were controlled chemically, a short wheat with upright leaves was the highest yielding in the test. At another location, with no herbicide treatment, the same entry yielded lowest. Weed growth between the rows of the upright leaf type was markedly more profuse than between the rows of the wide, floppy leaf types. The latter types established a canopy quickly enough to suppress weed growth. At one location the results indicated that the upright leaf type should be discarded, whereas results at the other location indicated good grain yielding ability (Tanner et al. 1966).

As a result of improved plant architecture coupled with rust resistance, the yield potential for wheat at the CIMMYT research stations in Mexico was raised in the mid-sixties from 4 t/ha to 7 and 8 t/ha (60 bu/ac to 105 and 120 bu/ac). Since then, the yield potential of wheat has gradually increased to the 8 to 9 t/ha (120 to 135 bu/ac) range but appears to be levelling off (Anonymous 1979).

A crop producer does not have to rely on plant breeders to supply morphological types but may search out various **phenotypes** from available cultivars and adopt his cropping program to utilize a particular feature. But producers must realize there is no one best architectural design. Donald (1962, 1968) and Donald and Hamblin (1976) have encouraged breeders to develop architecturally exciting plants and to use them to advantage. Examples of plants differing architecturally are shown in stylized drawings in Figure 6-2. Smaller plants with a leaf configuration to intercept more sunlight may increase photosynthesis [see Figure 6-2(b)]. Plants with upright leaves may require less physical space than plants with horizontal leaves (Figure 6-3) and may be adapted to narrow row widths. Perhaps an advantage may be achieved by combining all the seemingly advantageous features (Figure 6-4).

Leaf shape may be a morphological feature of advantage in light interception. For example, cotton plants with okra or superokra leaves illustrate a morphological feature of potential value in raising cotton fiber yields under management practices involving narrow rows and high populations (Figure 6-5).

Because of good light penetration properties and free air movement, a low incidence of boll rot was found in superokra types (Andries et al. 1969, 1970). Few vegetative branches were formed, resulting in reduced vegetative growth. Also, when planted in narrow rows (51 cm) (20 in) at Tuscon, Arizona, rates of photosynthesis were higher for superokra types than for normal leaf plants. Perhaps better air movement in the canopy and more advantageous leaf temperature due to better light penetration are the reasons for this higher rate of photosynthesis.

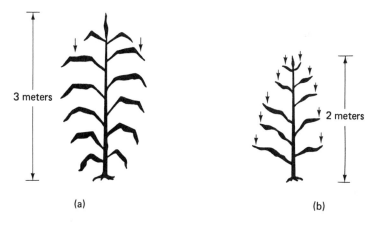

(a)   (b)

Figure 6-2. Diagrammatic sketches of various plant architecture. Big plants such as the one shown in (a) with broad, spreading leaves produce good yields in standard rows, but lower leaves cannot get adequate light. Such plants compete well with weeds, a feature that may no longer be of value in a weed-free environment. Smaller plants like that shown in (b) with a Christmas-tree shape intercept more sunlight. Lower leaves are photosynthetically active, and the entire plant may have a higher NAR value. Because the plant shown in (b) has fewer leaves than that in (a), it may have a shortened vegetative period and be capable of higher yields because of improved sunlight interception, more plants per unit area to reach optimum leaf area, and an extended filling period.

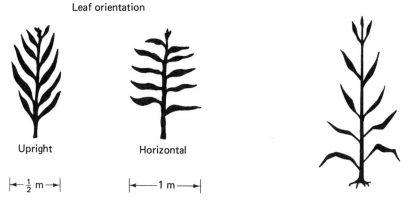

Figure 6-3. *(Left)* Graphic representation of two types of leaf orientation. The upright leaf type may require less physical space than the horizontal type, and a higher population may be possible. When both are planted at the same population, the upright leaf type plant should be more efficient in capturing the sun's energy than are conventional plants with horizontal foliage.

Figure 6-4. *(Right)* The "ideal" plant. The upper leaves are vertically oriented; the lower leaves are flat but short; the plant has ten leaves and is relatively short in height with a shortened vegetative and extended reproductive period; it is adapted to narrow rows and high plant populations because of its reduced height and width and its Christmas-tree shape.

Figure 6–5. Typical leaves of normal (center), okra (right), and superokra leaf (left) cottons. Could okra or superokra be used to advantage in high plant populations?

LEAF ANGLE AND UPRIGHT LEAVES

Upright leaves have been proposed (Warren Wilson 1960) as having an architectural advantage, perhaps because upright leaves are a discrete entity that can be observed, measured, and evaluated. There are several reasons why upright leaves are believed to be superior. The first is better light interception. The advantage of upright leaves can be understood by examining the implications of light striking a leaf surface. On a bright day at high noon, 10,000 to 12,000 foot-candles of light may be received by a leaf.[1] Leaves of most species cannot utilize this high an intensity; maximum rates of photosynthesis occur at a light saturation of 2000 to 3000 foot-candles. Light saturation is the point at which no increase in photosynthesis occurs with additional light intensity. By spreading the light over a greater leaf area, intensity can be reduced. Imagine a vertical beam of sunlight, 10,000 foot-candles in intensity falling on a horizontal leaf (Figure 6–6). Suppose then that the leaf is gradually inclined; the area illuminated will progressively increase, and the intensity of illumination within this area will decrease. At an inclination of about 80° from the horizontal, the intensity is such that the efficiency of light utilization is optimal as it is spread over a greater area. To determine the area illuminated, a calculation using trigonometry is employed:

$$\frac{1}{\text{cosine of the angle}}$$

$$\frac{1}{\text{cosine } 80°} = 3 \text{ cm}^2$$

$$\frac{10,000 \text{ foot} - \text{candles per cm}^2}{3 \text{ cm}^2} = 3300 \text{ foot candles}$$

[1]Although foot-candles are not an accurate measure of light useful to a plant, *foot-candles* is a simple, easily understood term.

This analysis cannot be applied directly to a field crop under all conditions. Although sunlight is strongly directional, diffuse light under a cloud cover comes almost equally from all directions. Also, some plants such as clover may orient their leaves to face the sun just as a sunflower follows the movement of the sun across the sky. The bending or turning response of plants to a light source is known as **phototropism.**

At 800 foot-candles, photosynthetic efficiency is greatest (Duncan 1965). Although the photosynthetic rate per leaf area is low, efficiency will be high in a plant population. Photosynthesis of the leaves in a plant population can usually be increased by as much as two times with upright leaves (Mitchell 1970). The light compensation point is around 300 foot-candles. This is the intensity at which photosynthesis is equal to respiration. Below this level a negative NAR occurs. Upright leaves allow light to penetrate into the leaf canopy of a crop so that lower leaves may receive adequate light.

Sunlight hitting a leaf is either reflected, absorbed, or transmitted. Horizontally disposed leaves reflect light back into the atmosphere, whereas vertically disposed leaves reflect more light into the leaf canopy to the advantage of lower leaves (Kriedeman, Neales, and Ashton 1964) (Figure 6–7). Light penetrating to the base of the plant may stimulate tiller development and promote thick-walled, lodging-resistant lower internodes.

Light penetration into a stand of plants with different leaf angles was demonstrated by Pearce, Brown, and Blaser (1967), who seeded barley at three rates into flats tilted at 0°, 30°, and 60° from the horizontal. Because barley plants are **geotropic,** (roots grow downward, leaves upward) the developing seedlings assumed a vertical stance regardless of the angle of the flat, and when the first leaf had developed, uniform leaf angles of 90°, 53°, and 18° at the three leaf areas were obtained. The vertical leaves had a higher NAR and allowed more light to penetrate at high LAI's than horizontal leaves. Leaves at 90° required a LAI of 11 to intercept 95% of the light, whereas LAI's of 7.0 and 4.5 were required for 95% light interception at 53° and 18°, respectively.

In a soybean crop with horizontal leaves, light penetration was limited into the fully developed canopy (Sakamoto and Shaw 1967).

Leaf angle was the main factor accounting for differences between three high- and three low-yielding cultivars of barley in Ontario (Gardener et al. 1964). The high-yielding cultivars had narrow, upright leaves, whereas the low-yielding cultivars had wide, floppy leaves.

Corn hybrids differing only in leaf angle (liguless versus normal) showed a 9.8% grain yield advantage for vertical leaf orientation over flat or horizontally disposed leaves when tested over a four-year period in Illinois (Lambert and Johnson 1978). Without a **ligule** at the top of the leaf sheath the leaf blade adopts an upright position. Improved light penetration to the lower leaves that produced a higher overall NAR in the upright leaf type was presented as the reason for the yield advantage.

In forages, Brougham (1960) found that in ryegrass with vertical leaves, 74%

of the incoming light energy was transmitted into the lower leaf canopy per unit leaf area, whereas in the horizontally disposed clover leaves, only 50% of the incoming light energy was transmitted per unit leaf area.

When yields of wild and cultivated sugar beets were compared, the yield advantage of the cultivated types was related to better light distribution due to upright leaves (Watson and Witts 1959). Flat leaf type sugar beets were superior in spaced tests, but higher yields were obtained with the erect leaf type under conditions of plant competition (Mitchell 1970).

UPRIGHT LEAF ADVANTAGES USING A COMPUTER

There is evidence to suggest that upright leaves offer crop producers an architectural advantage, yet breeding for upright leaf types has not resulted in overwhelming success. Various reasons may be cited:

1. The upright leaf feature may not endure for the life of the plant. In cereals, extremes in leaf angle were detectable during the vegetative period, but the leaves assumed a more horizontal position as the head emerged from the boot at the beginning of the important reproductive period. As the spike emerged from the boot, the upper part of the leaf sheath was spread and forced the attached flag leaf into a horizontal position. As the reproductive period progressed, drought may have reduced cell turgidity, and leaves flagged over.

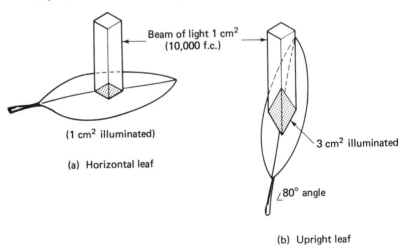

Beam of light 1 cm²
(10,000 f.c.)

(1 cm² illuminated)

(a) Horizontal leaf

3 cm² illuminated

∠80° angle

(b) Upright leaf

Figure 6-6. Light striking horizontal (a) and upright (b) leaves.

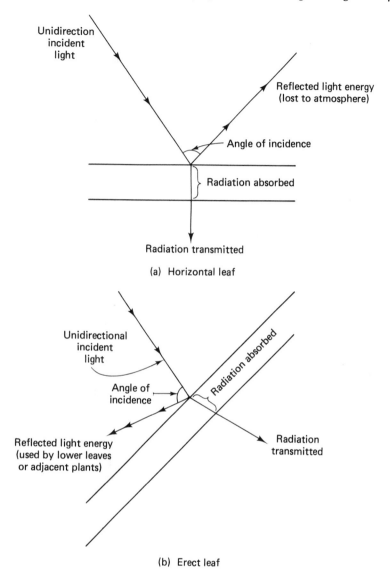

Figure 6-7.   A simple model of the fate of radiant energy falling upon the surface of a leaf. (After Kriedeman, Neales, and Ashton 1964.)

2. Genotypes rated as having upright leaves under one environmental regime often failed to exhibit upright leaf characteristics under another environment. Under high-nitrogen fertility levels, upright leaves often flagged over.

3. Perhaps plant breeding efforts to select for upright leaves were done at the expense of other features.

4. Perhaps conventional systems of evaluating selections failed to determine the merits of upright leaves.

5. Perhaps the advantages attributed to upright leaves were related to other factors.

6. Perhaps species variations can be assumed. Internode length, number of leaves, and amount of leaf flexibility could be factors (Graham and Lessman 1966).

To obtain a more definitive evaluation of the merits of upright leaves, Duncan (1965) and Duncan et al. (1967) used the computer and concluded that leaves at an 80° angle are associated with higher yields than at lesser angles in temperate regions. Theoretical differences in photosynthesis (expressed as dry-matter production in $g/m^2/h$) for leaves displayed at 0° and 80° and at leaf areas of 2, 4, and 8 are shown in Figure 6-8. The greatest advantage of upright leaves was obtained under the high light intensity at noon.

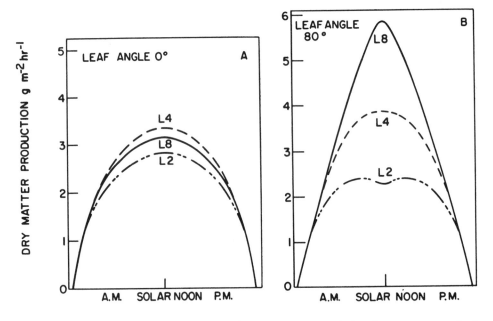

Figure 6-8. Theoretical values for dry-matter accumulation by corn at two leaf angles. (Duncan et al. 1967.)

In a subsequent study, Duncan (1971) showed that the highest photosynthetic rates were obtained using crop canopies with vertical leaves in the upper layers, with lower leaves becoming progressively horizontal. Upright leaves were beneficial only above a LAI of 2.0 (see Figure 6-4). The higher the LAI, the more layers of vertical leaves were required for maximum photosynthesis. In the computer study, Duncan assumed that the plant canopy consisted of ten layers of leaves with an equal leaf area in each layer. All leaves were assumed to transmit 5% and reflect 15% of the light striking them. The light climate was that of a clear July day at 30° and 40° north latitudes. In this study, leaf angle assumed a greater importance in the tropics and least importance in the far north. The advantage of upright leaves was greatest at the high LAI's (Duncan 1971; Monteith 1965).

CONCLUSIONS

Plant architecture is one more tool that the crop producer has at his disposal to increase yield. There is no one best plant shape to suit all crop production systems, and like all biological systems, any single modification in plant architecture such as upright leaves may not by itself result in a yield increase. Each modification must be coordinated with the many other physiological processes occurring in plants and must be coordinated into a suitable cropping system. The challenge to the crop producer—after identifying an architectural modification—is to be able to define the best cropping practice to take advantage of the modification.

QUESTIONS

1. The NAR values for two bromegrass cultivars A and B, expressed as NAR per LAI per day, are shown in Table 6-3. The crop growth rate can be calculated by multiplying LAI × NAR.

TABLE 6–3. NAR and LAI values for two bromegrass cultivars

LAI	NAR (cultivar A)	NAR (cultivar B)
1	12	15
2	11	13
3	10	11
4	9	9
5	8	7
6	7	5
7	6	3

(a) What is the optimum LAI of cultivar A?

(b) What is the optimum LAI of cultivar B?

    (c) What is the LAI of cultivar A when both cultivars are producing at the same crop growth rate?

    (d) What is the LAI of cultivar B when both cultivars are producing at the same crop growth rate?

    (e) Which cultivar has the greatest yield potential?

    (f) Sketch cultivars A and B to indicate morphological differences.

2. Are tall corn hybrids superior to shorter hybrids, or do they just look better? What is a more accurate method of evaluating the yield potential of a hectare of corn than plant height?

3. In Ontario, a final plant population for corn is 44,000 to 59,000 pph (17,800 to 24,000 ppa). Outline the conditions or circumstances under which the high or low rate would be chosen.

4. Sketch designs of plants of the following species that you think would be advantageous in crop production, and list reasons why each feature is considered desirable.

    (a) Sorghum.

    (b) Corn.

    (c) Cereals (spring).

    (d) Cereals (winter).

    (e) Soybeans (indeterminate).

    (f) Sugar beet.

    (h) Asparagus.

    (i) Rhubarb.

    (j) A forage legume.

    (k) A forage grass.

    (l) A lawn grass.

    (m) Cassava with palmately compound leaf.

    (n) A plant with pinnately compound leaves.

5. Which species listed in Question 4 do you feel may have the best architectural design now commonly associated with that plant and give reasons?

6. Construct a hypothetical plant and give reasons for each architectural inclusion.

7. What crop management practices might be associated with a sparse plant with upright leaves?

8. Name changes that have occurred in the past in relation to plant architecture that have resulted in crop management changes.

9. On the north side of corn planted in east-west rows, a large aluminium surface was placed to reflect sunlight onto the corn. Corn plants without a reflector yielded 12,475 kg/ha (200 bu/ac); corn plants beside the reflector yielded 15,711 kg/ha (250 bu/ac).

(a) What do these results suggest?

(b) Explain why plants at the end of the row or on the outside are often superior.

10. Soybeans and corn were planted in alternating strips of four rows each. The corn yields increased about 25%, but the soybean yields decreased by about the same percent. Explain why.

11. Yield per hectare (acre) can be expressed as the average yield per plant times the number of plants per hectare (per acre).

(a) Speculate on the yield per plant in corn as the number of plants per hectare (acre) is increased.

(b) Trace the yield of cereals per plant as the number of plants per hectare (acre) is increased.

(c) Sketch corn and cereal plants at wide spacings, without population stress, and at a close spacing with extreme population stress.

(d) Explain the yield differences found in (a) and (b) and explain the drawings in (c).

REFERENCES FOR CHAPTER SIX

Andries, J.A., et al. 1969. "Effects of Okra Leaf Shape on Boll Rot, Yield, and Other Important Characters of Upland Cotton, *Gossypium hirsutum* L.," *Crop Science* 9:705–710.

———. 1970. "Effects of Superokra Leaf Shape on Boll Rot, Yield, and Other Characters of Upland Cotton, *Gossypium hirsutum* L.," *Crop Science* 10: 403–407.

Anonymous. 1979. *Wheat Improvement.* Published by the International Maize and Wheat Improvement Centre, Mexico.

Black, J. N., and D. J. Watson. 1960. "Photosynthesis and the Theory of Obtaining High Crop Yields by A. A. Niciporovic. An Abstract with Commentary," *Field Crop Abstract* 13:169–175.

Brougham, R. W. 1956. "Effect of Intensity of Defoliation on Regrowth of Pasture," *Australian Journal Agricultural Research* 7:377–387.

———. 1960. "The Relationship between the Critical Leaf Area, Total Chlorophyll Content and Maximum Growth Rate of Some Pasture and Crop Plants," *Annals of Botany* 24:463–474.

Crookston, R. K., et al. 1978. "Agronomic Cropping for Maximum Biomass Production," *Agronomy Journal* 70:899–902.

Daynard, T. B., J. W. Tanner, and W. G. Duncan. 1971. "Duration of the Grain Filling Period and Its Relation to Grain Yield in Corn, *Zea mays* L.," *Crop Science* 11:45–48.

Donald, C. M. 1962. "In Search of Yield," *Journal Australian Institute of Agricultural Science* 28:171–178.

———. 1968. "The Breeding of Crop Ideotypes," *Euphytica* 17:385–403.

Donald, C. M., and J. Hamblin. 1976. "The Biological Yield and Harvest Index of Cereals as Agronomic and Plant Breeding Criteria," *Advances in Agronomy* 28:361–405.

Duncan, W. G. 1965. "A Model for Simulating Photosynthesis and Other Radiation Phenomena in Plant Communities," *Proc. 10th Intern. Grassland Congress,* pp. 120–125.

———. 1971. "Leaf Angles, Leaf Area and Canopy Photosynthesis," *Crop Science* 11:482–485.

Duncan, W. G., et al. 1967. "A Model for Simulating Photosynthesis in Plant Communities," *Hilgardia* 38:181–205.

Eik, K., and J. J. Hanway. 1966. "Leaf Area in Relation to Yield or Corn Grain," *Agronomy Journal* 58:16–18.

Gardener, C. J., et al. 1964. *A Physiological Basis for the Yield Performance of High and Low Yielding Barley Varieties. Agronomy Abstracts.* Published by American Society of Agronomy, Kansas City, Mo.

Graham, D., and K. J. Lessman. 1966. "Effect of Height on Yield and Yield Components of Two Isogenic Lines of *Sorghum vulgare* Pers.," *Crop Science* 6:372–374.

Kriedeman, P. E., T. F. Neales, and D. H. Ashton. 1964. "Photosynthesis in Relation to Leaf Orientation and Light Interception," *Australian Journal of Biological Science* 17:591–600.

Kuleshov, N. N. 1933. "World's Diversity of Phenotypes of Maize," *American Society of Agronomy Journal* 25:688–700.

Lambert, R. J., and R. R. Johnson. 1978. "Leaf Angle, Tassel Morphology, and the Performance of Maize Hybrids," *Crop Science* 17:499–502.

Mason, L., and M. S. Zuber. 1976. "Diallel Analysis of Maize for Leaf Angle, Leaf Area, Yield and Yield Components," *Crop Science* 16:693–696.

Milthorpe, F. L., ed. 1956. "The Growth of Leaves," *Proceed. Univ. Nottingham Third Easter School in Agr. Sci.* London: Butterworth Publ. (223 pp.)

Mitchell, R. L. 1970. *Crop Growth and Culture.* Ames, Iowa: Iowa State University Press. p. 349.

Monteith, J. L. 1965. "Light Distribution and Photosynthesis in Field Crops," *Annals of Botany* 29(113):17–37.

Pearce, R. B., R. H. Brown, and R. E. Blaser. 1967. "Photosynthesis in Plant Communities as Influenced by Leaf Angle," *Crop Science* 7:321–324.

Rutgers, J. N., C. A. Francis, and C. O. Grogan. 1971. "Diallel Analysis of Leaf Characteristics in Maize (*Zea mays* L.)," *Crop Science* 11:194–195.

Sakamoto, C. M., and R. H. Shaw. 1967. "Apparent Photosynthesis in Field Soybean Communities," *Agronomy Journal* 59:73–75.

Tanner, J. W., et al. 1966. "Some Observations on Upright Leaf-Type Small Grains," *Canadian Journal of Plant Science* 46:690.

Tanner, J. W., and N. C. Stoskopf. 1967. "The Plant Resource," *Agricultural Institute Review* 22:25–29.

Thorne, G. N. 1959. "Photosynthesis of Lamina and Sheath of Barley Leaves," *Annals of Botany* 23:365–370.

Troyer, A. F., and W. L. Brown. 1976. "Selection of Early Flowering in Corn: Seven Late Synthetics," *Crop Science* 16:767–772.

Warren Wilson, J. 1960. "Influence of Spatial Arrangement of Foliage Area on Light Interception and Pasture Growth," *Proceedings of the Eighth International Grassland Congress.* Paper 12 A/2, pp. 275–279.

Watson, D. J. 1947. "Comparative Physiological Studies on the Growth of Field Crops. I. Variation in Net Assimilation Rate and Leaf Area between Species and Varieties, and within and between Years. *Annals of Botany* 11:41–76.

———. 1956. "Leaf Growth in Relation to Yield," in F. L. Milthorpe ed., *The Growth of Leaves*, pp. 178–191. London: Butterworth & Co. (Publishers) Ltd.

Watson, D. J., and K. J. Witts. 1959. "The Net Assimilation Rates of Wild and Cultivated Beets," *Annals of Botany* 23:431–439.

Whyte, R. O. 1960. *Crop Production and Environment*, pp. 80–84. London: Faber & Faber Ltd.

Williams, W. A., R. S. Loomis, and C. R. Lepley. 1965. "Vegetative Growth of Corn as Affected by Population Density. II. Components of Growth, Net Assimilation Rate and Leaf Area Index," *Crop Science* 5:215–219.

chapter seven

Photosynthesis, Row Width, Plant Architecture, and Plant Population

Leaf area, NAR, leaf shape and angle are important considerations in management aspects such as plant population and row width. Rhodes (1971) indicated that the relationship between canopy type and yield is highly management-dependent. In practice, row widths and plant populations must be considered together. This chapter deals with integrating the variables related to photosynthesis into a field planting system and specifically, row widths.

ROW PLANTINGS

Cultivated crops have been seeded in rows ever since Jethro Tull (1674–1741) invented the seed drill to eliminate hand throwing of seed, a practice he considered wasteful and uncertain. The best row width for any crop was established through empirical observations and once established, changed little over the years. Sometimes specific reasons for selecting a row width existed. In crops requiring interrow cultivation, a standard width of 100 cm (40 in) was established because it allowed a horse and, later, powered machines to cultivate, spray, or harvest. Architecturally large plants performed well at these spacings, and the songwriters Rodgers and Hammerstein immortalized the image of corn in the lyrics "The corn is as high as an elephant's eye, and it looks like it's climbing clear up to the sky"—at least for Oklahoma.

Large plants required a lot of space, and at the low populations of 14,000 to 20,000 pph (5600 to 8000 ppa), secondary stalks or tillers were produced, which increased leaf area and eliminated attempts to control the leaf area by seeding rate. Tillering or stooling was soon recognized as undesirable for grain production because these secondary stalks produced a small ear or no ear at all. Tillers were referred to derogatorily as **suckers** because they used light and nutrients but contributed little or nothing to grain yield. Under a system of a fixed row width, efforts to increase plant populations were restricted to reducing spacings within the row, generally unsatisfactory for such architecturally large plants.

Producers didn't give up, however. A **check-row planting** or hill planting method was tried, which allowed for cultivation in more than one direction. Although weed control was improved and hand-hoeing was reduced, yield increases were minimal because excessive interrow cultivation caused damaging root pruning; competition among the two to four plants per hill was intense; sunlight interception was inadequate; and hill planting requiring great accuracy in seeding, was too laborious, time-consuming, and costly (Dungan, Lang, and Pendleton 1958). Some efforts were directed toward tests at wider row spacings to reduce lodging problems and to determine if a forage legume could be underseeded in corn. Rows 2 metres (6½ ft.) and wider were tried but showed reduced corn yields.

Increased plant populations in corn came when the tillering habit was bred out of corn, and when preemergence herbicides eliminated the need for interrow spraying and cultivation. Theoretically at least, row widths could be reduced. Conventional row widths could not be discarded, however, until changes in machinery were made and, equally important, changes in plant architecture. Changes in plant shape and size were slow to develop perhaps because tests continued to be conducted in standard 100-cm (39–40 in.) wide rows. The change is under way, and it is occurring in conjunction with the introduction of early maturing hybrids adapted to what was once considered the northern limit for corn production (Major and Hamilton 1978). Early maturing hybrids have a shortened vegetative period and a plant size much smaller than the corn that prompted Rodgers and Hammerstein's lyrics. Over the years, corn rows have become progressively narrower, and plant populations have been increased.

Success with small plants of corn established a trend toward architecturally smaller plants in other crops (Anonymous 1979). It was recognized that for grain-production purposes, small plants facilitate harvesting operations because of reduced biomass and less lodging. Sorghum traditionally was 180 cm (70 in) tall, and the release in 1956 of a dwarf sorghum, 100 to 125 cm (40–50 in) in height, was welcome, and today all grain sorghum hybrid cultivars have small stature (Karper 1949; Martin 1951; Anonymous 1979). Wheat was reduced 15 cm (6 in) (Colvin 1949), and further reductions may be desirable because dwarf cultivars require less photosynthetic energy to be used in producing straw. Of great importance, however, is the fact that dwarf cultivars may tolerate high plant populations (Singleton 1949).

Changes in row widths and plant populations have been made as indicated below.

ROW WIDTH RECOMMENDATIONS FOR CORN

The row width can be determined by the machinery available on each farm and will be a personal choice. Whether a population of 44,000 pph (18,000 ppa) or one up to 59,000 pph (24,000 ppa) is used as recommended in Ontario will depend on several considerations:

1. Where previous yield levels have been low and where plant lodging has been a problem, the low end of the recommended range (44,000 pph) (18,000 ppa) should be chosen. This population should also be chosen under conditions of late planting, on soils that are low in fertility, susceptible to drought, or imperfectly drained.
2. On farms with a history of high yields with no lodging problems, where soils are fertile and well-drained and with a high moisture-holding capacity, plant populations up to 59,000 pph (24,000 ppa) may be chosen.
3. In the short-season areas where adapted hybrids tend to be small, when a small hybrid cultivar can be deliberately chosen, and under conditions of early planting that tend to promote small plant size, the high plant population of 59,000 pph (24,000 ppa) is warranted.
4. Corn for whole plant silage production can be seeded at 57,000 to 62,000 pph (23,000 to 25,000 ppa) in an effort to increase stover production and total silage yields without concern about ear size and development. Increased populations will lead to greater stalk break-age for corn that is being produced for grain, but this can be avoided in silage production by the earlier date of harvest.

A corn producer, having established a row width will choose a plant population according to his objectives and farm conditions. The distance between seeds in the row at seven row spacings and the seeds recommended per hectare are shown in Table 7–1. At 59,000 pph (24,000 ppa) in 102-cm (40 in) wide rows, plants will be 15 cm (6 in) apart in the row (Table 7–1), a distance generally considered to be the minimal acceptable distance between plants in a row. Obviously a more uniform distribution of plants occurs at a fixed plant population when row width is reduced. Without the need to seed in rows, other plant arrangements may be possible that improve sunlight interception.

TABLE 7–1. Centimetres between seeds to achieve specific populations

Final Plants/Hectare*	Seed per Hectare**	Row Width—centimetres (inches)***						
		71(28)	76(30)	81(32)	86(34)	91(36)	97(38)	102(40)
40,000	43,900	32	30	28	26	25	24	22
44,000	49,400	28	27	25	24	22	21	20
49,000	54,900	26	24	22	21	20	19	18
54,000	60,400	23	22	20	19	18	17	16
59,000	65,900	21	20	19	18	17	16	15
64,000	71,400	19	18	17	16	15	14	13

*1 hectare = 2.47 acres.

**Based on 10% loss of plants. With planting after the optimum planting date, a loss of 5% of the plants may be more realistic.

***1 centimetre = 0.39 inches.

Source: Field Crop Recommendations, 1980 Publication 296. Ontario Ministry of Agriculture and Food.

HEXAGONAL ARRANGEMENT

There are 10,000 square metres in a hectare, and if a plant population of 10,000 was established on a square planting method, each plant would be 1 metre apart (Figure 7–1). If plants were seeded on the square ½ metre apart (50 cm), nine plants per metre or 90,000 pph could be seeded (Figure 7–2). Although the distance between plants A, B, C, and D in Figure 7–2 is 50 cm, the distance from A to C is 70.7 cm. If 90,000 pph could successfully be grown at ½-metre spacing, then by arranging plants in a hexagonal spacing (Figure 7–3), 103,300 could be seeded in a hexagonal arrangement with all plants 50 cm apart.

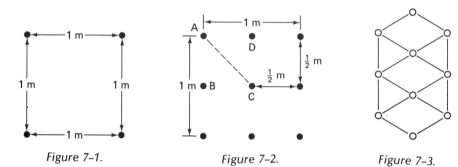

Figure 7–1.   Figure 7–2.   Figure 7–3.

The results of testing corn hybrids under an equidistant planting arrangement illustrates the need for a cautious approach. In 1966, at Guelph, highest yield was obtained at 69,900 pph (28,300 ppa) (Table 7–2). In 1967, the high population was superior (Table 7–3).

TABLE 7–2.   Grain yields of corn seeded in a hexagonal planting arrangement, Guelph, Ontario, 1966

Plant Population		Grain Yield		LAI
(pph)	(ppa)	(kg/ha)	(bu/ac)	
105,900	42,800	5100	81.3	4.2
91,300	36,900	5570	88.8	3.9
79,500	32,100	6020	96.0	3.4
69,900	28,300	6290	100.3	3.2

Source: Data based on personal communication from J. W. Tanner.

Equidistant planting arrangements have been successful in other crops. According to one source (Anonymous 1976) soybeans in 75-cm (30 in) wide rows, 7½ cm (3 in) apart in the row, compared with an equidistant 24-cm (9.4 in) grid, gave better light interception, which resulted in more symmetrical plants and a better grain-to-straw ratio. Narrow rows prevent inter-row

TABLE 7–3. Grain yields of corn seeded in a hexagonal planting arrangement, Guelph,
Ontario, 1967

| Plant Population | | Grain Yield | | LAI |
(pph)	(ppa)	(kg/ha)	(bu/ac)	
124,300	50,300	7110	113.4	5.1
79,500	32,100	6940	110.7	4.2
44,700	18,100	5430	86.6	2.2

Source: Data based on personal communication from J.
W. Tanner.

cultivation for weed control. Chemical weed control is essential therefore. Without inter-row cultivation the soil remains level and facilitates combining directly while standing in the row. More than 40 years ago Wiggans (1939) also concluded that "the nearer the arrangement of plants on a given area approaches a uniform distribution, the greater will be the yield" (p. 321).

Wilcox (1974) tested three soybean cultivars in approximately equidistant spacings at fourteen population densities ranging from 25,000 to 582,000 pph (10,100 to 235,600 ppa) at Lafayette, Indiana. The three cultivars differed in their response to population density; two cultivars were more responsive at 281,000 pph (113,700 ppa) and one at 456,000 pph (187,600 ppa).

PROBLEMS AND PROSPECTS FOR CORN ROW WIDTHS

The effect of plant density and distribution in corn has received much emphasis and was reviewed thoroughly by Dungan, Lang, and Pendleton (1958). With few exceptions, researchers have found that yield increases with higher populations up to an optimum, above which the yield declines due to a reduction in size or number of ears. Corn ears are located midway up the stalk and are subject to shading by upper leaves, and under conditions of high plant populations, shading may be too intense for silk development and to coincide with pollen release. For conventional hybrids, the maximum LAI has usually occurred in the narrow range of 3.3 to 4.0 (Eik and Hanway 1966; Nunez and Kamprath 1969; and Williams et al. 1968). For higher plant populations to be successful, the shade intolerance problem will have to be overcome. There is evidence that steady progress in this direction is possible. Change may not be abrupt, and it is in the short-season areas where small hybrids are adapted that the greatest thrust can be expected.

An apparent tolerance of five short-season commercial cultivars to population pressure led to studies in Ontario with 46- and 91-cm (18 and 36 in) rows at populations of 48,000, 62,000, and 72,000 pph (19,400, 25,100 and 29,100 ppa), (Hunter, Kannenberg, and Gamble 1970). All hybrids increased in grain yield with each increase in population and gave small but significant yield increases when row width was narrowed (Table 7-4).

Experienced corn producers in short-season areas and using architecturally small plants can increase plant populations above the 44,000 to 59,000 pph (18,000 to 24,000 ppa) recommendation. A 10% increase to a population range of 48,400 to 65,000 final plant population per hectare (20,000 to 26,000 ppa) may be in order.

ROW WIDTHS FOR SOYBEANS

Cultivars of soybeans can be grouped into ten maturity groups (00 to VIII). Groups 00, 0, I, and II can be grown in southern Canada and the U.S. Midwest, with groups III to VIII adapted to more southerly regions. Early soybean cultivars are architecturally smaller, shorter, and less leafy than later maturing soybeans and are well adapted to narrow row systems. Northern types (groups 00 to II) are indeterminate and achieve about half their vegetative growth when reproduction begins. Row widths narrow enough to allow the plants to close the leaf canopy between rows at the onset of flowering should have an advantage. In contrast, southern determinate soybeans achieve most of their vegetative growth before reproductive growth begins, and the canopy will have developed sufficiently to intercept all the light even in 100-cm wide rows.

TABLE 7–4. Grain yield of five corn hybrids grown at two row widths and three plant populations at Fullarton and Elora, Ontario, in 1967 and 1968

Hybrid	Row Width (cm)				Plant Population/ha (per acre)						mean	
	46 (18 in)		91 (36 in)		48,000 (19,400)		62,000 (25,000)		72,000 (29,100)			
	t/ha	bu/ac	t/ha	bu/ac								
A	7.45	119.7	7.28	117.0	6.96	111.8	7.31	117.4	7.82	125.6	7.36	118.2
B	7.36	118.2	7.21	115.8	6.81	109.4	7.45	119.7	7.58	121.8	7.28	117.0
C	7.29	117.1	7.16	115.0	6.94	111.5	7.23	116.1	7.51	120.6	7.23	116.1
D	7.90	126.9	7.42	119.2	7.22	116.0	7.60	122.1	8.16	131.1	7.66	123.1
E	6.68	107.3	6.45	103.6	6.01	96.6	6.67	107.2	7.01	112.6	6.56	105.4
mean	7.33	117.8	7.10	114.1	6.79	113.1	7.24	116.3	7.61	122.3	7.22	116.0

Source: Data from Hunter, Kannenberg and Gamble 1970.

A substantial variation in architectural size of soybean plants is found across the range of maturity groups. Plant height of six group VIII varieties averaged 120 cm (47.2 in); group VII, 100 cm; and group VI, 88 cm (34.6 in) when grown at Athens, Georgia (Boerma 1979). Height of six group II varieties grown at Columbia, Missouri averaged 74 cm (29.1 in); group III, 81 cm (31.9 in); and group IV, 96 cm (37.8 in) (Luedders 1977). At Guelph, four varieties from group 00 averaged 66 cm (26 in) in height.

Soybean production in North America can be divided into three zones. In the southern soybean producing area of the United States, architecturally large determinate soybean cultivars are well suited to conventional 100-cm

(39–40 in) wide rows. In the northern areas of Minnesota, Wisconsin, southern Manitoba, Ontario, and Quebec where maturity group 00 and 0 soybeans are grown, narrow rows or solid-seeded soybeans give top yields under weed-free environments.

Between the two extremes of soybean production, the response to row width is less predictable. Soybeans adapted to this region are intermediate in height, and the use of corn-planting equipment for soybeans, plus concern about adequate weed control, has delayed the use of narrow rows in this area. Much of the research in this region has involved row widths that will permit mechanical cultivation for weed control (Cooper and Lambert 1965; Cooper 1969; Lehman and Lambert 1960). In studies with narrow rows, the general observation was that yields were maximized at approximately the 50-cm (20 in) row widths (Hicks et al. 1969; Pendleton, Bernard, and Hadley 1960).

A common observation with narrow-row soybeans is that as populations are increased, less light penetrates to the base of the soybean, cells at the base of the stem become elongated and weaker than stems with adequate light, and lodging results. Under narrow-row conditions and at a fixed seeding rate, a higher percentage of the plants that emerged survive (Cooper 1970, 1971), and the higher final plant population results in greater lodging. Seeding rate for narrow-row soybeans is more critical than in wide rows.

To determine the best seeding rate, Cooper (1977) tested nine soybean cultivars in three different row widths at five populations in field plots in central Illinois over a six-year period. When seeding rate was low enough to prevent lodging, yield advantages of 10 to 20% (300 to 600 kg/ha) (268 to 536 lb/ac) were obtained from 17-cm (6.7 in) rows as compared to 50- or 75-cm (20 or 30 in) rows. Earlier maturing cultivars were more responsive to 17-cm (6.7 in) rows, and under high yield conditions and irrigation, 30 to 40% (900 to 1200 kg/ha) (892 to 1071 lb/ac) yield responses were observed with one of the more responsive cultivars (Table 7–5).

A small-seeded cultivar may have 18,000 seeds per kilogram (8,163 seeds per pound). At a target seeding rate of 37.5 viable seeds per square metre (31.5 seed/sq yard), the seeding rate would be:

10,000 m² (hectare) x 37.5 = 375,000 seeds/hectare (151,800 seeds/acre)

If 18,000 seeds are found per kilogram, this equals

$$\frac{375,000}{18,000} = 20.8 \text{ or } 21 \text{ kg/ha } (18\% \text{ lbs/ac})$$

Medium-sized soybeans have 6000 to 13,000 seeds per kilogram (2,721 to 5,897 seeds per pound), and the seeding rate would be 29 to 64 kg/ha (26 to 57 lb/ac). For large-seeded soybeans, cultivars with 2000 seeds per kilogram (907 seeds per lb), a seeding rate of 187.5 kg/ha (167 lb/ac) would be required to give 37.5 seeds per square metre (or 31.5 seeds per sq yd). Depending on germination, these seeding rates can be increased by about 10% to compensate for less than 100% germination.

TABLE 7–5. Soybean seed yields as influenced by cultivar and row width treatments, central Illinois, 1971

	Row Width (cm) (in)					
Cultivar	75 t/ha	(30) bu/ac	50 t/ha	(20) bu/ac	17 t/ha	(6.7) bu/ac
Corsoy	3.50	56.2	3.95	63.4	4.44	71.3
Amsoy 71	3.56	57.2	3.90	62.6	3.93	63.1
Beeson	3.69	59.3	3.70	59.4	4.26	68.4
Wayne	3.78	60.7	3.93	63.1	4.17	67.0
Calland	3.68	59.1	3.89	62.5	3.85	64.1
Williams	3.43	55.1	3.70	59.4	3.86	62.0
Miller 67	2.92	46.9	3.02	48.5	3.18	51.1
Clark 63	3.34	53.7	3.33	53.5	3.28	52.7
Cutler	3.78	60.7	3.88	62.3	3.44	55.3
Mean	3.52	56.5	3.70	59.4	3.82	61.4

Source: Data from Cooper 1971

Close spacing with its accompanying reduced light penetration to the base of the soybean plant causes lower internodes to elongate, thereby raising the lower pods of the ground to facilitate harvesting (Hoggard, Shannon, and Johnson 1978) (Table 7-6), or may result in fewer bottom pods because of flower abortion under low light intensity (Dominquez and Hume 1978).

TABLE 7–6. Means for several characteristics of three soybean cultivars at each of four plant populations, Portageville, Missouri, 1975 and 1976

Cultivar	Population of Plants per m of row	3 foot row	Lodging Score*	Height (cm)	(in)	Lower Internode Length (cm)	(in)	Yield (t/ha)	(bu/ac)
Essex	23	21	1.3	69	27	3.1	1.2	2.77	44.5
	33	30	1.4	71	28	3.3	1.3	2.68	43.1
	43	39	1.6	72	28	3.3	1.3	2.64	42.4
	53	48	2.0	75	30	3.8	1.5	2.69	43.2
Mean			1.6	72	28	3.4	1.3	2.70	43.4
Forrest	23	21	2.1	95	37	3.5	1.4	2.87	46.1
	33	30	2.3	95	37	3.8	1.5	2.78	44.7
	43	39	2.3	93	37	3.8	1.5	2.60	41.8
	53	48	2.7	91	36	3.9	1.5	2.69	43.2
Mean			2.3	93	37	3.7	1.5	2.74	44.0
Mack	23	21	2.3	89	35	3.2	1.3	2.66	42.7
	33	30	2.7	89	35	3.6	1.4	2.45	39.4
	43	39	2.9	90	35	3.6	1.4	2.50	40.2
	53	48	3.3	90	35	4.0	1.6	2.48	39.8
Mean			2.8	90	35	3.6	1.4	2.52	40.5

Source: Hoggard, Shannon and Johnson 1978.
*1.0 Indicates no lodging

Donovan, Dimmock, and Carson (1963) found highest grain yields at a 17-cm (6.7 in) wide row spacing with plants 7.5 cm (3 in) apart in the row, but highest oil content was obtained at the widest row and widest plant spacing. A 17-cm (6.7 in) row with a 10-cm (4.0 in) in-row spacing was suggested as a compromise to help reconcile the differential response of oil and yield.

Differences in soybean plant architecture are evident in Figure 7–4.

Figure 7–4. The row width used for soybeans depends on plant architecture. Small stature soybeans can be planted at higher plant populations than can large statured plants. (Courtesy Ontario Ministry of Agriculture and Food.)

ROW WIDTH FOR DRY OR WHITE BEANS

White beans are commonly grown in rows 60 to 90 cm (24 to 36 in) apart to facilitate mechanical pulling at maturity, for interrow cultivation, and to allow for free air movement to reduce the incidence of white mold. Effective herbicides and the prospects for more flexible harvesting equipment may allow for a reduction in row widths. In seven of nine experiments in New York State conducted from 1966 to 1977, narrow rows produced 7 to 48% higher yields (Kueheman et al. 1979). At a given density, plants with more equidistant arrangements outyielded the more rectangular arrangements. Plants spaced at 25 x 25 cm (10 in x 10 in) yielded 13% more than those at 76 x 8 cm (30 x 3) spacings, plants spaced at 20 x 20 cm (8 x 8 in) cm yielded 12% more than at 76 x 5 cm (30 x 2 in) and yields averaged across five cultivars were 48% higher at 30 x 10 cm (12 x 4 in) than those at 60 x 5 cm (24 x 2 in).

Seeding rates and row-width recommendations for field beans are a compromise between the harvesting equipment, disease, LAI, and yield. In Ontario a 60- to 70-cm (24 to 28 in) row width with 16 to 20 seeds per metre (17 to 18 seeds per yard) of row is recommended. In a 70-cm (28 in) row, the seeding rate is 16 to 18 kilograms per hectare (14 to 16 lb/ac).

Work in Idaho with two determinate bush snap beans and two indeterminate semivining types at plant populations of 107,600 to 968,700 pph (43,560 to 392,180 ppa) indicated that seed yields of determinate (but not indeterminate) bean plants could be increased by higher plant populations in equidistant plant arrangements (Westermann and Crothers 1977). This appeared to be related to a tolerance of determinate types to high plant populations, a constant number of seeds per pod, and grams per seed.

ROW WIDTH IN CEREALS

Cereals generally are seeded in 17- to 22-cm (7 to 9 in) wide rows and are responsive to planting pattern or spatial arrangement but are unaffected over a wide range of seeding rates. This fact is related to the tillering habit of most cereals. At a low seeding rate, tillering is encouraged, whereas at a high seeding rate, tillering is restricted.

Row-width studies have produced remarkably consistent results, and a study with three spring barley genotypes over a three-year period seeded at 134 kg/ha (120 lb/ac) (Brinkman, Luk, and Rutledge 1979) (Table 7–7) is typical. Narrow rows produced a 12% increase over wider rows at a constant seeding rate. Similar results have been produced at diverse areas: higher yields at row spacings 7.5- to 22-cm (3 to 9 in) apart have been obtained than for rows 30 cm (12 in) or wider.

In Europe, Holliday (1963) concluded that cereals drilled in rows narrower than 15 cm (6 in) gave a yield increase of 2 to 10%; similar increases were

TABLE 7–7.   Percentage means across three genotypes and three environments for five traits over three years with three row spacings at Madison, Wisconsin

Row Spacing (cm)	Grain Yield	Straw Yield	Spikes per m²	100 kernel wt.	Plant Height
30	100	100	100	100	100
15	108	109	110	99	100
7.5	112	114	118	100	102

Source: Data from Brinkman, Luk, and Rutledge 1979.

Note: Means are expressed as a percentage of the means for the 30-cm row spacing.

obtained for fall- and spring-sown small grains in rows less than 15 cm (6 in) wide in England (Baldwin 1963); in Michigan, Foth, Robertson, and Brown (1964) reported a 23% increase in oat yields with 9-cm (3.5 in) over 18-cm (7 in) rows in a two year study; in Ontario, Stoskopf (1967) found that upright leaf winter wheat cultivars yielded an average 9% more grain in 9- and 11-cm (3.5 and 4 in) rows over 18- and 23-cm (7 and 9 in) rows, and five barley cultivars yielded 5% more in 11- (4 in) than 18-cm (7 in) rows (Finlay, Reinbergs, and Daynard 1971); in Manitoba, Siemens (1963) reported highest yields in the narrowest row spacings when spring barley, oats, wheat, and flax were seeded in rows 8 to 38 cm (3 to 15 in) apart; and in Alberta, Briggs (1975) found a small increase in 15-cm (6 in) over 30-cm rows with three spring wheat cultivars. Siemens (1963) reported similar findings in the Netherlands and Australia.

Note the "waste" space between the rows of upright-leaved oat plants (Figure 7–5).

Figure 7–5. The five rows in this oat plot have upright leaves on short-statured plants. In standard 18-cm wide rows, sunlight interception may be inadequate for maximum photosynthesis for this cultivar. The architecture of this oat cultivar has allowed the underseeded forages to benefit from sunlight. Oat yields may be increased if this cultivar is seeded in rows closer than 18 cm (7 in). (Courtesy Ontario Ministry of Agriculture and Food.)

FORAGES

Forages are normally seeded in broadcast stands, and from a light interception standpoint this is ideal. In low rainfall areas, however, trials with alternate-row seedings of grasses and legumes 46 cm (18 in) apart showed a yield advantage. At Swift Current, Saskatchewan, the yield advantage has been as high as 25% or 2800 kg/ha (2500 lb/ac) and in very dry years 560 to 1700 kg/ha (500 to 1518 lb/ac) (Anonymous 1977). The theory is that competition for moisture is reduced and during a dry season, alfalfa roots penetrate deeply for moisture and allow adjacent grass rows to use more shallow water sources.

Forages planted in rows in higher rainfall areas are less satisfactory than broadcast seedings because leaf area and light interception become the limiting factors (Rhodes 1971).

QUESTIONS

1. Explain the term *management-dependent*.
2. The results of a corn yield trial are shown in Table 7–8 under weed-free conditions and adequate moisture and nutrient fertility.

TABLE 7–8. Grain yield (kg/ha) (bu/ac) of two corn hybrid cultivars in 1-meter wide rows at six plant populations

Plant Population		Hybrid A		Hybrid B	
(pph)	(ppa)				
28,000	11,300	3763	60	2509	40
33,600	13,600	5018	80	4077	65
38,400	15,500	5958	95	5331	85
43,200	17,500	6586	105	6272	100
48,000	19,400	5645	90	6899	110
52,800	21,400	5018	80	7213	115

   (a) Speculate why the yield of hybrid A declined above 43,200 pph (17,500 ppa).
   (b) Speculate why hybrid B continued to give an increase in yield with increased population.
   (c) What advice would you give a farmer who is considering growing one of these hybrid cultivars?
3. Row widths suitable for soybeans can vary between 15 cm to 102 cm (6 to 40 in). What row width would you select for a soybean from group 00, group IV, and group VIII (see text) to be grown at three different latitudes? Suggest reasons for your selection of row width.
4. How many centimetres (inches) space will be found between soybean plants seeded at a recommended seeding rate in 71- and 18-cm (28 and 7 in) wide rows? Could a better seeding arrangement be developed? On what factors will a modified plant arrangement depend?
5. What are the sources of information in your region for seeding rate and row width recommendations? Give the recommended row widths and seeding rates for selected crops in your area.

6. Why is one rate of seeding often given in a recommendation booklet? How should such a recommendation be treated? For whom is such a general recommendation provided? Why are more specific recommendations not available? How can you decide on the best recommendation for your particular situation?

7. Interpret Table 7-6, paying close attention to how cultivar height influenced lodging, lower internode length, and grain yield. What is the best number of plants per metre of row? Are these results applicable over a wide geographic area?

8. The work of Westermann and Crothers (1977) with determinate and indeterminate beans indicated that the greatest potential for seed yield increases in high plant populations is with determinate cultivars. Explain why.

9. Why is increased plant population often associated with a decrease in quality factors?

10. Discuss the virtues of reducing row width under conditions of a fixed plant population. Under what conditions might this be most advantageous?

REFERENCES FOR CHAPTER SEVEN

Anonymous. 1976. "Soybeans Planted Equidistantly Increase Yields," *Crops and Soils* 29(3):26.

Anonymous. 1977. "New Planting Patterns for Forages," *The Furrow* 82(1):24–25.

Anonymous. 1979. "Crop Plants: Short is Beautiful," *The Furrow* 84(1):2–5.

Anonymous. 1979. *Field Crop Recommendations, 1980.* Publication 296. Ontario Ministry of Agriculture and Food.

Baldwin, J. H. 1963. "Closer Drilling of Cereals," *Agriculture* 70:414–417.

Boerma, H. R. 1979. "Comparison of Past and Recently Developed Soybean Cultivars in Maturity Groups VI, VII, VIII," *Crop Science* 19:611–613.

Briggs, K. G. 1975. "Effects of Seeding Rate and Row Spacing on Agronomic Characteristics of Glenlea, Pitic 62, and Neepawa Wheats," *Canadian Journal of Plant Science* 55:363–367.

Brinkman, M. A., T. M. Luk, and J. J. Rutledge. 1979. "Performance of Spring Barley in Narrow Rows," *Agronomy Journal* 71:913–916.

Colvin, W. S. 1949. "New Short Wheats May Double Yields in Soft Wheat Belt," *Crops and Soils* 1(7):18–19.

Cooper, R. L. 1969. "What's New in Soybean Row Spacing and Populations," *Soybean News* 20(3):1–2.

———. 1970. "Early Lodging—A Major Barrier to Higher Yields," *Soybean Digest* 30:12–13.

———. 1971. "Influence of Soybean Production Practices in Lodging and Seed Yield in Highly Productive Environments," *Agronomy Journal.* 63:490–493.

————. 1977. "Response of Soybean Cultivars to Narrow Rows and Planting Rates under Weed-Free Conditions," *Agronomy Journal* 69:89–92.

Cooper, R. L., and J. W. Lambert. 1965. "Narrow Rows: How Much Do They Increase Soybean Yields in Minnesota?" *Minnesota Farm Home Science* 22(4):5–7.

Dominquez, C., and D. J. Hume. 1978. "Flowering, Abortion, and Yield of Early-Maturing Soybeans at Three Densities," *Agronomy Journal* 70:801–805.

Donovan, L. S., F. Dimmock, and R. B. Carson. 1963. "Some Effects of Planting Pattern on Yield, Percent Oil and Percent Protein in Mandarin (Ottawa) Soybeans," *Canadian Journal Plant Science.* 43:131–140.

Dungan, G. H., A. L. Lang, and J. W. Pendleton. 1958. "Corn Plant Population in Relation to Soil Productivity," *Advances in Agronomy.* 10:435–473.

Eik, K., and J. J. Hanway. 1966. "Leaf Area in Relation to Yield of Corn Grain," *Agronomy Journal* 58:16–18.

Finlay, R. C., E. Reinbergs, and T. B. Daynard. 1971. "Yield Response of Spring Barley to Row Spacing and Seeding Rate," *Canadian Journal Plant Science* 51:527–533.

Foth, H. D., L. S. Robertson, and H. M. Brown. 1964. "Effect of Row Spacing Distance on Oat Performance," *Agronomy Journal* 56:70–73.

Hicks, D. R., et al. 1969. "Response of Soybean Plant Types to Planting Patterns," *Agronomy Journal* 61:290–293.

Hoggard, A. L., J. G. Shannon, and D. R. Johnson. 1978. "Effect of Plant Population on Yield and Height Characteristics in Determinate Soybeans," *Agronomy Journal* 70:1070–1072.

Holliday, R. 1963. "The Effect of Row Width on the Yield of Cereals," *Field Crop Abstracts* 16:71–81. (Review article)

Hunter, R. B., L. W. Kannenberg, and E. E. Gamble. 1970. "Performance of Five Maize Hybrids in Varying Plant Populations and Row Widths," *Agronomy Journal* 62:255–256.

Karper, R. E. 1949. "The Rise of Combine Sorghums," *Crops and Soils* 1(5): 14–16.

Kueneman, E. A., et al. 1979. "Effect of Plant Arrangements and Densities on Yields of Dry Beans," *Agronomy Journal* 71:419–424.

Lehman, W. F., and J. W. Lambert. 1960. "Effects of Spacing on Soybean Plants between and within Rows on Yield and Its Components," *Agronomy Journal* 52:84–86.

Luedders, V. D. 1977. "Genetic Improvement in Yield of Soybeans," *Crop Science* 17:971–972.

Major, D. J., and R. I. Hamilton. 1978. "Adaptation of Corn for Whole-Plant Silage in Canada," *Canadian Journal Plant Science* 58:643–650.

Martin, J. H. 1951. "Tailor-Made Sorghum for Industry and Agriculture," *Crops and Soils* 3(5):20–23.

Nunez, R., and E. Kamprath. 1969. "Relationships between N Response, Plant Population, and Row Width on Growth and Yield of Corn," *Agronomy Journal* 61:279–282.

Pendeton, J. W., R. L. Bernard, and H. H. Hadley. 1960. "For Best Yields Grow Soybeans in Narrow Rows," *Illinois Research* 2(1):3–4.

Rhodes, I. 1971. "The Relationship between Productivity and Some Components of Canopy Structure in Ryegrass (*Lolium* spp.).II: Yield, Canopy Structure and Light Interception," *Journal of Agricultural Science* (Cambridge). 77:283–292.

Siemens, L. B. 1963. "The Effect of Varying Row Spacings on the Agronomic and Quality Characteristics of Cereals and Flax," *Canadian Journal Plant Science* 43:119–130.

Singleton, W. R. 1949. "Short Corn May Be Good Corn," *Crops and Soils* (7): 22–24.

Stoskopf, N. C. 1967. "Yield Performance of Upright-Leaved Selections of Winter Wheat in Narrow Row Spacings," *Canadian Journal Plant Science* 47:597–601.

Westermann, D. T., and S. E. Crothers. 1977. "Plant Population Effects on the Seed Yield Components of Beans," *Crop Science* 17:493–496.

Wiggans, R. G. 1939. "The Influence of Space and Arrangement on the Production of Soybean Plants," *Journal of the American Society Agronomy* 31: 314–321.

Wilcox, J. R. 1974. "Response of Three Soybean Strains to Equidistant Spacings," *Agronomy Journal* 66:409–412.

Williams, W. A., et al. 1968. "Canopy Architecture at Various Population Densities and the Growth and Grain Yield of Corn," *Crop Science* 8:303–308.

chapter eight

Economic Yield, Biological Yield, and Harvest Index

If a crop production scientist spots a cultivar believed to have the potential for yield increases through selective use of morphological features, how would he best evaluate this "ideal" cultivar? If tests are designed to include the many variables that might influence performance, extensive and expensive research tests would be required. Such tests may need to include several seeding rates and plant populations, a number of fertility levels, and possibly variations in planting patterns. On the other hand, if a standardized testing program is adopted, the unique feature(s) of the cultivar may not be expressed. Standardized tests involve comparisons with regionally adapted **check** cultivars with proven performance grown under cultural methods considered typical of good local practice. Results of such tests are more useful for recommendation purposes and for providing assurance of performance. They tend to maintain the *status quo*. And finally, if a modified and limited testing program is adopted with selected treatments only, there is concern that maybe one key factor was overlooked, possibly the missing link to greater productivity. Also, plant response is unpredictable, and preguessing the response may be risky. Therefore, rather than rely on scientists at research stations, a farmer may wish to evaluate a specific treatment on his own farm using a cultivar he believes to have yield potential. Field-scale tests are often the final evaluation of a new production recommendation.

Whatever the testing program, it is customary to emphasize economic, grain, or dry-matter yield, depending on the crop, as the measure of success or failure of the cultivar. Although disease, lodging, vigor, height, protein or oil content, or other criteria may be measured, the "bottom line" invariably becomes yield. However, performance based on yield alone provides incomplete information. A cereal cultivar tested under a specific regime may not produce more grain perhaps because unproductive or low-yielding tillers were promoted, and a high leaf area and poor partitioning of food energy into grain and straw resulted in an increase in straw yield. In a forage cultivar, the treatment(s) may have increased dry-matter yield but may be accompanied by reduced forage quality, shorter stand longevity, or decreased seasonal distribution. If a number of variables are included in the testing program, interpretation of the results is often very complex and requires the use of mathematical interpretation involving statistical analysis, a computer, and extensive know-how.

If crop production practices are to contribute toward increasing yield, and also keeping pace with and effectively utilizing the new cultivars released, additional information and understanding are needed on the overall performance that can be understood readily by the farm producer. In the hands of trained farm personnel, a high-yielding cultivar may become an exceptional cultivar if standarized test results provide information for additional insight into performance. Imaginative new approaches to crop production may be encouraged if sufficient research data are available to allow plant responses to be understood.

The purpose of this chapter is to give an insight into plant responses to various treatments by common measurements supplementary to yield performance. The end result is to help reduce the perplexity frequently associated with performance data when several treatments are applied because often as one component is increased, another is decreased. When several variables are being measured, a single response measure is much easier to comprehend than two or more. A single **productivity score** is proposed, therefore, with the personal view that it provides a simple method of evaluating a complex set of interacting data in an easy-to-understand manner adaptable to most crops. Of greater importance than the mechanics of calculating a productivity score is the easy interpretation afforded complex plant responses. Logical conclusions can be drawn that lead to understandable steps in crop production. Productivity scores, therefore, are used throughout this chapter as a key to a better understanding of crop production. This method allows yield and quality to be advanced on an equal basis. The background to the development of the scoring is discussed, and productivity scores are applied to examples of various crops and treatments taken from the literature.

PAST APPROACHES TO SCORING PERFORMANCE

Scoring criteria have been sought for many years, and various measurements have been proposed, but few have been adopted. As early as 1914, Beaven (1947), as reviewed by Donald and Hamblin (1976), proposed an "efficiency of grain production" or "migration coefficient" to show the distribution of assimilate into grain and nongrain parts. Both grain and straw dry-matter yields were required measurements. An index value was given as the proportion by weight of grain to total above-ground produce, today referred to as the **harvest index.** If the weight of grain and straw totaled 100 units and the weight of grain was 50 units, then the harvest index would be calculated in percent as follows:

$$\frac{50 \text{ units of grain}}{100 \text{ units of grain and straw}} \times 100 = 50\%$$

In an attempt to determine optimum seeding rates, Holliday (1960*a* and *b*) developed mathematical formulae including a "coefficient of multiplication" (yield divided by seed weight). This approach was of some value in determining population density in crops where vegetative growth was measured; but where grain yield was measured, the formula provided an approximation only.

Niciporovic (Black and Watson 1960) emphasized the distinction between economic yield (grain, tubers, fiber, oil, seed) and the total dry-matter or **biological yield**. The term *biological yield* or *biomass* should include the weight of the roots; but since roots are not readily or usually recovered, the term has come to mean the weight of tops only, cut at ground level at harvest or maturity. Biological yield is the net result of photosynthesis, respiration, and nutrient uptake.

Donald (1962) and Donald and Hamblin (1976) strongly advocated that biological yield should be a standard measurement along with grain yield to allow for the calculation of a harvest index. They emphasized that the harvest index is an appropriate measure because it does not overlook the important yield aspect; it can be expressed as a simple percentage; it requires no physiological explanation; it is free of **teleology;** and although economic yield is used to calculate the harvest index, biological yield and harvest index are unrelated. The measurements of economic yield, biological yield, and harvest index express the sum-total relationships and interactions found in the field. Donald and Hamblin (1976) concluded:

Though biological yield and the linking ratio of grain yield, harvest index, are extremely simplified statements of multiple and complex growth processes, they nevertheless permit a far more analytical interpretation of environmental and genotypic influences than is possible from grain yields alone" [p. 363].

But instead of providing one measurement, three values (economic yield, biological yield, and harvest index) added to the complexity because Donald and Hamblin failed to show how to use these values equally or effectively.

Several other workers measured the effect of different planting dates, population densities, fertility levels, or other environmental variables on harvest index (Johnson and Major 1979; DeLoughrey and Crookston 1979; Gardener and Rathjen 1975) but were unable to use the information to effectively demonstrate improved production practices.

An example of the difficulties involved in relating the various measurements is strikingly demonstrated below—as taken from the works of Gardener and Rathjen (1975), to show the complexity of biological data.

Twelve barley cultivars, comprised of equal numbers of two- and six-row types and ranging in maturity from exceptionally early (Bankuti) to very late (Maraini), were grown under nine nitrogen fertility levels to determine the best cultivar, and the results are graphically presented in Figure 8–1. The results are perplexing! What fertility level or cultivar is best? It is tempting to base this decision on grain yield alone, which showed cultivar differences.

Velvon and Maraini gave the highest grain yield at 0 kg N/ha; C.I.3576 showed a small increase as nitrogen fertility increased; others (Excelsior, Vaughn) showed a typical parabolic response (See Figure 8-1). But every producer knows that straw is an important factor in harvesting and lodging. Most cultivars showed an increase in the weight of straw produced with increasing nitrogen levels; some, like Excelsior and Proctor, showed a very marked increase. The harvest index percentage declined for all cultivars except BR1239, but grain yield was below the yield of other cultivars. Precise values from the work of Gardener and Rathjen are not available, and without being able to score each cultivar at each level of fertility, it is impossible to determine the superior treatments or cultivars.

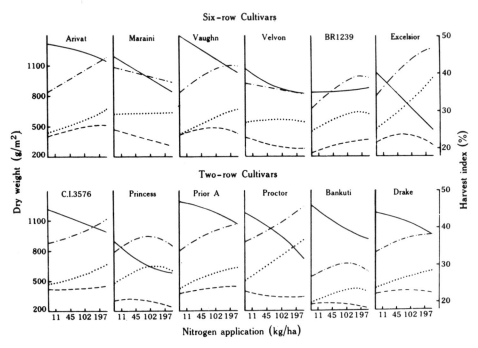

Figure 8-1.   Relationship of grain yield, straw yield, total yield, and harvest index in twelve cultivars at nine levels of nitrogen application. Fitted curves show harvest index as percentage (____), total dry weight per m² (.....), straw dry weight per m² (......), and grain dry weight per m² (..........).

PRODUCTIVITY SCORE

A productivity score, obtained by summing the grain yield, biological yield, and harvest index, gives a single rating on the reactions of these variables to a treatment as shown in a study with wheat of varying height (Table 8-1) grown at a standardized row width at Guelph, Ontario (Singh and Stoskopf 1971). Based on grain yield alone, tall wheats appear superior; based on harvest index, dwarf wheats appear best; based on biological yield alone the tall wheats appear superior; but based on productivity scores, the medium and dwarf height groups were superior to the check and the tall cultivars. The question is raised whether increased fertility could be used to further improve the performance of the medium and dwarf wheats.

TABLE 8–1.   Performance of tall, medium and dwarf wheat in Guelph, Ontario compared to check cultivars, averaged over two years, in 1966 and 1967

Cultivar	Height		Grain Yield		Biological Yield		Harvest Index	Productivity Score
	(cm)	(in)	(t/ha)	(tons/ac)	(t/ha)	(tons/ac)	(%)	
Genesse and Talbot checks	113	44	5.03	2.26	12.65	5.69	36.0	53.7
10 tall	86	34	5.04	2.27	11.24	5.06	38.0	54.3
10 medium	78	31	4.63	2.08	10.07	4.53	40.0	54.7
10 dwarf	58	23	4.74	2.13	9.49	4.27	42.0	56.2

Source: Data from Singh and Stoskopf 1971.

Evidence does exist that semidwarf wheats can outyield tall (standard height) cultivars (Vogel, Allan, and Peterson (Table 8-2) 1963). The moderate fertility level reported was comparable with local practice and usually sufficiently high to differentiate tall cultivars for lodging resistance. The high nitrogen fertility level averaged 224 kg N/ha (200 lb N/ac). Semidwarf wheats showed higher and increasing productivity scores compared to the declining scores of standard tall cultivars with increasing nitrogen fertility.

TABLE 8–2.   Agronomic responses in terms of four components of semidwarf and standard-height commercial winter wheat cultivars, under moderate and high fertility levels at Pullman, Washington, 1960

Fertility Level	Height		Agronomic Component		
	(cm)	(in)			
A. Three efficient semidwarf winter wheats					
Moderate	74	29.1	Grain yield (t/ha)	4.3	1.94 tons/ac
			Biological yield (t/ha)	11.0	4.95 tons/ac
			Harvest index (%)	38.5	
			Productivity score	53.8	
High	77	30.3	Grain yield (t/ha)	4.9	2.21 tons/ac
			Biological yield (t/ha)	12.6	5.67 tons/ac
			Harvest index (%)	38.5	
			Productivity score	55.0	

TABLE 8–2.   *(Continued)*

B. Five standard-height commercial wheats					
Moderate	117	46.1	Grain yield (t/ha)	3.3	1.49 tons ac
			Biological yield (t/ha)	10.7	4.82 tons/ac
			Harvest index (%)	30.5	
			Productivity score	44.5	
High	122	48.0	Grain yield (t/ha)	3.4	1.53 tons/ac
			Biological yield (t/ha)	12.5	5.63 tons/ac
			Harvest index (%)	27.6	
			Productivity score	41.5	

Source: Data from Vogel, Allan, and Peterson 1963.

An even more striking situation is shown with rice cultivars (Table 8–3) from the work of Langfield and reported by Donald (1968) and Donald and Hamblin (1976). At low fertility levels, the short rice gave a 23.7% lower grain yield and a 8.1% lower productivity score than that of the tall rice cultivars. As nitrogen fertility was increased, grain yield of both types increased; but the tall rice cultivars produced more straw as reflected in biological yield, and as a result the harvest index declined. With the short rice cultivars, biological yield increased; but the increase in straw was less than in the tall cultivar, the harvest index increased, and the productivity score increased.

TABLE 8–3.   Agronomic performance of tall and short rice as related to nitrogen fertility levels in northern Australia, 1961

Description	Agronomic Component	Low Nitrogen		High Nitrogen	
Tall rice	Grain yield (t/ha)	3.8	(1.71 tons/ac)	4.5	(2.03 tons/ac)
	Biological yield (t/ha)	17.5	(7.88 tons/ac)	28.6	(12.87 tons/ac)
	Harvest index (%)	22.0		16.0	
	Productivity score	43.3		49.1	
Short rice	Grain yield (t/ha)	2.9	(1.31 tons/ac)	5.4	(2.43 tons/ac)
	Biological yield (t/ha)	11.7	(5.27 tons/ac)	18.2	(8.19 tons/ac)
	Harvest index (%)	25.0		30.0	
	Productivity score	39.6		53.6	

Source: Donald 1968; Donald and Hamblin 1976

RESPONSES OF THE CORN CROP

Plant Population

Subjecting corn to high plant populations in association with narrow rows generally has failed to show grain yield gains because of shading on the ears, which develops as leaf area is increased. Under such stress conditions nubbins are formed. Productivity scores are especially useful to show the most responsive corn cultivars as indicated in corn trials in Minnesota (DeLoughery and Crookston 1979). Under nonstress conditions (Table 8-4), yield and productivity scores generally increased up to a population of 100,000 pph (40,500). The cultivar with a relative maturity rating of 105 was most productive. Under a stress environment (Table 8-5), stover yield was restricted, and best yields and productivity scores were achieved with the earliest maturity ratings and at low plant populations. Under stress conditions grain yield alone did not indicate superior performance.

TABLE 8-4. Performance of five corn hybrid cultivars differing in maturity ratings grown at five plant densities under a nonstress environment at Waseca, Minnesota in 1976

Relative Maturity Rating	Agronomic Component*	Population Density (pph) or (ppa)					
		12,500 (5,060)	25,000 (10,100)	50,000 (20,250)	100,000 (40,500)	200,000 (81,000)	Mean
75	A	3.1 1.4	5.0 (2.3)	5.6 (2.5)	6.1 (2.7)	4.3 (1.9)	4.8 (2.2)
	B	6.5 (2.9)	11.5 (5.2)	12.3 5.4	14.1 6.3	12.6 5.7	11.4 5.1
	C	47.7	43.5	45.5	43.3	34.1	42.1
	D	57.3	60.0	63.4	63.5	51.0	58.3
90	A	4.1 1.8	5.4 2.4	6.8 3.1	7.6 3.4	5.9 2.7	6.0 2.7
	B	9.8 4.4	14.8 6.7	16.1 7.2	15.7 7.1	15.9 7.3	14.5 6.5
	C	41.8	36.5	42.2	48.4	37.1	41.4
	D	55.7	56.7	65.1	71.7	58.9	61.9
105	A	4.5 2.0	7.2 3.2	7.5 2.5	8.7 3.9	6.1 2.7	6.8 3.1
	B	9.0 4.1	15.7 7.1	16.3 7.3	19.5 8.8	16.2 7.3	15.3 6.9
	C	50.0	45.9	46.0	44.6	37.6	44.4
	D	63.5	68.8	69.8	72.8	59.9	66.5
120	A	4.2 1.9	6.9 3.1	7.4 3.3	8.5 3.8	8.1 3.6	7.0 2.2
	B	9.3 4.2	15.2 6.8	15.4 6.9	19.2 8.6	18.9 8.8	15.6 7.0
	C	45.2	45.4	48.7	44.7	42.9	44.9
	D	58.7	67.5	71.5	71.9	69.9	67.5
135	A	5.0 2.3	7.4 3.3	7.7 2.7	7.7 3.5	7.0 3.2	7.0 3.2
	B	11.5 5.2	17.7 8.0	18.0 8.5	18.3 8.2	19.0 8.6	16.9 7.6
	C	43.4	41.8	42.7	42.1	36.8	41.4
	D	59.9	66.9	68.4	68.1	62.8	65.3
Mean	A	3.6 1.6	3.3 1.5	2.6 1.2	1.1 0.5	0.7 0.3	2.3 1.0
	B	8.1 3.6	8.5 3.8	8.3 3.7	7.7 3.5	8.5 3.8	8.2 3.7
	C	44.2	39.1	30.9	14.1	8.4	28.0
	D	55.9	50.9	41.8	22.9	17.6	38.5

Source: Data from DeLoughery and Crookston 1979.

*A = Grain yield (t/ha) (tons/ac).
 B = Biological yield (t/ha) (tons/ac).
 C = Harvest Index %.
 D = Productivity Score.

TABLE 8–5.   Performance of five corn hybrid cultivars differing in maturity ratings grown at five plant densities under a stress environment at St. Paul, Minnesota in 1976

Relative Maturity Rating	Agronomic Component*	Population Density (pph) or (ppa)											
		12,500 (5060)		25,000 (10,100)		50,000 (20,250)		100,000 (40,500)		200,000 (81,000)		Mean	
75	A	3.2	1.4	3.3	1.5	3.2	1.4	1.5	0.7	1.1	0.5	2.5	1.1
	B	6.6	3.0	7.6	3.4	7.7	3.5	6.9	3.1	10.5	4.7	7.9	3.6
	C	48.5		43.4		41.6		21.7		10.5		31.6	
	D	58.3		54.3		52.5		30.1		22.1		42.0	
90	A	3.9	1.8	3.7	1.7	3.7	1.7	2.2	1.0	1.2	0.5	2.9	1.3
	B	8.5	3.8	9.1	4.1	8.8	4.0	8.3	3.7	7.9	3.6	8.5	3.8
	C	45.9		40.6		42.0		26.5		15.2		34.1	
	D	55.3		53.4		54.5		37.0		24.3		45.5	
105	A	3.7	1.7	3.5	1.6	2.5	1.1	0.9	0.4	0.7	0.3	2.3	1.0
	B	8.3	3.7	9.0	4.1	8.0	3.6	7.4	3.3	7.8	3.5	8.1	3.6
	C	44.6		38.9		31.3		12.2		9.0		28.4	
	D	56.6		51.4		41.8		20.5		17.5		38.8	
120	A	3.8	1.7	3.3	1.5	2.1	0.9	0.7	0.3	0.6	0.3	2.1	0.9
	B	9.0	4.1	8.8	4.0	7.8	3.5	7.9	3.6	8.7	3.9	8.4	3.8
	C	42.2		37.5		26.9		8.9		7.9		25.0	
	D	55.0		49.8		36.8		17.5		17.2		35.5	
135	A	3.2	1.4	2.9	1.3	1.3	0.6	0.1	0.05	0.0	0.0	1.5	0.7
	B	7.9	3.6	8.2	3.7	9.1	4.1	7.9	3.6	7.8	3.5	8.2	3.7
	C	40.5		35.4		14.3		1.3		0		18.3	
	D	51.6		46.5		24.7		9.3		7.8		28.0	
Mean	A	3.6	1.6	3.3	1.5	2.6	1.2	1.1	0.5	0.7	0.3	2.3	1.0
	B	8.1	3.6	8.5	3.8	8.3	3.7	7.7	3.5	8.5	3.8	8.2	3.7
	C	44.2		39.1		30.9		14.1		8.4		28.0	
	D	55.9		50.9		41.8		22.9		17.6		38.5	

Source: Data from DeLoughery and Crookston 1979.

*A = Grain yield (t/ha) (tons/ac).
B = Biological yield (t/ha) (tons/ac).
C = Harvest Index %.
D = Productivity Score.

Moisture Stress and Irrigation

High plant population can cause stress to a crop under dry conditions, but associated yield reductions are dependent on when the stress occurs during the development of the plant. Under semiarid conditions in Colorado (Fairbourn, Kemper, and Gardner 1970), sufficient moisture during the vegetative period caused biological yields to increase with increasing plant population (Table 8-6). Drought stress during reproductive growth may have limited ear development at 42,000 pph (17,000 ppa) so that productivity score at this population was substantially reduced. The superior performance at 32,100 pph (13,000 ppa) is made clear by the productivity score.

TABLE 8–6.   Agronomic performance of two corn hybrid cultivars under semiarid conditions in Colorado grown at three plant populations in 1967 and 1968

Agronomic Component	Plant Population (pph) or (ppa)		
	21,000 (8,500)	32,110 (13,000)	42,000 (17,000)
Grain yield (t/ha) (tons/ac)	3.2 1.4	4.3 1.9	4.3 1.9
Biological yield (t/ha) (tons/ac)	5.3 2.4	7.3 2.3	9.7 4.4
Harvest index (%)	59.1	59.0	44.7
Productivity score	67.6	70.6	58.7

Source: Data from Fairbourn, Kemper, and Gardner 1970.

In contrast, a stress during vegetative growth may prove beneficial if vegetative growth is restricted under conditions of a high plant population. Irrigation applied so as to reduce moisture stress entirely produced a biologically similar grain yield to stress conditions during vegetative growth (Downey 1971) at a high fertility level and at 59,000 pph (24,000 ppa). Late stress conditions reduced grain yield 50% (Table 8-7). Although small yield differences between the no stress and early stress conditions were found, the high productivity score for the early stress period suggests that the moderate plant population of 59,000 pph (24,000 ppa) did not produce an adequate LAI and a further grain yield increase may be possible with a higher plant population.

Cultural practices such as high plant populations, narrow rows, and irrigation that promotes stover growth may be unaccompanied by a comparable increase in grain yield. High stover production is often associated with

TABLE 8–7.   Agronomic performance of corn under three water regimes in New South Wales

Agronomic Component	No Stress	Early Stress	Late Stress
Grain yield (t/ha) (tons/ac)	3.6 1.6	3.8 1.7	1.9 0.9
Biological yield (t/ha) (tons/ac)	12.7 5.7	9.1 4.1	8.9 4.0
Harvest index (%)	28.3	41.8	21.3
Productivity score	44.6	54.7	32.1

Source: Data from Downey 1971.

Figure 8-2. The grain of sorghum is produced on a terminal inflorescence. Reproductive organs are not shaded under high plant populations. (Courtesy Ontario Ministry of Agriculture and Food.)

shading of the ear, causing delayed silk development, accompanying poor pollination, and subsequent nubbin development. Part of the problem with corn development is the location of the ear, midway up the plant. Other crops which have their reproductive organs located at the top of the plant such as sorghum (Figure 8-2), cereals and sunflowers, are not subject to shading and will be examined.

SUNFLOWER

Like sorghum, the reproductive organs of sunflowers develop on the top of the plant, and like sorghum and unlike corn, no shading stress on reproductive development appeared to exist as population density increased. Under dry-land conditions at Hyderabad, India and Swift Current, Saskatchewan, sunflower crops were successfully grown over a wide range of plant populations and row spacings (Vijayalakshmi et al. 1975). At Hyderabad (Table 8-8), wide rows that increased competition among plants within a row (intrarow competition) had little influence on productivity scores. In other tests at Hyderabad, yield plateaus of approximately 900 and 1350 kg/ha (803 and 1,205 lb ac) existed over population ranges of 18,000 to 32,000 (7,200 to 13,000 ppa) and 56,000 to 98,000 (22,600 to 39,600 ppa) plants/ha, respectively. On the basis of these tests it is recommended that for dry-land production, sunflowers be grown at populations of 60,000 to 75,000 pph (24,200 to 30,300 ppa) at row spacings of 35 to 60 cm (14 to 24 in) but extrapolation to North American conditions must be done cautiously. Performance trials are needed to verify this conclusion.

TABLE 8-8. Agronomic performance of Armaveric sunflower at Hyderabad at a plant population of 75,000 pph (30,300 ppa) during the period July to October, 1971.

Agronomic Component	Row-to-row and plant-to-plant spacing (cm)				
	37 x 37 (15 x 15 in)	45 x 30 (18 x 12 in)	60 x 22 (24 x 9 in)	90 x 15 (35 x 6.0 in)	135 x 10 (53 x 4 in)
Grain yield (t/ha)	0.87 0.39	0.91 0.41	0.91 0.41	0.87 0.39	0.89 0.40
Biological yield (t/ha)*	2.23 1.00	2.46 1.11	2.22 1.00	2.18 0.98	2.17 0.98
Harvest index (%)	39.00	37.00	41.00	40.00	41.00
Productivity score	42.10	40.37	44.13	43.05	44.06

Source: Data from Vijayalakshmi et al. 1975.

CEREALS

The application of nitrogen to cereals (wheat, oats, barley, rice, rye) commonly promotes an increase in biological yield that may lead to lodging, parasitic leaves, and reduced grain yields. Under adequate moisture conditions this situation is exaggerated. These observations were found to be true with spring wheat (McNeal et al. 1971), with the highest productivity score at 44.8 kg N/ha (40.2 lb/ac) (Table 8-9), a level that did not produce the highest grain yield. At 89.7 kg N/ha (80 lb/ac), the highest grain yield was obtained but was accompanied by a 6700 kg/ha (5,882 lb/ac) straw increase. Without a productivity score, this fact was overlooked and subsequent work was directed at the 89.7 kg N/ha (80 lb/ac) rate.

TABLE 8-9. Agronomic response of five spring wheat cultivars to nitrogen fertility, at Belgrade, Montana, 1969

Agronomic Component	Nitrogen fertility level (kg/ha) or lb/ac									
	Check		22.4 (20)		44.8 (40.2)		67.3 (60)		89.7 (80)	
Grain yield (t/ha)	1.9	0.9	2.2	1.0	2.6	1.2	2.9	1.3	3.1	1.4
Biological yield (t/ha)	4.5	2.0	5.2	2.3	6.3	2.8	7.3	3.3	13.0	5.9
Harvest index (%)	42.9		42.3		41.6		39.7		24.1	
Productivity score	49.4		49.8		50.6		49.9		40.2	

Source: Data from McNeal et al. 1971.

The value of productivity scores to show the potential of semidwarf wheat and rice cultivars has been examined in this chapter (Tables 8-3 and 8-4), but in these cases the advantages of dwarf cultivars were obvious from grain yield alone. The results of McNeal et al. (1971) may be more typical (Table 8-10). On the basis of grain yield alone, little differences among the cultivars are apparent. Based on productivity scores, the medium and short cultivars are superior.

TABLE 8-10. Agronomic response of five spring wheat cultivars when averaged over all fertility levels, at Belgrade, Montana, 1971

Agronomic Component	Cultivar and height (cm)									
	Fortuna 101.0(40 i)		Centana 110.3(43 i)		Tall Centana 109.2(43 i)		Medium Centana 82.2(32 i)		Short Centana 58.6(23 i)	
Grain yield (t/ha) (tons/ac)	2.7	1.2	2.6	1.2	2.4	1.1	2.6	1.2	2.4	1.1
Biological yield (t/ha) (tons/ac)	6.7	3.0	6.8	3.1	6.4	2.9	6.1	2.7	5.3	2.4
Harvest index (%)	40.7		38.2		37.8		42.8		45.8	
Productivity score	50.1		47.6		46.6		51.5		53.5	

Source: Data from McNeal et al. 1971.

Relating leaf area and grain yield provided a physiological explanation of barley responses to nitrogen levels under moisture stress (Luebs and Laag 1967). A level of 68 kg N/ha promoted vegetative growth, and LAI reached a maximum of 4.8 at late straw elongation and became associated with high transpiration rates that led to moisture stress that reduced grain yields. Productivity scores clearly showed the advantage of low nitrogen levels under such conditions (Table 8-11) as reported by Luebs and Laag 1967.

Since moisture can promote excessive vegetative growth in cereals, it can be assumed that moisture deficiency could be used to restrict vegetative growth, a situation that was found to be true in corn (see Table 8-7). Under a system of

TABLE 8-11. Agronomic performance of barley at two moisture and nitrogen fertility levels under dry-land conditions at Riverside, California, 1964

Agronomic Component	Treatments*							
	M_1N_1		M_1N_2		M_2N_1		M_2N_2	
Grain Yield (t/ha) (ton/ac)	0.93	0.42	0.81	0.36	1.53	0.69	1.33	0.60
Biological yield (t/ha)	3.83	1.72	3.89	1.75	4.89	2.20	4.98	2.24
Harvest index (%)	24.28		20.82		31.29		26.71	
Productivity score	29.04		25.52		47.71		33.02	

Source: Data from Luels and Laag 1967.
*M_1= low moisture level (15 cm) (6 in)
M_2 = high moisture level (19 cm) (7.5 in)
N_1 = 17 kg N/ha (6.7 in)
N_2 = 68 kg N/ha (26.8 in)

controlled irrigation, Storrier (1965) was able to demonstrate that one irrigation at ear emergence (treatment 6) was nearly as effective as five irrigations (treatment 9) (Table 8-12).

In cereals, plant population is an important consideration; but because of tillering, seeding rate offers little population control. At a recommended seeding rate of 130 kg/ha (116 lb/ac) for wheat, assuming 35 seeds per gram (15,876 seeds per lb) and with 90% germination, a plant population of 4,100,000 per hectare (1,659,900 acre) could result. Lower rates would encourage tillering. Productivity scores are most useful to determine the response of cereal crops to seeding rates.

SOYBEAN

Corn, sorghum, sunflower, and cereals are determinant plants with a clearly defined vegetative and reproductive growth. Northern-type soybeans are indeterminate and produce about half of their vegetative growth after reproductive development begins. The effect of seeding dates on vegetative growth, biological yields, and productivity scores is of considerable interest. Ten soybean cultivars representing a wide range of maturity (two from each of maturity groups I to V) were tested at five seeding dates at Columbia, Missouri (Johnson and Major 1979). Possibly because of a strong photoperiodic response, biological yields did not increase, and harvest index values remained high at all seeding dates and remained consistent (Table 8-13). The early maturing group gave the highest grain yield and productivity score when seeded June 1, but for later cultivars, earlier seedings were required for top yields and scores. The mean productivity score of 46.9 for maturity group II suggests these two cultivars are best suited for conditions at Columbia, Missouri.

In soybeans and in other crops, protein, oil content, or kilograms of nitrogen fixed per hectare (acre) may constitute economic yield, with biological yield comprised of total kilograms (pounds) of beans and residue. Economic conditions may dictate this choice. Productivity scores can determine management practices to promote high protein, oil, or sugar content, or other factors for crops marketed on this basis. Canola (rapeseed) has a photosynthetic area consisting mainly of seed pods and stem sections (Figure 8-3). Productivity scores on canola (rapeseed) are not available. The crop may tolerate high plant populations if suitable environmental conditions exist.

FORAGES

Unlike most annual crops in which management is directed toward maximizing the production at one harvest date, perennial forage crop production is concerned with a multiplicity of interacting factors, which makes this crop

TABLE 8–12.   Agronomic response of wheat watered at different stages of development at Wagga Wagga, New South Wales, 1961

Agronomic Component*	Treatment **									
	1	2	3	4	5	6	7	8	9	10
A	1.8 0.8	1.3 0.6	1.7 0.8	1.2 0.5	1.7 0.8	2.2 1.0	2.0 0.9	2.1 0.9	2.9 1.3	1.6 0.7
B	7.8 3.5	9.3 4.2	8.6 3.9	8.2 23.7	8.4 3.8	9.1 4.1	10.2 4.6	8.4 3.8	11.3 5.1	7.9 3.6
C	22.8	14.0	19.5	15.0	20.0	24.7	19.8	25.4	25.7	20.0
D	32.4	24.6	29.8	24.4	30.1	36.0	32.0	35.9	39.9	29.5

Source: Data from Storrier 1965.

*A = Grain yield (t/ha)(tons/ac).
B = Biological yield (t/ha)(tons/ac).
C = Harvest index (%).
D = Productivity score.

**1 = A single irrigation at 4-to-5 leaf stage.
2 = A single irrigation at jointing stage.
3 = Treatments 1 + 2.
4 = A single irrigation at flag leaf stage.
5 = Treatments 1 + 2 + 4.
6 = A single irrigation at ear emergence.
7 = Treatments 1 + 2 + 4 + 6.
8 = A single irrigation at milk stage.
9 = Treatments 1 + 2 + 4 + 6 + 8.
10 = Check or control.

TABLE 8–13.   Average agronomic performance of two soybean cultivars from each of five maturity groups planted at five dates in 1971 and 1973 at Columbia, Missouri

Maturity Group	Agronomic Component*	Late April		Mid-May		June 1		June 21		Mid-July		Mean
I	A	2.2	1.0	2.3	1.0	2.5	1.1	1.8	0.8	1.3	0.6	
	B	5.8	2.6	6.5	2.9	6.7	3.0	5.5	2.5	3.7	1.7	
	C	37.9		35.4		37.3		32.7		35.1		
	D	45.9		44.2		46.5		40.0		40.1		43.3
II	A	2.6	1.2	3.2	1.4	2.8	1.3	2.3	1.0	1.5	0.7	
	B	6.4	2.9	8.1	3.6	7.7	3.5	6.1	2.7	4.0	1.8	
	C	40.6		39.5		36.4		37.7		37.5		
	D	47.6		50.8		46.9		46.1		43.0		46.9
III	A	3.4	1.5	2.8	1.3	2.6	1.2	2.4	1.1	1.5	0.7	
	B	8.4	3.8	8.5	3.8	8.4	3.8	6.9	3.1	4.7	2.1	
	C	40.5		32.9		30.9		34.8		31.9		
	D	52.3		44.2		41.9		44.1		38.1		44.1
IV	A	3.0	1.4	2.5	1.1	2.2	1.0	2.2	1.0	1.3	0.6	
	B	8.7	3.9	8.5	3.8	7.5	3.4	6.4	2.9	4.3	1.9	
	C	34.5		29.4		29.3		34.4		30.2		
	D	46.2		40.4		39.0		43.0		35.8		40.9
V	A	2.5	1.1	2.2	1.0	2.5	1.1	2.4	1.1	1.8	0.8	
	B	8.4	3.8	8.4	3.8	8.5	3.8	8.0	3.6	5.5	2.5	
	C	29.8		26.2		29.4		30.0		32.7		
	D	40.7		36.8		40.4		40.4		40.0		39.7

Source: Data from Johnson and Major 1979.
*A = Economic or grain yield (t/ha)(tons/ac).
 B = Biological yield (t/ha) (tons/ac).
 C = Harvest index.
 D = Productivity score.

the most complex and challenging to produce. Some crop producers have abandoned the highly complex forage production system for a simpler program of cash cropping and on livestock farms for an increased hectarage (acreage) of whole plant corn silage to meet forage needs. Forage crop production, nevertheless, remains attractive because it offers a sound soil management system, can be a source of high-quality protein feed, and provides the greatest flexibility in livestock programs. High nitrogen fertilizer prices and soil management problems associated with monoculture and cash cropping have placed nitrogen-fixing forage legumes in a new perspective as a soil-holding and soil-building rotational crop even on nonruminant livestock farms.

To be a successful crop, forages must be profitable and the management problems must be tackled in a clear, logical, and precise manner. Productivity scores can be used to greater advantage in forage management than in most other crops. For productivity scores to be useful, forage crop production objectives need to be defined more precisely and explicitly than they have in the past and perhaps more carefully than in other crops where a management treatment that produces as little as 50 to 100 kg/ha (one or two bushels per

Figure 8–3. The photosynthetic area of the canola crop (rapeseed) is comprised mainly of seed pods and stem sections. High plant populations do not result in an excessive lower "leaf area." The reproductive pods at the top half of the plant effectively intercept sunlight but with good light penetration to all pods. (Courtesy Ontario Ministry of Agriculture and Food.)

acre) extra yield may be considered a worthy goal. In forage crop production, yield increases of small proportions are of minor consequence when the possible impact on other factors such as longevity, aftermath growth, seasonal distribution, quality, weather patterns, time of fertilizer or manure application, or irrigation are to be considered. Because explicit objectives are so hard to define for a forage program, a general objective to maximize the yield of livestock products (meat, milk, eggs, wool) has been accepted. But such a general objective gives no criteria or guidance to the many possibilities and alternatives a forage producer has in reaching the general objective. Specific objectives must be determined for each operation, for unlike a universally accepted objective such as gluten strength in wheat for bread-making, a forage producer may have a different objective of forage production to meet the needs on each farm. For example, forage programs for a high-producing dairy herd may differ from that for overwintering beef cattle or heifers, for grazing of horses and sheep, or for low-value "rough lands" compared to high-value first-class land, according to protein feed supplement prices or according to market demands. Productivity scores can be a useful management tool if the economic yield component can be defined accurately.

The following should be considered in relation to forage crop management and productivity when defining economic yield:

1. In forages, total dry-matter production or biological yield is the economic yield, and this single measurement characteristically has been the "bottom line" for evaluating production systems or plot treatments. Dry-matter production is an inadequate measure because it gives no indication of quality, livestock productivity, or guide to crop production practices. In many areas the bulk of preserved forage is harvested from the first crop, but this is not the "bottom line" because regrowth and number of subsequent cuts must be considered. Seasonal distribution may be of greater consequence than total seasonal yield, especially on pasture or dry-lot feed programs. Also, longevity and yields in subsequent years may be of greater significance than a very high yield in any one year. Thus, calculation of productivity scores using a clearly defined economic yield rather than total dry-matter productivity can lead to more meaningful productivity scores.

2. Under a cash crop program, financial gain from the sale of forages per hectare (acre) may comprise the economic yield component.

3. For livestock producers, quality aspects undoubtedly offer the best criteria for economic yield. Protein feeds often represent the highest single cost in animal production. Under such circumstances, protein produced may be a wise measure of economic yield. Cutting forages early and frequently may lead to greatest protein yields per hectare (acre), but low dry-matter yields and productivity scores indicate clearly the preferred treatment. In addition to protein, forage digestibility may serve as a measure of economic yield. The development of an in vitro fermentation technique in the laboratory using a small forage sample in conjunction with an extract of rumen liquor removed directly from a ruminant animal actively digesting roughage has replaced guesswork as to forage quality (Mowat et al. 1965). Forage analyses giving protein, digestibility, or other nutritional aspects are available from regional laboratories for a nominal cost.

Because high dry-matter yields are not compatible with high digestibility or protein level (Fulkerson et al. 1967), a compromise between quality and dry-matter yield must be considered. Productivity scores can make this perplexing problem clear.

Producers are not in a position to run elaborate farm trials to evaluate the parameters needed to calculate productivity scores; but work in conjunction with local research stations and agricultural advisors, along with a greater availability of research station results, can make productivity scores useful in determining how to meet a specific farm need.

4. Productivity scores based on quality aspects may be tempered by economic considerations and the feasibility of implementation. A forage producer, for example, accustomed to maintaining a forage field at a reasonably productive level for five years, may wish to consider the merits of "mismanaging" a field to obtain the highest economic yield (assume it to be protein per hectare or acre), and instead of harvesting at the recommended one-tenth bloom on alfalfa to allow for three or four cuttings, he may cut the field six or eight times. Such mismanagement may result in the field being plowed under at the end of two years. If such a system fits a rotation and is economically feasible, and if the field can be safely plowed without subjecting it to undue erosion, such a practice may be attractive.

By using milk production as the economic yield, and with meaningful "on farm" comparisons as a guide, productive management programs will evolve. Seemingly "unsound management" may become accepted practice. An open-minded willingness to change and to face peer criticism is a needed producer attribute.

5. In pasture programs, economic yield may be measured by an aspect of livestock productivity such as kilograms of beef produced per hectare (acre). A producer concerned about losses from bloat may switch from the high-bloat potential of alfalfa or red clover to a grass-legume mixture with less bloat potential or to a no-bloat pasture of birdsfoot trefoil or pure grass. Although an alfalfa or an alfalfa-grass mixture may outyield other mixtures or species, the decision of which to use can be based on productivity scores.

6. The contribution of nitrogen fixed by a legume, once taken for granted, has received such prominence that this factor should be carefully considered in a productivity score. On some farms, especially cash crop farms, legumes may be included as a source of nitrogen, and ability to fix nitrogen may take priority over other factors. Subterranean clover was described by Donald (1962) as the wonder plant of southern Australian pastures because of its ability to fix nitrogen and to thrive on the extensive poor soils of the region. However, as a forage species it is not fully satisfactory because the seed is hard to harvest and it can be toxic to livestock, the clover is shaded out by grasses, and it does not respond to summer rain, which results in a low yield and poor seasonal distribution.

With high nitrogen fertilizer prices, the role of legumes in a forage mixture or in a rotation may become jealously guarded, with some considerable sacrifice to total dry-matter yields.

APPLICATION OF PRODUCTIVITY SCORES TO FORAGES

Most forage crop cutting and management recommendations represent a comprise among the many factors associated with forage production. The one-tenth bloom stage of alfalfa is recommended as a guide for cutting because it gives the highest dry-matter yield in combination with protein and disgestibility, allows for two or three additional cuts, and encourages longevity. This recommendation is based primarily on dry-matter yield. Application of productivity scores involving forage crops (from the scientific literature) often gives startling results. Many of these studies suggest that productivity scores are higher when cutting deviates from the recommendation. Although management systems that produce high productivity scores need a comprehensive examination, the evidence points strongly in this direction.

An example of an unexpected result can be shown by applying productivity scores to the forages listed in Table 3-5 (in Chapter Three). The 30 forages were arranged in descending order of dry-matter yield. Forage entry No. 1 yielded 13,926 kg/ha (12,434 lb/ac), whereas entry No. 15 yielded 6895 kg/ha (6,156 lb/ac), or less than half when based on 22 cuts in one season, with a 23rd cut occurring the following spring to determine winter hardiness. When economic yield was taken as digestible organic matter, the highest mean productivity score of 60.6 was obtained for entry No. 15, the highest of all 30 entries, marginally ahead of entry No. 1 with a productivity score of 59.5 (Table 8-14). Seasonal distribution for entry No. 15 was superior and dropped below entry No. 1 only in the fall cutting period. Under the conditions of Swift Current, however, entry No. 15 was not considered to have sufficient winter hardiness. Based on dry matter yields alone, entry No. 15 would be overlooked as having agricultural potential.

In a study of cutting dates of first cut legume and grasses, Fulkerson et al (1967) found that dry matter yields increased as cutting date was delayed (Table 8-15, note B or biological yield). When crude protein was considered as economic yield, the rapid decline in crude protein with delays in cutting date invariably pointed to the desirability of early cuts. This fact is convincingly and clearly demonstrated by the productivity scores.

Table 8-15 merits careful attention. Unlike productivity scores that coincide with maximum economic yield in other crops, the highest productivity scores in Table 8-15 occur at the lowest dry-matter (biological) yield. Percent crude protein in legumes declined from 32.6% on May 14 to 14.2% on July 16. In grasses the protein decline was even more dramatic. If yield of crude protein is the economic yield objective of a forage management program, then a major review of management practices is in order. The effects of doubling or tripling the cutting frequency on longevity, the fertility program, financial advantage, the cost of harvesting in relation to the returns obtained, the need for bulk or cellulose in the ration, and other forage production factors need to be examined. The conclusion reached by Fulkerson and his associates from the results of the three-year study in Table 8-15, obtained without benefit of a

TABLE 8–14.   Seasonal distribution of dry matter and digestible organic matter yields of two entries from Table 2–5 with the average productivity score for the season when grown at Swift Current Saskatchwan. Results are for the first harvest year 1971.

Agronomic Component*	Mean values of 23 cuts over the season											
	Spring		Summer		Late Summer		Fall		Over Winter		Mean	
Entry 1:												
A	1.6	0.7	2.1	0.9	1.9	0.8	1.3	0.6	0.9	0.4	1.6	0.7
B	2.5	1.1	3.6	1.6	3.5	1.6	2.4	1.1	2.0	0.9	2.8	1.3
C	65.0		58.0		53.0		53.0		48.0		57.1	
D	69.1		63.7		57.4		56.7		50.9		59.5	
Entry 15:												
A	0.8	0.4	1.4	0.6	1.1	0.5	0.6	0.3	0.3	0.1	0.8	0.4
B	1.2	0.5	2.2	1.0	2.0	0.9	1.0	0.5	0.6	0.3	1.4	0.6
C	67.0		63.0		56.0		54.0		52.0		58.4	
D	69.0		66.6		59.1		55.6		52.9		60.6	

Source: Data from Lawrence 1978.
*A = yield of digestible organic matter (t/ha) (tons/ac)
 B = biological yield or total dry matter (t/ha) (tons/ac)
 C = harvest index %
 D = productivity score

productivity score, was to "cut alfalfa at the very first-flower-stage to permit three cuttings before September to allow for maximum yields of nutrients. Cutting the grasses early encourages a more rapid aftermath recovery by taking advantage of more favorable moisture conditions" (Fulkerson et al. 1967, p. 689). Indeed, the decline in productivity scores was much more pronounced in grasses than in legumes. A forage producer who cut his grass hay on June 11, according to recommendations, would obtain over 50% less protein than if he cut on May 14.

To amplify the suggestion that frequent cutting may produce more seasonal protein, the results of a frequency of cutting test on ryegrass (Reid 1978) were examined (Table 8-16). Nitrogen fertility levels were applied at increments up to 897 kg N/ha (800 lb/ac), but yield levels above 448 kg N/ha (181 lb/ac) levelled off according to the law of diminishing returns (see Figure 3-3). On the basis of productivity scores using crude protein yields as economic yield, in every case ten seasonal cuts was superior to three or five cuts even though dry-matter yields were always substantially reduced as cutting frequency increased. Increasing nitrogen fertility had the effect of increasing dry-matter yields and crude protein percent in all cutting regimes, but economic considerations may determine the nitrogen level used.

TABLE 8-15.   Productivity scores calculated on forages cut at selected dates and averaged over three years—1961, 1962, and 1963

		*Date of First Cut										
Forage	Agronomic Component*	May 14		May 28		June 11		June 25		July 9		July 16	
Legume**	A	0.31	0.14	0.79	0.36	0.89	0.40	1.00	0.45	1.12	0.50	1.06	0.48
	B	0.94	0.42	2.63	1.18	4.48	2.02	5.87	2.64	7.20	3.24	7.49	3.37
	C	32.55		29.85		19.95		17.00		15.50		14.20	
	D	33.80		33.27		25.32		23.87		23.82		22.76	
Grass***	A	0.25	0.11	0.43	0.19	0.53	0.24	0.53	0.24	0.51	0.23	0.47	0.21
	B	1.02	0.46	2.82	1.27	4.98	2.24	6.39	2.88	7.56	3.40	7.75	3.49
	C	24.58		15.25		10.68		8.33		6.75		6.12	
	D	25.86		18.51		16.20		15.26		14.82		14.34	

Source: Data from Fulkerson et al 1967.

*A = economic yield (yield of crude protein (t/ha) (tons/ac).
B = biological yield (t/ha) (tons/ac).
C = harvest index (percent crude protein).
D = productivity score.

**Mean of two alfalfa cultivars, Vernal and Dupuits.

***Mean of three species each represented by two cultivars: bromegrass, cultivar Saratoga and Canada; orchard grass, cultivar Frode and Ottawa; timothy, cultivar Climax and Essex.

TABLE 8-16.   Productivity scores on perennial ryegrass averaged over a three-year period (1967, 1968, and 1969) at five selected fertility levels and subjected to three, five, and ten cuts per year.

N Fertility Applied in kg/ha (lb ac)		Agronomic Component*	Number of Cuts					
			3		5		10	
0	0	A	0.28	0.13	0.25	0.11	0.20	0.09
		B	3.49	1.57	2.69	1.21	1.55	0.70
		C	8.10		10.13		14.20	
		D	11.87		13.07		15.95	
112	100	A	0.62	0.28	0.64	0.29	0.49	0.22
		B	7.89	3.55	6.42	2.89	3.13	1.41
		C	7.83		10.20		16.07	
		D	16.34		17.26		19.69	
224	200	A	1.06	0.48	1.12	0.50	0.01	—
		B	11.83	5.32	9.94	4.47	5.10	2.30
		C	9.17		11.40		17.97	
		D	22.06		22.46		23.98	
336	300	A	1.61	0.72	1.66	0.75	1.39	0.63
		B	14.08	6.34	12.72	5.72	6.93	3.12
		C	11.53		13.20		20.23	
		D	27.22		27.58		28.55	
448	400	A	1.88	0.85	2.03	0.91	1.89	0.85
		B	14.07	6.33	13.21	5.94	8.63	3.88
		C	13.53		15.50		21.97	
		D	29.48		30.74		32.49	

Source: Data from Reid 1978.

*A = economic yield—crude protein (t/ha) (tons/ac).
B = biological yield—total dry-matter yields (t/ha) (tons/ac).
C = harvest index—percent crude protein.
D = productivity score.

CONCLUSIONS

The assessment of crop management practices by dry-matter yield alone, along with the help of statistical analysis, has contributed much to management practices. But "cut and weigh" experiments leave a lot to be desired because the results may be misinterpreted or the most successful treatments or practices overlooked.

The application of productivity scores to data available from the literature reporting the necessary parameters has left no doubt about the usefulness of productivity scores in understanding crop responses. As production techniques become more sophisticated, crop producers need more detailed information collected at research institutions. Unlike a statistical analysis, productivity scores can be easily comprehended by the layman, and can provide him with an insight into management practices.

Many cropping programs are undergoing extensive modifications, notably forage production and management. The result is that farm research will become more important, general recommendations will be less meaningful, maximum and efficient use of resources will be demanded, and "accepted" practices will be challenged. General procedures will be replaced with recommendations specific to each farm.

The productivity score may be a tool in this transition. The limitation to this method is that total dry-matter or biological yield must be taken as well as the standard economic yield. A surprising number of references contain the necessary determinations. In any research test involving field plots, the effort of seeding and managing plots is substantial, and the results are diminished if biological yield is not measured. The argument that productivity scores add to the complexity of reporting data is not valid in the hands of capable, well-trained farmers, although they may not be versed in statistical methods. In an age of pocket calculators and minicomputers, calculation of productivity scores offers no great difficulty.

The very useful and comprehensive review of the virtues of harvest indices and biological yield measurements of Donald and Hamblin (1976) is acknowledged. It was this work that led to the productivity score method.

QUESTIONS

1. Holliday (1960*a* and *b*) presented data on the influence on spring and winter wheat yields of seeding rates at Leeds University, England (Table 8–17).

TABLE 8–17. Grain and straw yields as influenced by four seeding rates over a three- and five-year period in England 1956 to 1960

Seed		Winter Wheat (5-year mean)				Spring Wheat (3-year mean)			
Rate		Grain		Straw		Grain		Straw	
(kg/ha)	(lb/ac)	(kg/ha)	(lb/ac)	(kg/ha)	(lb/ac)	(kg/ha)	(lb/ac)	(kg/ha)	(lb/ac)
62	55	3524	3146	2316	2068	2587	2310	2612	2332
123	110	4103	3663	2600	2321	2871	2563	2858	2552
246	220	4768	4257	3092	2761	2784	2486	3067	2738
493	440	4521	4037	3388	3025	2476	2211	3351	2992

Source: Data from Holliday 1960*a* and *b*.

(a) Determine the best seeding rate for winter wheat at Leeds, England from Table 8-17. On what do you base your conclusion?

(b) What seeding rate gave the lowest productivity score for winter wheat? Suggest reasons why.

(c) What conclusions can be drawn regarding grain-straw relationships in winter wheat from the Leeds data?

(d) What cultural practices might favor a seeding rate of 62 kg/ha (55 lb/ac) for winter wheat at Leeds?

(e) Is the best seeding rate for winter wheat also the best for spring wheat? Give reasons why.

(f) On the basis of the data presented, give reasons why winter wheat outyielded spring wheat.

(g) Would your conclusions regarding the best and the poorest seeding rates reached on the basis of a productivity score be the same if you had not calculated a productivity score?

(h) How do the best seeding rates at Leeds compare with recommended seeding rates for winter and spring wheat in your area?

(i) Why might seeding rates differ from area to area?

(j) Speculate on the morphology of an individual winter wheat plant at the high- and low-seeding rates in terms of the number of vegetative and fertile tillers per plant together with the number and weight of seed per ear.

(k) Locate the references of Holliday (1960*a* and *b*) in the library and determine what conclusions he reached regarding seeding rates of winter and spring wheat.

2. A situation that producers may encounter and must learn to deal with was described by Akinola and Whiteman (1975). These workers observed the widespread local use of pigeon pea in India and Africa, and commercial production and use of the crop in Puerto Rico. They wished to evaluate the potential of this crop, untested under Australian conditions. They knew that the crop was sensitive to photoperiod and that the sowing date would be important in influencing vegetative and reproductive processes. Four cultivars of pigeon pea were seeded at eight dates with the results as shown in Table 8-18. The four cultivars gave strikingly similar performance, and so the mean values are given.

(a) What is the best seeding date by a visual observation only of Table 8-18?

(b) Calculate productivity scores and compare the results with those for (a).

(c) Discuss the merits of a late December seeding.

(d) Describe the photoperiodic response of pigeon pea from the data in Table 8-18.

TABLE 8–18. The effects of sowing dates on straw and seed yields of pigeon pea grown from 1970 to 1972 in the subtropical region of southeastern Queensland.

Sowing Date	Straw Yield		Seed Yield	
	(kg/ha)	(lb/ac)	(kg/ha)	(lb/ac)
Sept. 1, 1970	14,270	12,741	4350	3884
Sept. 29	12,770	11,402	4070	3,634
Oct. 27	11,920	10,643	3240	2,893
Nov. 24	9,430	8,420	4900	4,375
Dec. 22	6,880	6,143	3710	3,313
Jan. 19, 1971*	2,410	2,152	2,940	2,625
Feb. 16	1,820	1,625	1190	1,063
Mar. 16	1,670	1,491	940	839

Source: Data from Akinola and Whitman 1975.

*Some leaf and pod losses occurred at this date and could not be included in the straw yield.

(e) What conclusions regarding seeding dates were reached by the authors?

3. Define with reasons the best time(s) to irrigate a corn crop, a cereal crop, a soybean crop, and a watermelon crop.

4. From the material presented on forages in this chapter, what lesson is dramatically emphasized by the productivity scores regarding the time of cutting forages?

5. What factors must be considered in addition to the time of cutting in your area?

6. In Table 8-15, explain why the yield of crude protein increased from the May 14 to July 9 cutting dates and then declined, whereas the biological yield continued to increase.

7. What prevents forage producers from harvesting at the dates showing the highest harvest indices?

8. Following the law of diminishing returns, herbage dry-matter yields of perennial ryegrass plateaued as nitrogen fertilizer rates were increased from 0 to 897 kgN/ha (800 lb/ac)(see Figure 3-3). Productivity scores increased as nitrogen levels increased at five selected nitrogen levels ranging from 0-448 kgN/ha (0-400 lb/ac). What effect did nitrogen levels above 448 kgN/ha (400 lb/ac) have on productivity scores when the economic yield is crude protein? Total yields of herbage dry matter and crude protein for six levels of nitrogen are given in Table 8-19. Does the law of diminishing returns apply to protein yield levels? In addition to economic considerations, what other factors would a forage producer have to consider when selecting the nitrogen fertility level to be applied?

9. List economic yield objectives that various forage producers may seek.

TABLE 8-19.   Total yields of herbage dry matter (t/ha) and crude protein (t/ha) of perennial ryegrass at nitrogen levels ranging from 504 to 896 kgN/ha averaged over a three-year period (1967–1969) in Scotland

N Applied (kg/ha)	Herbage Dry Matter						Herbage Crude Protein					
	3 cuts		5 cuts		10 cuts		3 cuts		5 cuts		10 cuts	
	(t/ha)	(tons/ac)	(t/ha)	(tons/ac)	(t/ha)	(tons/ac)	(t/ha)	(tons)	(t/ha)	(tons/ac)	(t/ha)	(tons/ac)
504	14.44	6.50	13.76	6.19	9.31	4.19	2.09	0.94	2.21	0.99	2.12	0.95
560	14.63	6.58	13.90	6.26	9.65	4.34	2.26	1.02	2.33	1.05	2.27	1.02
616	14.77	6.65	13.89	6.25	9.87	4.43	2.40	1.08	2.50	1.13	2.39	1.08
672	14.30	6.44	13.71	6.17	10.25	4.61	2.40	1.08	2.56	1.15	2.55	1.15
784	13.89	6.25	13.78	6.20	10.61	4.77	2.54	1.14	2.74	1.23	2.74	1.23
896	14.55	6.55	13.87	6.24	10.54	4.74	2.73	1.23	2.96	1.33	2.82	1.27

Source: Data from Reid 1978.

REFERENCES FOR CHAPTER EIGHT

Akinola, J. O., and P. C. Whiteman. 1975. "Agronomic Studies of Pigeon Pea (*Cajanus cajan* (L.) Millsp.). I. Field Responses to Sowing Time," *Australian Journal Agricultural Research* 26:43-56.

Beaven, E. S. 1947. *Barley: Fifty Years of Observation and Experiment.* London: Gerald Duckworth Co. Ltd. (394 pp.)

Black, J. N., and D. J. Watson. 1960. "Photosynthesis and the Theory of Obtaining High Crop Yields, by A. A. Niciporovic. An Abstract with Commentary," *Field Crop Abstracts* 13:169-175.

DeLoughery, R. L., and R. K. Crookston. 1979. "Harvest Index of Corn Affected by Population Density, Maturity Rating and Environment," *Agronomy Journal* 7:577-580.

Donald, C. M. 1962. "In Search of Yield," *Journal Australian Institute of Agricultural Science* 28:171-178.

———. 1968. "The Breeding of Crop Ideotypes," *Euphytica* 17:385-403.

Donald, C. M., and J. Hamblin. 1976. "The Biological Yield and Harvest Index of Cereals as Agronomic and Plant Breeding Criteria," *Advances in Agronomy* 28:361-405.

Downey, L. A. 1971. "Effect of Gypsum and Drought Stress on Maize (*Zea morp.* L.). I. Growth, Light Absorption and Yield," *Agronomy Journal* 63:569-572.

Fairbourn, M. L., W. D. Kemper, and H. R. Gardner. 1970. "Effects of Row Spacing on Evapo-Transpiration and Yields of Corn in a Semiarid Environment," *Agronomy Journal* 62:795-797.

Fulkerson, R. S., et al. 1967. "Yield of Dry Matter. *In vitro* Digestible Dry Matter and Crude Protein in Forages," *Canadian Journal Plant Science* 47:683-690.

Gardener, C. J., and A. J. Rathjen. 1975. "The Differential Response of Barley Genotypes to Nitrogen Application in a Mediterranean-Type Climate," *Australian Journal of Agricultural Research* 26:219-230.

Holliday, R. 1960a. "Plant Population and Crop Yield, Part I," *Field Crop Abstracts* 13(3): 159-167.

———. 1960b. "Plant Population and Crop Yield, Part II," *Field Crop Abstracts* 13(4): 247-254.

Johnson, D. R., and D. J. Major. 1979. "Harvest Index of Soybeans as Affected by Planting Date and Maturity Rating," *Agronomy Journal* 71:538-541.

Lawrence, T. 1978. "An Evaluation of Thirty Grass Populations as Forage Crops for Southwestern Saskatchewan," *Canadian Journal Plant Science* 58:107-115.

Luebs, P. E., and A. E. Laag. 1967. "Nitrogen Effect on Leaf Area, Yield, and Nitrogen Uptake of Barley under Moisture Stress," *Agronomy Journal* 59:219-222.

McNeal, F. H., et al. 1971. "Productivity and Quality Response of Five Spring Wheat Genotypes, *Triticum aestivum* L., to Nitrogen Fertilizer," *Agronomy Journal* 63:908-910.

Mowat, D. N., et al. 1965. "The in vitro Digestibility and Protein Content of Leaf and Stem Portions of Forages," *Canadian Journal Plant Science* 45:321-331.

Reid, D. 1978. "The Effects of Frequency of Defoliation on the Yield Response of a

Perennial Ryegrass Sward to a Wide Range of Nitrogen Application Rates," *Journal of Agricultural Science (Cambridge)* 90:447-457.

Singh, I. D., and N. C. Stoskopf. 1971. "Harvest Index in Cereals," *Agronomy Journal* 63:224-226.

Storrier, R. R. 1965. "The Influence of Water on Wheat Yield, Plant Nitrogen Uptake and Soil Mineral Nitrogen Concentration, *Australian Journal Agriculture and Animal Husbandry* 16:310-316.

Vijayalakshmi, K., et al. 1975. "Effects of Plant Population and Row Spacing on Sunflower Agronomy," *Canadian Journal Plant Science* 55:491-499.

Vogel, O. A., R. E. Allan, and C. J. Peterson. 1963. "Plant and Performance Characteristics of Semidwarf Winter Wheats Producing Most Efficiently in Eastern Washington," *Agronomy Journal* 55:397-398.

chapter nine

Carbon Dioxide and Yield

Photosynthesis is dependent on carbon dioxide in the atmosphere. On the average, carbon dioxide (CO_2) constitutes about 0.03% or 300 parts per million (ppm) of the atmosphere, and it is a well-known fact that if CO_2 concentrations were increased, many plants would photosynthesize at a higher rate. But CO_2 fertilization of crops under field conditions is not a common practice. Has this aspect of crop production been overlooked? Is the importance of CO_2 in crop production recognized by crop producers? Can crop producers modify the crop environment to improve the CO_2 concentration? These and other questions form the basis of this chapter.

HISTORICAL PERSPECTIVE

In the seventeenth century, the Dutch chemist Jan Baptista van Helmont (Cannon 1963) attempted to find the substance from which plants developed. He planted a 2270 gram willow branch in an earthen vessel with 90.8 kilograms of oven-dried soil. The vessel was carefully guarded by a perforated lead shield to keep out any foreign matter, and only water could be added as needed. At the end of five years, the tree was removed and weighed. Although the tree had increased by 74.46 kilograms, the soil in the container appeared to have lost 57 grams, which van Helmont attributed to loss when washing off the roots. He concluded that growth was due to the water alone.

Today we understand that the loss of weight in the soil was due to the uptake of fertilizer nutrients, and the realization that the majority of dry-matter increase came from carbon dioxide had to await the experiments of the Swiss scientist Nicolas Theodore de Saussure around 1790. To discover the "food" of plants, de Saussure enclosed leaves, branches, roots, and the flowering parts of plants in flasks. He was able to determine that water and air contributed more to the formation of plant dry matter than the minerals absorbed through the roots. His careful work led to the observation that plants in sunlight absorb carbon dioxide through their leaves, and when the carbon is used in photosynthesis with water absorbed through the root system, it forms the first simple sugars. He observed that at night, carbon dioxide was given off and pointed out that this was true respiration, much like the breathing of animals.

The discovery that carbon dioxide was essential to plant growth raised the question in de Saussure's mind as to whether more carbon dioxide would be better for plant growth. He then conducted the first carbon dioxide fertilization tests by filling the flasks containing intake leaves with more concentrated CO_2 mixtures, but perhaps he did not understand the relatively low levels involved. He reported that plants died in an air mixture containing 50% CO_2. In 1804 de Saussure published his classical work on agricultural chemistry ("Chemical Researches on Vegetation") (Cannon 1963).

It may have been de Saussure's experiments that eventually led to modern trials with CO_2 enrichment. Early studies gave variable results probably because of contaminants in the CO_2 released. A detailed report of outstanding results from CO_2 enrichment in greenhouses and growth chambers at Michigan State University was given by Wittwer and Robb (1964). Several cultivars of lettuce, tomatoes, and cucumbers were subjected to CO_2 concentrations, initially at 500 to 600 ppm and subsequently to 1200 to 1500 ppm, for the period January through April. Light intensities were 500 to 1500 footcandles.

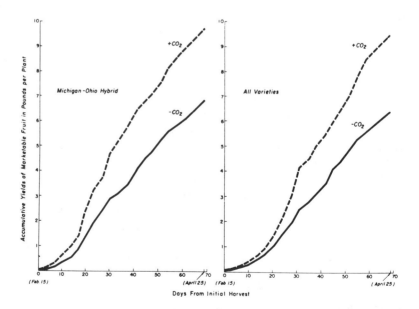

Figure 9–1. Accumulative yields of Michigan-Ohio hybrid and the mean accumulative yields of 22 cultivars and selections of tomatoes subjected to 125 to 500 ppm. ($-CO_2$) and 800 to 2000 ppm. ($-CO_2$) of carbon dioxide during daylight hours in a greenhouse atmosphere. (From Wittwer and Robb et al. 1964, p. 42.)

A fresh weight increase of 30% was found in seven lettuce cultivars grown over a 39-day period at CO_2 levels double the normal air level. At CO_2 levels of 800 to 2000 ppm, average yield increases of 70% expressed as fresh weights were observed (Table 9-1).

In tomatoes, yields of marketable fruit during the first 110 days of harvest for nine cultivars at 800 to 2000 ppm of CO_2 showed a mean yield increase of 43%. Yield increases varied between 25 and 70% depending on the cultivar. Accumulated yields became progressively greater with an extension of the harvest period. Yield increases were associated with larger fruit size and improved color, flavor, and sweetness, all of which resulted in an increase in the percentage of top-grade tomatoes. Vegetative growth and reproductive development were accelerated by increased CO_2 concentrations; but effects were more pronounced on reproductive development, and it appeared that tomato fruit yields were significantly increased at the expense of the vegetative growth (Figures 9-1 and 9-2).

Vegetative growth increases in cucumber plants from added CO_2 were equally as great as those for lettuce and tomatoes. The most striking

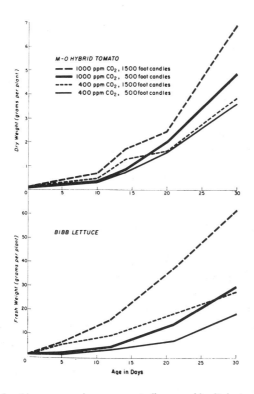

Figure 9-2. Yields of lettuce and tomatoes influenced by light intensity and carbon dioxide level of the atmosphere in controlled environment growth chambers. (From Wittwer et al. 1964, p. 37.)

difference, however, was that during the first 60 days of growth, the number of pistillate flowers (fruit-bearing flowers) was approximately doubled. No differences occurred in the number of staminate flowers.

It is suggested that most plants grow far more rapidly and luxuriantly in an atmosphere that contains more than the normal CO_2 level (Lake 1963; Plass 1959). The statement that CO_2 stimulation of plant growth is "an abandoned gold mine" was endorsed by Wittwer and Robb (1964).

Increases in the yield of marketable fruit in tomato of 40% by enrichment with 900 ppm of CO_2 led Hicklenton and Joliffe (1978) to conclude that the benefits of CO_2 enrichment in terms of increased yield result not only from increased photosynthesis and growth but also from changes in the innate capacity of photosynthetic systems to utilize CO_2. The effects of enriching greenhouse atmospheres with CO_2 are clearly beneficial to the grower and may result in greater modifications to the physiology of crop plants than were previously imagined.

TABLE 9–1. Yields of lettuce cultivars influenced by carbon dioxide added to the greenhouse atmosphere

Lettuce Cultivar	Normal CO_2 Level (g)	(lb)	Enriched with CO_2 (g)	(lb)
Bibb	544.8	1.2	862.6	1.9
Cheshunt No. 5B	726.4	1.6	1271.2	2.8
Grand Rapids H-54	590.2	1.3	1180.4	2.6
Average	620.5	1.4	1104.7	2.4

Source: Data from Wittwer and Robb 1964.

CARBON DIOXIDE AND CROP PRODUCTION

The results with increased CO_2 concentrations in enclosed environments suggest that normal CO_2 levels in the air may limit photosynthesis and yield. It is logical to conclude that higher CO_2 levels under field conditions would lead to higher yields. Under field conditions, however, the direct addition of CO_2 seems impractical because of the volume of air involved, and winds and air turbulence limit control, although they serve to replenish depleted supplies of CO_2. Tisdale and Nelson (1966) nevertheless suggested that it may not be unreasonable to suppose that atmospheric enrichment of CO_2 for certain field crops in certain locations may someday be practiced. Whether it is feasible or not, CO_2 is a key element in photosynthesis, and if crop producers understand its importance, they may be able to effect changes and practices that might result in an improved carbon dioxide environment. To understand the implications and extent of factors associated with carbon dioxide enrichment in the field, factors of the carbon cycle are presented here.

In ancient times, when oil and coal fields were formed by the products of photosynthesis, levels of carbon dioxide in the atmosphere may have been 10 to 100 times that of today's levels (Plass 1959). The carbon locked in fossil fuels millions of years ago is now being released by combustion and may be contributing to an increasing carbon dioxide concentration in the atmosphere today.

On a world basis, photosynthesis now uses about 20 to 30 billion tonnes (22 to 33 billion tons) of carbon annually, which may be stored as food and agricultural products in forest trees and various forms of life and then released by respiration, combustion, or decay of organic matter. On land, the greatest absorbers of carbon are the forests, which use about 14 billion tonnes (15.4 billion tons) a year and store about 400 to 500 billion tonnes (440 to 550 billion tons).

Harvesting forest products with the eventual release of carbon dioxide, along with 3 to 6 billion tonnes (3.3 to 6.6 billion tons) of CO_2 released from combustion of fuels in industry, homes, and transportation systems, may cause a slow increase in atmospheric carbon dioxide levels. Rapid carbon dioxide build up in the atmosphere is prevented by the buffering action of oceans. If CO_2 levels increase in the earth's atmosphere, they may cause an alteration of future weather conditions through the "greenhouse" effect, a situation where a CO_2 mantle prevents reradiated heat to be trapped within the CO_2 mantle, and temperature increases occur. A carbon dioxide increase would have a direct effect (increase in photosynthesis) and an indirect effect (increase in air temperature) on crop production.

Atmospheric carbon dioxide measurements taken at the Mauna Loa Observatory in Hawaii suggest that an annual increase in atmospheric carbon dioxide levels is occurring: From 1960 to 1965, measured carbon dioxide increased slightly less than 1%; between 1965 and 1970, the increase was slightly more than 1%; and between 1970 and 1975, the increase was 1.2%. Similar results were obtained at the South Pole; Barrow, Alaska; American Samoa; and in Sweden and Australia (Anonymous 1978).

Estimates of carbon dioxide concentrations when the Industrial Revolution began in 1860 ranged from 265 to 305 ppm. If present levels of CO_2 lie between 330 and 368 ppm, then the increase since 1860 is between 8.2 and 13.8%.

A warm trend in the ocean resulting from the "greenhouse effect" could result in a release of carbon dioxide into the atmosphere. The ocean is a vast reservoir of carbon; there is about 30 times as much carbon dissolved in the oceans as occurs in the atmosphere and living matter combined. It is impossible to determine the portion of carbon dioxide released by fossil fuel burning that has remained in the atmosphere and that has been released or absorbed from the oceans. Carbon dioxide concentrations in the atmosphere have increased and may increase further. An increase would have a desirable effect on photosynthesis.

OTHER SOURCES OF CARBON DIOXIDE FOR CROP USE

Soil respiration is possibly an important source of carbon dioxide for most plant communities and results from aerobic metabolism of soil microorganisms, respiration of soil-burrowing animals, and respiration of plant roots with some CO_2 release by chemical action. The soil is a mecca of living organisms (Table 9-2).

TABLE 9–2.   Average number of soil microflora and live weight per hectare to plow depth

Group	Average Number per Gram of Soil	Live Weight per Hectare to Plow Depth (kg)	(lb)
Bacteria	1 billion	560	255
Actinomycetes	10 million	840	382
Fungi	1 million	1200	545
Algae	100 thousand	168	76

Source: Data from Logsdon 1975, p. 91.

Under favorable temperature conditions, organic matter is continuously decomposing, giving rise to carbon dioxide gas, which may escape into the atmosphere or dissolve in the soil water to form carbonic acid. Roots in the soil are living organisms and through the process of respiration give off carbon dioxide.

Within forests, on grasslands, and in river bottom lands, carbon dioxide concentrations at or near the soil level may be two or three times above the normal 300 ppm present in the atmosphere. (Fuller 1948).

Levels of carbon dioxide over muck soils, which consist almost entirely of decomposing organic matter, have appreciably higher CO_2 levels than over mineral soils (Hopen and Ries 1962). Onions, lettuce, and celery grow better in muck (organic) than mineral soils, which has been attributed to a better moisture relationship, nitrogen availability, and more favorable soil density and root media; but perhaps some of the advantage of a muck soil lies in the fact that it releases more CO_2 than does a mineral soil (Figure 9-3).

CO_2 from the soil was found to contribute significantly to photosynthesis of a corn crop under conditions of cloudy weather when total photosynthesis was low; but under bright, sunny conditions, the CO_2 supply from the soil was only a fraction of the total assimilated by the crop (Moss, Musgrave, and Lemon 1957).

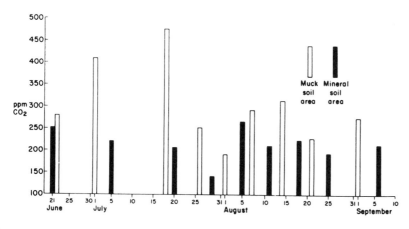

Figure 9–3. Average CO_2 level, at sunrise (5 cm above soil surface) at mineral and muck soil areas 18 km apart during the summer of 1961. (From Hopen and Ries 1962, p. 366.)

If a humus content (organic matter) of 3% is assumed to exist in a given soil, one hectare (2.47 acres) of soil therefore will contain about 200,000 kg (36,800 lb/ac) organic matter, the equivalent of about 300,000 kg (52,200 lb/ac) CO_2 after oxidization occurs. One hectare (acre) of soil produces about 10,000 kg (1065 lb/ac) of CO_2 a year (Lundegardh 1927).

Carbon dioxide evolution rates can vary widely. In forest soils, variations of 100 to 700 mg per square metre per hour were measured (Ellis 1969), with the high rate occurring in hot climates and the low rate in cold climates.

At night, in the absence of photosynthesis, plant and soil respiration will cause CO_2 levels to increase. In a timothy-meadow fescue forage crop, CO_2 levels exceeding 430 ppm were observed (Monteith 1962) during the night. The greater density of carbon dioxide will cause it to remain within the crop canopy, and under conditions of still morning air, CO_2 levels within the canopy will be high when photosynthesis begins. As photosynthesis depletes the CO_2 supply within the crop canopy, air turbulence and wind currents are essential to renew and maintain carbon dioxide levels. Niciporovic (Black and Watson 1960) suggested that the main source of CO_2 for photosynthesis within a crop canopy must come from atmospheric air. He reasoned that if 20% of the total weight of a previous crop remained in the soil as roots, and if a further 20% was returned as organic matter (crop residue at harvest or applied as manure), the return of 40% organic matter from a previous crop(s) would satisfy only a small part of a subsequent crop's CO_2 requirements. In a commentary on this paper, Black and Watson (1960) criticized the suggestion that CO_2 was replenished by convection currents and suggested that CO_2 is replenished by downward eddy-diffusion from air moving horizontally, and, except in very calm conditions, atmospheric transport processes are sufficiently active to maintain the CO_2 concentration near 300 ppm above the crop canopy. High rates of air flow cannot increase it further.

Photosynthesis within a corn crop canopy was measured (Wright and Lemon 1966) on a clear day, August 1, 1961 at various heights in a densely planted four-hectare (10 acre) field at Ithaca, New York. The crop was 220 cm (87 in) tall and had not reached the tasseling or reproductive stage. At noon when light penetration into the crop was greatest, the lowest level of CO_2 was 260 ppm measured at 130 cm (51 in) above the soil. CO_2 levels in this zone were restored between 0500 and 0600 hours (Table 9-3). Lemon and Wright stressed that wind to cause CO_2 movement is essential; that the percentage CO_2 fixed in photosynthesis closely paralleled wind intensity; and that in this study the maximum hourly change of CO_2 was less than 10 ppm under conditions of moderate winds. The low rate of CO_2 replenishment was attributed to the fact that such a tall crop resisted air movement within the canopy. Between 53 and 89% of CO_2 fixation, depending on the time of day, was accounted for in the 175- to 200-cm (69-79 in) zone probably because this represented the zone of greatest light interception and photosynthetic surface, although at noon, more CO_2 was fixed in the 20- to 175-cm (8-69 in) zone as light penetrated the canopy. Highest CO_2 concentrations were found at the 20-cm (8 in) height, but it is uncertain if this was due to low photosynthetic usage, effective transfer of CO_2 into this zone by air currents, or release of CO_2 from the soil. Much speculation remains on the significance of CO_2 release from the soil for photosynthesis.

TABLE 9-3.   Concentrations of carbon dioxide at several heights within and above an immature corn crop, 220 cm (87 in) tall, August 1, 1961

Height		CO_2 Concentration (ppm) EST. hours			
(cm)	(in)	1800–1900	1900–2000	2000–2100	0500–0600
350	138	304	348	374	312
260	102	300	350	376	313
200	79	295	349	376	314
175	69	298	355	379	314
135	53	300	354	378	317
75	30	305	356	381	326
20	8	305	358	383	327

Source: Data from Lemon and Wright 1966.

MANAGING CARBON DIOXIDE FOR CROP PRODUCTION

Manure and Crop Residues

To encourage early development of transplants for use in the farm garden, a "hot bed" was commonly used and consisted of an outdoor bin with a base of rapidly decomposing manure, covered with soil and glass or plastic facing, and angled to face south. The decomposing manure provided warmth and nutrients, regarded as the key for a successful hot bed, but the release of CO_2 into this enclosed bin undoubtedly was of significance in explaining the rapid growth that occurred.

The success of hot beds led to the use of huge amounts of animal manures to be employed in greenhouses, and many historical reports relate successful cucumber culture under this system. As much as 448 t/ha (200 tons/ac) of manure has been reportedly applied for the culture of greenhouse cucumbers, and as much as 5000 ppm of CO_2 have been measured in closed greenhouses in Denmark in midwinter that evolved from the manure, peat, or straw used.

Successful potting materials for greenhouse crops contain high levels of organic materials. In contrast, it is noted in hydroponic plant culture, i.e., growing plants in nutrient solution, that little if any organic matter is present, and the lack of release of CO_2 may be a factor explaining the varying and often disappointing results obtained with nutrient culture.

Many farmers insist that manure offers an advantage over commercial inorganic fertilizers. Although a plant does not distinguish between a molecule of nitrogen from an organic as opposed to an inorganic source, manure does offer the advantage of adding valuable organic matter, which loosens soil, releases nutrients, increases water-holding capacity, and releases CO_2. Incorporation of green manure (plowing under an actively growing crop), especially a legume, will produce comparable results. If a nonlegume crop is incorporated, the addition of inorganic fertilizers is needed to aid decomposition.

Incorporating material with a high carbon content such as straw, corn stover, wood chips, or sawdust for purposes of promoting CO_2 release without the addition of adequate nitrogen to promote decomposition may result in a crop showing nitrogen deficiencies. Nitrogen is essential for the decomposition of organic matter as it permits and promotes the growth of microorganisms that attack and break down the organic materials. If the organic matter contains a low level of nitrogen in proportion to the carbon present, the microbes will utilize any nitrogen in the soil for their own use in decomposing organic matter. Until decomposition is completed and the microbes decay, thereby releasing the nitrogen, a growing crop may be unable to obtain sufficient nitrogen for adequate growth, and high levels of carbon dioxide are of little or no benefit because the nitrogen is the limiting factor.

When organic materials with a 30:1 C:N ratio (**carbon:nitrogen ratio**) are added to soil, shortages of soil nitrogen during the initial decomposition process can be expected. Since straw, stover, and wood chips may have a C:N ratio of 90:1, adequate nitrogen should be incorporated to prevent crop nitrogen deficiencies.

Carbon Dioxide Fertilization

One of the obvious methods to increase the CO_2 level in a crop is to supply dry ice to a field. Dry ice is composed of gaseous carbon dioxide, which becomes a solid when cooled at atmospheric pressure to $-78.5°C$ $(-173.3°F)$. If applied to the soil in chunks, dry ice does not melt to form a liquid but sublimes directly back to a gas. Various unofficial and unsubstantiated claims that cannot be duplicated suggest that yields can be increased significantly with dry-ice applications in low-lying protected areas. The idea does not appear practical because of the volume of air involved and because of the large amount of CO_2 required to produce a crop.

How much CO_2 is required to produce a grain yield of corn of 6200 kg/ha (100 bu/ac)? This can be calculated as follows. The amount of CO_2 required to produce a molecule of sugar ($C_6H_{12}O_6$) can be determined from the balanced photosynthetic equation.

$$6CO_2 + 12H_2O \rightarrow C_6H_{12}O_6 + 6H_2O + 6O_2$$

The atomic weights of the elements involved are:

Carbon — 12 grams
Oxygen — 16 grams
Hydrogen — 1 gram

Six molecules of carbon dioxide weigh $6(12 + 32) = 264$ grams, and a mole of glucose sugar weighs $(6 \times 12) + (12 \times 1) + (6 \times 16) = 180$ grams. If 264 grams of CO_2 produce 180 grams of glucose, then to produce a grain yield of 6200 kg/ha (100 bu/ac) corn crop, 20,650 kilograms (9386 lb) of CO_2 are required. This agrees with the figure presented by Norman (1962) that a 6200 kg/ha (100 bu/ac) grain corn crop will require 22,800 kg/ha (10,360 lb) of CO_2 and at 300 ppm of CO_2 in the atmosphere, the plant must process 30,500 tonnes (33,516 tons) of air to procure the carbon for the crop. Thus, the logistics of increasing CO_2 levels by the release of dry ice are so unmanageable as to be impractical.

Windbreaks

If carbon dioxide fertilization is not practical on a field scale, crop producers may have to rely on CO_2 replenishment by air currents. Based on diffusion theory, Waggoner, Moss, and Hesketh (1963) predicted an increase in photosynthesis by increasing the turbulence of air and hence of CO_2 movement, even though they observed that according to previous reports still air was not a common phenomenon outdoors. On a macro or large-scale basis, at a 50-cm (20 in) height above a forage crop, calm air occurred less than 1% of all daylight hours, and half of the hours had wind speeds in excess of 133 cm per sec (3 mph). On a micro basis, air turbulence around the leaf also affects the rate at which CO_2 diffuses into a leaf by modifying leaf resistance to CO_2 movement. A sugarcane leaf in very still air at CO_2 concentrations of 200 and 300 ppm showed differences in the rates of photosynthesis (Figure 9-4). Stirring the air containing 200 ppm increased photosynthesis to a level equal to that of the upper curve, i.e., as fast as increasing CO_2 levels to 300 ppm in the quiet air. Turbulence in the leaf microclimate occurs when solar radiation is absorbed, and heat is produced, which heats the air layer near the leaf, causing the air in this microclimatic area to rise and be replaced by heavier and cooler air containing CO_2. On a microclimatic basis, wind speed generally increases as the sun warms the plants and air associated with a field crop. Sunlight serves the dual role, therefore, of supplying the energy for photosynthesis and also causing turbulent air that helps replenish the supply of CO_2 to the leaf surface.

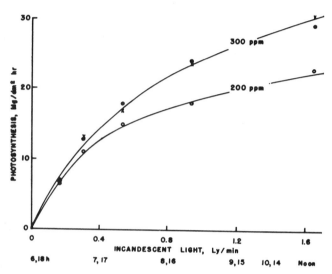

Figure 9–4.   Response of the photosynthesis of cane to increasing light in calm (o) or turbulent (*) air containing 200 ppm CO_2 and in calm air containing 300 ppm CO_2 (X). (From Waggoner, Moss, and Hesketh 1963.)

Within a well-developed crop canopy, which may hinder air currents, Musgrave (reviewed by Newman 1963) reported rather violent reductions from the 300 ppm value for CO_2, especially during calm early morning hours. CO_2 levels fluctuated according to the rate of photosynthesis during a low level or complete lack of turbulence and changes in a crop canopy to promote air turbulence may be useful in replenishing CO_2 levels. Strong hot winds may be detrimental in causing increased water losses and may not be as effective in increasing carbon dioxide within a crop canopy which results in more turbulent air currents.

Generally, field crops are more productive when protected by a windbreak (Bagley and Gowen 1960; Rosenberg 1966*a* and *b*); but yield increases were attributed to increased relative humidity and increased average air and soil temperature, due to reduced windspeed, although CO_2 levels were not measured in these studies. Windbreaks reduced wind speeds 50 to 70% within a distance of eight times the height of the windbreak (Bagley and Gowen 1960). Similar results were obtained by Rosenberg (1966*b*), but he indicated that the air in the sheltered area was not still or stagnant. During daylight hours, radiant energy caused strong buoyancy effects and a greater wind shear (gradient between any two levels), which resulted in greater turbulence. In total, a more strongly unstable atmosphere was found to exist in the lee of a shelterbelt during the daytime and a more intensely stable atmosphere existed at night. The literature on the effects of shelterbelts on CO_2 levels is sparse.

Small temporary windbreaks consisting of corn rows or snow fences were spaced at various intervals in soybean fields to study the effects of windbreaks on CO_2 levels and crop yields. Data from eleven tests showed that soybeans sheltered by temporary corn windbreaks grew taller, produced more dry weight, had a larger LAI, and produced higher grain yields (Radke and Burrows 1970) (Table 9-4).

TABLE 9-4.   Grain yields (t/ha)(bu/ac) of soybeans as affected by small temporary field windbreaks in Minnesota in the years 1963 to 1968 inclusive

Year*	Open Area		Corn Windbreaks		Percentage
	t/ha	bu/ac	t/ha	bu/ac	increase
1963	1.82	29.0	2.19	34.9	20.3
1964	1.06	16.9	1.33	21.2	25.5
1964	1.28	20.4	1.43	22.8	11.7
1965	1.26	20.1	1.43	22.8	13.4
1966	0.72	11.5	0.92	14.7	27.8
1967	1.93	30.8	1.92	30.6	−0.5
1967	1.50	23.9	1.47	23.4	−2.0
1968	1.79	28.5	2.03	32.4	13.4
1968	1.25	19.9	1.38	22.0	10.4
1968	2.25	35.9	2.37	37.8	5.3
1968	2.15	34.3	2.40	38.3	11.6
Snowfence windbreaks					
1963	1.82	29.0	1.82	29.0	0.0
1964	1.28	20.4	1.41	22.4	10.2

*Several sites explain why more than one result is given in some years.

Source: Data from Radke and Burrows 1970.

Soybean rows adjacent to corn windbreaks did not do as well as the rest of the windbreak-sheltered soybeans because of shading effects and root competition from the corn. East-west planted windbreaks, which were more nearly perpendicular to the prevailing winds, were more effective than those oriented in a north-south direction. Windbreak spacings between six times and fifteen times the windbreak height have been effective, and it is possible that positive results could be obtained with windbreak spacings greater than fifteen times the height. Radke and Burrows (1970) admit that the causes for increased soybean plant response have not been definitely established but suggest air turbulence effects on micrometeorological elements were responsible, including more favorable CO_2 replenishment. It is not impossible, however, that the results were due to light stress. Soybean leaves, to a limited extent, tend to orientate their leaves toward the sun, and under conditions of strong direct winds, this may be prevented and the more reflective underside of the leaves may be intercepting the light.

In another study on the Great Plains (Hagen and Skidmore 1974), under limited precipitation conditions, water-use efficiency was increased by reducing turbulent transfer by wind barriers or interspersed tall plants at least 20 cm (8 in) above the sheltered crop. The Great Plains area is characterized by high and nearly constant winds. These authors considered high wind speeds of up to 400 cm/sec (9 mph) and reported that in 1970, three weeks of warm southerly winds caused sheltered winter wheat to grow 11 cm (4 in) taller and to have 44 and 10% greater LAI's of the uppermost and all leaves, respectively, than wheat in an open field. The increase in crop water-use efficiency was considered the key cause, however.

Concern has been raised that windbreaks to increase water-use efficiency will impede CO_2 movement under calmer air conditions. Hagen and Skidmore (1974) reject this on the basis that resistances are greater in the CO_2 exchange path than in the transpiration path, and the percentage reduction in transpiration is greater than the percentage reduction in CO_2 exchange. The net result would be increased water-use efficiency with decreased wind speed but with a minor effect in relatively strong winds.

Application of Lime

The largest storehouses of carbon are not the trees in the vast forests nor the oceans, but rather limestone rocks. Over 99% of the earth's carbon is locked in these deposits. A common and recommended agricultural practice for successful crop production is control of soil acidity by the application of limestone ($CaCO_3$) directly to the soil. Generally limestone or agricultural lime is applied to soils below a pH of 6.5, unless crops favored by an acid reaction are to be grown. Most soils used for the production of field crops will have a pH of 4.5 to 8.5. The pH scale of soil reaction is as follows:

> pH 7.5 to 8.5 — moderately **alkaline**
> pH 7.1 to 7.4 — slightly alkaline
> pH 7.0 — neutral
> pH 6.6 to 6.9 — very slightly acid
> pH 6.0 to 6.5 — slightly acid
> pH 5.5 to 5.9 — moderately acid
> pH below 5.5 — strongly acid

Liming means the addition and incorporation into the soil of any calcium or calcium and magnesium compounds capable of reducing acidity. In addition to calcium oxide (CaO), lime includes such materials as calcium hydroxide, calcium carbonate, calcium-magnesium carbonate, and calcium silicate slags.

The beneficial effects of liming are often dramatic and produce striking increases in plant growth. The benefits of liming are attributed to (1) the value of calcium or magnesium as a plant nutrient, (2) the improvement of phosphorus availability, (3) enhanced decomposition of plant residues, and (4) increased availability of nitrogen for plant growth. Nitrogen fixation in legumes is enhanced, and soil physical condition is improved. The possible benefit of limestone as a source of CO_2 has apparently not been recognized because it has not been mentioned in textbooks. The general reaction of limestone in acid soil is:

$$CaCO_3 + 2HCl \rightarrow CaCl_2 + H_2O + CO_2$$

(calcium carbonate)   (soil acids)

$$or\ MgCO_3 + 2HCl \rightarrow MgCl_2 + H_2O + CO_2$$

(magnesium carbonate)   (soil acids)

The release of carbon dioxide when lime is applied may be an unrecognized benefit to crop production. Carbon dioxide concentrations were sampled at depths of 5, 10, 20, and 30 cm within the soil of fallow (uncropped) and cropped plots (Pritchard and Brown 1979). In all plots, CO_2 concentrations increased with depth. It is impossible to determine the extent that limestone contributed to the CO_2 levels; it was greater in the cropped than in the fallow plots, probably due to root respiration.

The amount of limestone needed to correct soil pH is substantial (Table 9-5) and varies according to the crop to be grown, the pH level of the soil, and the soil type and texture.

TABLE 9–5.   Estimates of limestone needed to correct soil pH for some common crops (t/ha) (tons/ac). Calculations determined at Guelph, Ontario

	Fine texture soil pH values			Medium texture soil pH values		
	5.0 and less	5.1 to 5.5	5.6 to 6.0	5.0 and less	5.1 to 5.5	5.6 to 6.0
Legume forages, small grains, and soybeans	9 (8)	7 (6)	4.5 (4)	7 (6)	4.5 (4)	2 (1.8)
Corn	9 (8)	4.5 (4)	0	7 (6)	4.5 (4)	0
Potatoes	4.5 (4)	0	0	2 (1.8)	0	0
Tobacco				4.5 (4)	2	0
Vegetables	9 (8)	7 (6)	4.5 (4)	7 (6)	4.5 (4)	2 (1.8)

Source: Data from Bates, Lane and Heeg 1977.

CONCLUSIONS

The extent to which a crop producer can "manage" the carbon dioxide levels for field crop production either through the addition of manure or crop residues, carbon dioxide fertilization, windbreaks, or the application of lime is a much debated, important, and unresolved question. No one solution at any one time may be economical, practical, or effective; but if CO_2 levels could effectively be increased, the results might be spectacular and the potential should not be underestimated.

Considerations of increasing CO_2 concentrations must recognize the law of the minimum (see Chapter Three). CO_2 enrichment can be effective only if another factor(s) does not limit photosynthesis. Gaastra (1962) stressed this fact, and in addition to the obvious factors such as light, temperature, moisture, and fertility, factors that influence water uptake by roots such as the rate of root extension, soil moisture stress, soil aeration, root temperature plus water transport in the upper part of the plant, transpiration rate, absorbed radiation, leaf tissue temperature, wind speed, humidity in the air, etc.—all of these must be taken into consideration. In addition to all these factors resistance to CO_2 movement in the external air near the leaf surface as well as within the leaves themselves may limit CO_2 intake.

Although Wittwer and Robb (1964) mentioned European reports of crops responding to CO_2 levels ranging from 20,000 to 30,000 ppm, the possibility of CO_2 toxicity cannot be overlooked. Early studies with increased levels of CO_2 in greenhouses reported evidence of toxicity possibly due to impurities in the CO_2 released. The human safety limit for CO_2 is set at 5000 ppm (Anonymous 1960), although much higher levels are safely tolerated for short periods.

It is also known that above-normal levels of CO_2 do reduce the size of stomatal openings and hence restrict transpiration and CO_2 intake. In the presence of decaying manures, however, small amounts of ammonia are released, which have the effect of keeping the stomata open. Because CO_2 at high concentrations can close stomata, this has been proposed as a means of increasing water-use efficiency (Begg and Turner 1976). The complexities of this subject are great and more research is needed.

Rather than providing a producer with a clear guide to increase crop yields by effective CO_2 management, this chapter has attempted to show the importance of CO_2 and its complexity in management, as well as to illustrate that crop production is an art as well as a science.

QUESTIONS

1. What practices do farmers employ in your area that might influence carbon dioxide levels in the field?

2. Speculate on why the practice of CO_2 enrichment in greenhouses is so slow to develop.

3. If increased food production was needed on a field scale and economic considerations were of no consequence, how might CO_2 enrichment on a field scale occur?

4. Apart from the unmanageable nature of CO_2 enrichment on a field scale, what other barriers to enrichment exist?

5. Suggest reasons why the growth of a crop is restricted for several metres along the edge of a dense woodlot.

6. A two-row barley cultivar was subject to CO_2 enrichment in the field in enclosed containers throughout the entire life cycle. Continuous carbon dioxide enrichment increased grain yield by 70% with much of the benefit accruing from enrichment before anthesis. Explain.

    The results of CO_2 enrichment are presented in Table 9-6. What impact did CO_2 enrichment have on productivity scores? What yield component contributed most to the increased yield?

TABLE 9–6.   Yield and yield component data from barley enriched with CO_2 in the field in enclosed containers

Component	Check	Enriched with CO_2
Grain yield (g/m²)	277	471
Straw and chaff (g/m²)	511	967
Weight per grain (mg)	38.8	40.9
Grain no./ear	11.9	11.8
Ears/m²	616	973

Source: Data from Gifford, Bremner, and Jones 1973.

REFERENCES FOR CHAPTER NINE

Anonymous. 1960. "The Safety Limits of Hazardous Chemicals," *Safety Maintenance Magazine*, pp. 37–39.

Anonymous. April–May 1978. "Atmospheric Carbon Dioxide Continues to Increase," *Crops and Soils*, p. 29.

Bagley, W. T., and F. A. Gowen. 1960. "Growth and Fruiting of Tomatoes and Snap Beans in the Shelter Area of a Windbreak," *Fifth World Forestry Congress Proceedings* 3:1667–1670.

Bates, T. E., T. H. Lane and T. J. Heeg. 1977. *Soil Acidity and Liming*. Ontario Ministry of Agriculture and Food. Fact Sheet No. 534.

Begg, J. E., and N. C. Turner. 1976. "Crop Water Deficits," *Advances in Agronomy*. 28:161–217.

Black, J. N., and D. J. Watson. 1960. Photosynthesis and the Theory of Obtaining

High Crop Yields, by A. A. Niciporovic. An Abstract with Commentary, *Field Crop Abstracts* 13:169-175.

Cannon, G. G. 1963. *Great Men of Modern Agriculture*, pp. 125-134. New York: Macmillan Publishing Co., Inc.

Ellis, R. C. 1969. "The Respiration of the Soil Beneath Some Eucalyptus Forest Stands as Related to the Productivity of the Stands," *Australian Journal of Soil Research.* 7:349-357.

Fuller, H. J. 1948. "Carbon Dioxide Concentrations of the Atmosphere above Illinois Forest and Grassland," *American Midland Naturalist* 39:247-249.

Gaastra, P. 1962. "Photosynthesis of Leaves and Field Crop," *Netherlands Journal of Agricultural Science.* 10:311-324.

Gifford, R. M., P. M. Bremner, and D. B. Jones. 1973. "Assessing Photosynthetic Limitation to Grain Yield in a Field Crop," *Australian Journal of Agricultural Research.* 24:297-307.

Hagen, L. J., and E. L. Skidmore. 1974. "Reducing Turbulent Transfer to Increase Water-Use Efficiency," *Agricultural Meteorology.* 14:153-168.

Hicklenton, P. R., and P. A. Jolliffe. 1978. "Effects of Greenhouse CO_2 Enrichment on the Yield and Photosynthetic Physiology of Tomato Plants," *Canadian Journal Plant Science.* 58:801-817.

Hopen, H. J., and S. K. Ries. 1962. "Atmospheric Carbon Dioxide Levels over Mineral and Muck Soils," *Proceedings of the American Society of Horticultural Sciences.* 81:365-367.

Lake, J. V. 1963. "Carbon Dioxide Is a Basic Need of Plants," *The Grower* 59(14): 687-688.

Logsdon, G. 1975. *The Gardener's Guide to Better Soil.* Emmaus, Pennsylvania: Rodale Press. 246 p.

Lundengardh, H. 1927. "Carbon Dioxide Evolution of Soil and Crop Growth," *Soil Science.* 23:417-451.

Monteith, J. L. 1962. "Measurement and Interpretation of Carbon Dioxide Fluxes in the Field," *Netherlands Journal of Agricultural Science.* 10:334-346.

Moss, D. N., R. B. Musgrave, and E. R. Lemon. 1961. "Photosynthesis under Field Conditions III. Some Effects of Light, Carbon Dioxide, Temperature, and Soil Moisture on Photosynthesis, Respiration and Transpiration of Corn," *Crop Science* 1:83-87.

Newman, J. E. 1963. "Carbon Dioxide in Crop Environments by R. B. Musgrave," in "Symposium: Responses of Field Crops to Environmental Factors Summary Statements," *Agronomy Journal.* 55:31.

Norman, A. G. 1962. "The Uniqueness of Plants," *American Scientist.* 50:436-449.

Plass, G. N. 1959. "Carbon Dioxide and Climate," *Scientific American* 201:41-45.

Pritchard, D. T., and N. J. Brown. 1979. "Respiration in Cropped and Fallow Soil, *Journal of Agricultural Science.* 92:45-51.

Radke, J. K., and W. C. Burrows. 1970. "Soybean Plant Response to Temporary Field Windbreaks," *Agronomy Journal.* 62:424-429.

Rosenberg, N. J. 1966a. "Influence of Snow Fence and Corn Windbreaks on Microclimate and Growth of Irrigated Sugar Beets," *Agronomy Journal.* 58: 469-475.

————. 1966*b*. "Microclimate, Air Mixing, and Physiological Regulation of Transpiration as Influenced by Wind Shelter in an Irrigated Bean Field," *Agricultural Meteorology.* 3:197–224.

Tisdale, S. L., and W. L. Nelson. 1966. *Soil Fertility and Fertilizers,* 2nd ed. New York: Macmillan Publishing Co., Inc. 694 pp.

Waggoner, P. E., D. N. Moss, and J. D. Hesketh. 1963. "Radiation in the Plant Environment and Photosynthesis. *Agronomy Journal.* 55:36–39.

Wittwer, S. H., and W. Robb. 1964. "Carbon Dioxide Enrichment of Greenhouse Atmospheres for Food Crop Production," *Economic Botony.* 18:34–56.

Wright, J. L., and E. R. Lemon. 1966. "Photosynthesis under Field Conditions. IX. Vertical Distribution of Photosynthesis within a Corn Crop," *Agronomy Journal.* 58:265–268.

chapter ten

Application of Photosynthetic Concepts

MANAGEMENT OF PERENNIALS AND WINTER ANNUALS

Fall Management

In temperate zones, fall-seeded **winter annuals** and perennial crops are often subjected to severe winter conditions, which may result in the loss of plants. Winterkill may be caused by low temperature conditions without a protective snow cover; by the freezing and thawing action of warm days and cold nights in spring; by smothering caused by ice-sheet formation and the lack of oxygen required in respiration; by a build-up of toxic gases; or by desiccation associated with frozen soil and drying winds whereby the plant cannot obtain moisture. Crop producers in temperate zones must take precautions to help assure winter survival, although under extreme conditions loss of plants may occur even with the best management program.

A major factor associated with winter plant survival is the build-up of adequate plant food reserves to withstand stress conditions. Adequate food reserves are essential for winter survival, for differentiation of stress-resistant cells and **crown buds,** as well as for energy for regrowth in spring of the crown buds. A lack of proper management that does not provide for food reserve build-up may be associated with not only poor winter survival and thin forage stands but also weakened plants, slow and sparse spring growth, and subsequent low seasonal yields.

Well-defined management practices designed to assure adequate build-up of food reserves have been established. The general recommendation is to avoid any clipping or grazing treatments for a six-week period in the fall, referred to as the **critical fall harvest period,** which begins three weeks prior to a selected calendar date for a given region and continues for three weeks after the date. Such a recommendation will assure adequate food reserves in the plant but gives no explanation of why it is effective, how it was developed, or if it can or should be modified under specific circumstances.

Understanding Fall Management Recommendations. Fall management practices are based on the fact that photosynthesis and growth processes occur within reasonably well-defined environmental limits (Bidwell 1974). Seasonal temperature variations differentially affect photosynthesis and growth; photosynthesis is less subject to temperature variations than growth is. As temperatures decline in the fall, growth is slowed up, and photosynthesis proceeds at a relatively high rate because respiration losses under cool conditions are reduced. The calendar date selected represents a date when temperature and photoperiodic conditions, based on long-term observations and study, will reach a point when further growth is restricted. Not cutting for a three-week period prior to this date allows for growth to occur and for the development of a leaf area sufficient for a high rate of photosynthesis. During this period, photosynthetic food energy produced in the leaves is used for growth, respiration, and normal plant functions. Following the calendar date, less food energy is needed for growth, respiration rate is reduced, and surplus food energy results. The unused or surplus food energy is stored and forms the reserve food supply. Slow growth conditions and surplus food energy are associated with cell differentiation, resulting in the development of crown buds and thick-walled cells resistant to stress. A period of three weeks following the selected calendar date is required for reserves to build up. Usually by the end of the critical fall harvest period, severe temperatures restrict photosynthesis. Grazing may take place after the critical fall harvest period has ended without a loss in food reserves.

Weather conditions during the critical fall harvest period can be expected to vary from year to year, but the recommendation has proven reliable. Perhaps this is the result of a predictable temperature pattern, because photoperiod as well as temperature are related to plant dormancy and growth reductions.

Indirect evidence is available that serves as a guide to understanding the plant processes occurring during the critical fall harvest period. In perennial ryegrass, a continuous day and night temperature of 25°C (77°F), causes rapid growth. A LAI of 6.25 was produced under such conditions in a ten week period (Beevers and Cooper 1964*a* and *b*). At 12°C (54°F) continuous temperature, a LAI of 2.88 was produced, suggesting that growth was less than half that at the high temperature. At 12°C (54°F) ryegrass plants were found to have higher carbohydrate levels and a higher nitrogen (protein content) than the plants grown at 25°C (77°F), suggesting that more food energy was produced than was needed for growth.

In alfalfa, greatest dry-matter production occurred between 15 and 20°C (59 and 68°F) (Christie and McLaughlin 1979). At lower temperatures, dry-matter yields could be expected to be less, but surplus food energy may result in higher quality as measured by protein content. At higher temperatures, respiration could be increased, resulting in a low net assimilation rate and possibly reduced yield. High temperature conditions are associated with high **evapotranspiration** rates, and under droughty conditions, moisture stress may reduce the rate of photosynthesis. Extreme conditions of this nature are unlikely to occur during the critical fall harvest period.

The lower limit for growth to occur was found to be at an air temperature of 8°C (46°F) for three temperate forage species—orchard grass, Kentucky bluegrass, and Canada bluegrass—according to a review by Wilsie (1962). Food energy produced at or below this temperature is used primarily for food reserves.

Surplus food energy is stored in the basal stem of orchard grass, in the spreading underground stem of bromegrass, or in the basal stem area of timothy known as a corm. The swollen corm of timothy (Figure 10-1) can readily be observed following the critical fall harvest period. In legume crops, food reserves are stored in the taproot or **crown** (Figure 10-2).

Figure 10-1. Perennial grasses and legumes require food reserves to withstand low winter temperatures and to supply energy for regrowth the following spring. The photo shows a timothy corm, the swelling at the base of the plant that serves as the storage organ. (Courtesy Ontario Ministry of Agriculture and Food.)

Winter annuals such as fall-seeded cereals must be planted in time to allow for the development of a sufficient leaf area and root growth, and for the differentiation of tiller buds. If seeded too early, excessive growth may result in smothering when snow and ice flatten the leaves against the crown. If seeded too late, winter survival is jeopardized, and grain yields are almost surely reduced.

In plants without an opportunity for root reserve build-up or cell differentiation to occur, winter temperature conditions may cause ice crystals to form in the cells, which lacerate and damage the cell walls and other structures. As thawing occurs, the plant has a watery appearance as water escapes from the cells. Eventually, the cells become desiccated and the plant dies.

Figure 10-2. Food reserves are essential to winter survival of perennial legumes. Reserves are stored in the crown and taproot. (Courtesy Ontario Ministry of Agriculture and Food.)

Although plants during winter months are dormant, they are living, breathing organisms. Without any food energy being produced, the energy required for respiration must come from stored food reserves. During the winter, stored food reserves may be depleted by respiration. Inadequate reserves may mean that although the plants survive the severe midwinter conditions, they succumb to spring conditions of night frosts.

During the critical fall harvest period, photosynthesis can be enhanced by fertilizer applications of phosphorus and potash with rates determined by a soil test and applied in late summer or early fall. Fall-applied nitrogen is not recommended because it stimulates growth that is not compatible with root storage. For fall-seeded cereals, the same principle relative to fertilizer application applies. However, a starter fertilizer applied with the seed and containing some nitrogen is recommended to stimulate leaf growth as quickly as possible. Excessive nitrogen should be avoided. Spring-applied nitrogen was found to be more effective than fall-applied nitrogen as measured by grain yield (MacLeod and MacLeod 1975).

sacrificing population management control. Prolific corn hybrids have the genetic ability to increase sink size, but unless provision is made to fill this sink through an increased source size or efficiency, prolific corn hybrids may offer little advantage at recommended plant populations.

Increasing the quantity of reflected light in spring oats, winter wheat, and corn by placing white, opaque reflective polyethylene film on the ground between the rows was investigated (Pendleton, Brown, and Weibel 1965; Pendleton, Peters, and Peek 1966). Since such a plastic ground surface cover might alter soil moisture, black polyethylene film of the same thickness was included as part of the test. The reflectivity rate of the black material was less than 4% compared to over 80% for the white plastic. Cereals were planted in 20 cm (8 in) rows, and plastic strips were fitted when the plants were 4 cm (1½ in) tall. Corn was planted in 102 cm (40 in) rows at populations of 40,000 and 60,000 pph (16,100 and 24,200 ppa), and plastic strips were applied when the plants were 30 cm (12 in) tall. Tests were conducted in Illinois for a two-year period. An average yield increase of 12% was recorded for cereals (Table 10-2). In corn, the average yield increases for 40,000 and 60,000 pph (16,200 and 24,200 ppa) were 12% and 7%, respectively, whereas the nonreflective ground cover increased yields 5% or less.

TABLE 10-2. Effects of reflective and non-reflective plastic ground covers on grain yield of spring oats and winter wheat at Urbana, Illinois

Ground Cover	Spring Oats					Winter Wheat		
	1963		1964		Percent	1964		Percent
	(kg/ha)	(bu/ac)	(kg/ha)	(bu/ac)	of check	(kg/ha)	(bu/ac)	of check
None (check)	2272	59.7	2964	77.9	100.0	3642	54.2	100.0
Black plastic	2448	64.3	3007	79.0	104.2	3931	58.5	107.9
White plastic	2713	71.2	3194	83.8	112.8	4106	61.0	112.9

Source: Data from Pendleton, Brown and Weibel 1965.

The relatively small increases suggest that at recommended plant populations and row widths of cereals and corn, most of the light is effectively captured by the leaf canopy, essentially by upper leaves, which clearly indicates their importance. The main benefits associated with the use of polyethylene mulches have been increased soil temperature (Andrew, Schlough, and Tenpas 1976; Harris 1965).

Plastic ground covers at northern locations may be practical for horticultural crops where early marketing and quality demand a premium. Based on the Illinois work (Pendleton, Brown, and Weibel 1965; Pendleton, Peters, and Peek 1966), a highly reflective plastic surface would be preferable. In New Hampshire, the use of clear polyethylene mulch reduced time to maturity and increased yield of sweet corn in an early planting; in a normal or late planting, clear polyethylene mulch enhanced maturation by two to five days, but without a yield increase (Lee, Estes and Wells 1978). For commercial field crop producers, the value of such tests may be of purely academic interest in helping to explain the environmental impact on crop yield.

Another example of light management is grape production, which has spread to northern limits previously considered unsuitable, with satisfactory yield and quality due to the following factors:

(a) Utilization is made of southern slopes to maximize sunlight interception.

(b) Production is enhanced by light reflected off an adjacent lake or river.

(c) Warm temperature-retaining and reflective shale on the surface is not removed because it benefits the crop and reduces the risk of late spring and early fall frosts.

As crop production is pushed by housing and industrial developments and food needs into areas less than ideal for production, effort may be directed toward utilizing more of the sun's energy.

Row Direction

Narrow rows, high plant populations, and the desire to maximize sunlight interception have raised the question of a possible advantage in row direction. There are few reports on the effect of row direction on yield, but what reports do exist generally indicate higher yields in north-south than in east-west rows. Such is the case for tests in Australia, India, U.S.S.R., Illinois, and Saskatchewan with corn, wheat, oats, barley, and pearl millet (Austenson and Larter 1970).

In western Canada, based on three years of testing, two barley cultivars in 30-cm wide north-south rows averaged 9% and two oat cultivars averaged 4% higher in grain yield (Austenson and Larter 1970). Yield increases were only half as great when the rows were 15 cm wide. Four wheat cultivars gave similar results.

Several reasons for the advantage of north-south rows over east-west rows may be suggested:

1. There is better light interception during the morning and afternoon hours for north-south rows because the sun travels east to west.

2. **Magnetic response** of roots may be a factor. Some plants are known to orient their roots in a north-south direction (Pittman 1962, 1964, 1970). Directional root orientation for wild oats was measured at 27 locations across the vast length of Canada, but the advantage of north-south root orientation is still not clear. Possibly under conditions of interrow cultivation, north-south rows with roots

parallel to the row direction may be less subject to root pruning damage than east-west oriented rows with roots oriented north-south. The reaction of roots to the earth's polarity is termed a magnetotropic response.

3. There are conflicting reports on the advantage of north-south orientation in cotton (Anonymous 1969). Rows 53- and 106-cm (20 in and 40 in) wide were compared, and the results may emphasize the importance of north-south orientation for wide rows over that of narrow rows where row direction may be of less consequence.

4. Rows planted perpendicularly to prevailing winds reduced wind erosion and produced higher sorghum yields than rows parallel to the wind. Under conditions of prevailing westerly winds, a north-south orientation may be superior (Skidmore, Nossaman, and Woodruff 1966).

5. Less water was lost to evapotranspiration in 30-cm (12 in) wide north-south rows with the peanut crop than in 90-cm rows, or than in east-west rows of these spacings. Peanut yields were enhanced by narrow rows, but no row-direction effect was noted (Chin Choy, Stone, and Garton 1977).

6. Phototropic plant responses or the movement of plants to directional light may be related to lodging in sunflowers. Leaves and heads of sunflowers are phototropic from emergence through flowering; i.e., leaves are oriented to the direction of the sun so that leaves and heads face east in the morning and west in the evening. North-south rows appeared to have better light penetration to the base of the plant, which increased cell differentiation during vegetative growth and reduced lodging over that of east-west oriented rows. With the onset of reproductive growth, the fully developed flowers ceased photo-tropic movements and most heads faced the east, and no further influence of row direction on grain yield was detected (Robinson 1975).

For farmers whose fields are laid out to facilitate seeding in north-south rows and where north-south directional planting can be done with no added cost or inconvenience, the advantage of north-south plantings should be considered.

SEEDLING LEAF DAMAGE AND YIELD

Before effective herbicidal weed control was possible in soybeans, and under conditions of inadequate preplanting weed control, a recommended practice was to cultivate soybeans with a rotary hoe, harrow, or finger weeder following emergence. Willard (1947) suggested cultivation every five days until the soybeans attained a height of 15 cm (6 in) and when the weeds were very small and until the soybeans were of adequate size to be competitive with weeds through shading. The method was widely adopted (Weiss 1949; Hughes and Metcalfe 1972). Soybeans subjected to this rough treatment suffered mechanical leaf loss or damage, but the benefits of weed control overrode any possible adverse effects.

Hail damage may cause tearing and loss of leaf. Frost can damage the sensitive seeding of corn or soybeans and can cause producers concern as to whether damaged plants will recover, what effect leaf loss will have on yield, and the advisability of reseeding. Decisions to reseed must be made with haste because each day's delay reduces the duration of photosynthesis and increases respiration in C_3 crops with a low NAR; long-day crops may also be subjected to a drastically reduced vegetative period. Producers, deciding on the proper course of action to follow, should consider the number of live plants remaining, leaf loss, weed control, and calendar date (Crookston, Hicks, and Miller 1976).

Live Plant Population

Plants can recover from leaf damage provided the growing point is intact. In corn, the growing point pushes its way from its place of origin in the seed to the soil surface, to eventually terminate in the tassel. In the seedling, the growing point is encased by the leaf sheaths (Figure 10-3), and from the time that the first leaf emerges and until two to three leaves are developed (two to three weeks), the growing point is below the soil surface. Except on sandy soils, on black muck soils which freeze to a greater depth, or under a very severe spring freeze, the growing point will be undamaged by frost. One month after seeding and when the seedlings have four to five leaves, the growing point may be above the soil surface, but unless the entire sheath and stalk are destroyed, an undamaged growing point will cause the plant to recover. The growing point is not likely to be destroyed by a single night frost because the leaf canopy prevents heat loss from the soil. However, repeated night frosts may cause damage to the growing point.

To determine if the growing point has been frost-damaged, cut the corn "stalk" (a tight **whorl** or bundle of leaves) down its length, unroll the protective leaf sheaths, and check if the growing point is fresh and turgid and hence capable of recovery. If the growing point is not damaged, the field should not be reworked, but continued close scrutiny is required because bacterial decay of damaged tissue above the growing point may lead to further damage under cool, damp conditions. Under favorable growing conditions, mechanically cut corn seedlings showed 2.5 cm to 4.0 cm (1 in to 1½ in) of regrowth within 24 hours. If a damaged crop has not shown evidence of recovery within this time period, the field should be reworked. Recovered plants sometimes have their leaves bound in a whorl called *ties* or *buggy whips*, which prevent tassel emergence. Damaged plants can form an ear and contribute to yield.

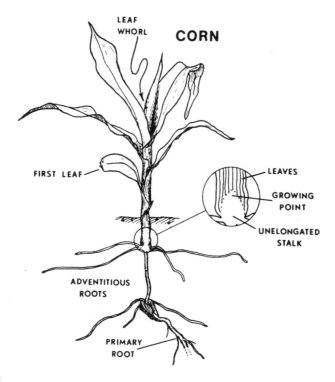

Figure 10-3. The development of a corn seedling showing the position of the encased growing point at the five leaf stage. (Courtesy American Society of Agronomy, Crops and Soils Magazine.)

A decision not to rework a field recognizes the advantage of an established root system and the delay caused by reworking. If a damaged field of corn is to be used for silage and if there is some question about the recovery of injured seedlings, a new corn crop could be planted close to the original rows without further tillage. At harvest, the **ensilage** would consist of plants from both seedlings, differing from each other perhaps in development but satisfactory for silage.

Damaged leaves do not grow when an injured plant recovers; new growth arises entirely from the undamaged nodes or **axillary buds**. A corn hybrid normally with sixteen leaves will produce eleven if five are destroyed. Provided an adequate leaf area can be achieved, damaged seedlings can produce a satisfactory grain yield at harvest. If no more than five or six leaves are destroyed in the seedling stage, maturity will be delayed by about three days, which may not affect grain yield under suitable fall conditions. When eight to ten leaves are destroyed, it is most unlikely that sufficient leaf area can be recovered to produce a satisfactory yield.

It was the recognition that an excessive leaf area may develop during the vegetative period and that some leaf loss during early development need not be detrimental to yield that led to machine weeding of soybeans or rolling cereals up to a 15-cm height. It is possible that some leaf loss may be beneficial, that the function of new leaves in helping the young plant to become established has been met, and that leaves lost to a frost in the seedling stage may have normally turned yellow and senesced by grain filling time. If a rapid and steady recovery results in an optimum LAI coinciding with the onset of reproductive development, a higher NAR without lower parasitic leaves could favor damaged plants.

To test this hypothesis, workers at the University of Minnesota defoliated corn plants at the five leaf stage over a three-year period, using an early and a late maturity hybrid (Hicks, Nelson, and Ford 1977). Complete leaf removal at the five leaf stage caused an average grain yield increase of 48% in the early maturing corn and an average 8% grain yield reduction in the late maturing corn in Minnesota (Table 10-3). The defoliation treatment resulted in a 16% increase of the early over the late hybrid. At harvest, ear moisture and maturity were not affected.

TABLE 10-3.   Yield of a full-season and a short-season corn hybrid with and without defoliation at the five leaf stage (Lamberton, Minnesota) at 60,000 pph planted in 76 cm wide rows

	Full-Season Hybrid					Short-Season Hybrid				
	Control		Cut		Change	Control		Cut		Change
Year	(t/ha)	(bu/ac)	(t/ha)	(bu/ac)	(%)	(t/ha)	(bu/ac)	(t/ha)	(bu/ac)	(%)
1973	6.4	102.8	6.1	198.0	−5	4.5	72.3	7.1	114.1	+59
1974	3.0	48.2	2.9	46.6	−3	2.0	32.1	3.6	57.8	+80
1975	6.8	109.2	5.8	93.2	−14	6.2	99.6	8.1	130.1	+30
Mean	5.4	86.8	4.9	78.7	−8	4.2	67.5	6.3	101.2	+48

Source: Data from Hicks, Nelson, and Ford 1977.

In a subsequent study (Crookston and Hicks 1978), twelve early corn hybrid cultivars were defoliated at the five leaf stage in 1975. Average yield increased 1.3% from 5.54 to 6.24 t/ha (89.0 to 100.2 bu/ac). The same twelve hybrid cultivars showed no change at one site, and reductions of 15% and 22% at two other sites.

It is unlikely that clipped plants can invariably recover an optimum LAI coinciding with reproductive development. Clipping of 28 single-cross corn hybrids near the soil surface at the four, six, and eight leaf stages reduced yields 11, 38, and 46%, respectively, apparently due to growing point damage as indicated by reduced survival (Cloninger, Zuber, and Horrocks 1974). When clipped at the four leaf stage, 5 of the 28 hybrids outyielded the unclipped control plots.

A 10 to 13% yield reduction was found in all hybrids clipped at the five leaf stage in Illinois (Johnson 1978) when corn was grown in 75-cm (30 in) wide rows at 61,200 pph (24,700 ppa) in 1976 and 54,400 pph (22,700 ppa) in 1977.

When a June 23rd frost at Madison, Wisconsin in 1972 damaged six or more leaves on corn plants 30 cm to 90 cm (12 in to 35 in) high with ten to eleven fully emerged leaves, (Arny and Upper 1973), plants could not recover; photosynthetic area was reduced and was accompanied by a 30% grain yield reduction from 8500 to 6000 hg/ha (136.6 to 96.4 bu/ac).

Clipping of corn is not a recommended practice. The lesson to be learned is that defoliation at an early stage by frost, hail, or mechanical damage may not reduce yield or may result in less yield reduction than reseeding. If more than five leaves are removed, the growing point is apt to be damaged and yield reductions are likely.

In the case of soybeans, the growing point is at the top of the plant, and it is subject to damage or loss (Figure 10-4). Vegetative buds in leaf axils (axillary buds) can serve as growing points and will develop into branches if flowering

SOYBEAN

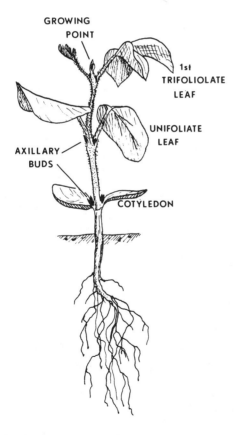

Figure 10-4. The development of a soybean seedling showing the position of the growing point when the first triplicate leaf has developed. (Courtesy American Society of Agronomy, Crops and Soils Magazine.)

has not begun. Normally these axillary buds are suppressed by the dominance of the apical bud, but apical dominance is lost if the apical bud is destroyed. Usually a frost will destroy axillary buds because soybeans are very tender and are subject to frost damage. Soybeans will not recover if destroyed to a level below the cotyledons, but if leaf loss is minimal and fresh axillary buds remain, recovery is possible.

EARLY SEEDING OF CEREALS

In contrast to tender crops such as corn and soybeans, cereals are not likely to be frost-damaged and can be seeded as early as the land can be prepared. However, the likelihood of subsequent cold wet weather has discouraged farmers from taking advantage of a very early seeding opportunity. To test if early seedings produced a yield advantage, cereals were broadcast on the frozen soil surface for a three-year period at Guelph (Stoskopf, Reinbergs, and Jones 1967; Stoskopf et al. 1968). Although the seed remained virtually uncovered with soil, the yield advantage of early seeding was strikingly demonstrated (Table 10-4).

TABLE 10-4.   Yield and date of anthesis of frost-seeded barley in 1967 as compared to a conventional seeding at Guelph, Ontario.

	Frost Seeding		Conventional	
	April 4	April 12	April 24	May 10*
Yield (kg/ha)	3003	2691	2308	2137
(bu/ac)	55.9	50.1	42.9	39.8
Date of anthesis	June 29	July 3	July 9	July 8

Source: Data from Stoskopf et al 1968.
*The first opportunity for conventional seeding.

To determine just how early seeding could occur safely, frost seedings were started in early January and were repeated at intervals into April, when suitable weather and soil conditions permitted field seeding. Generally it appeared that seedings as early as the last week in March and during the first week in April were satisfactory even though subsequent cold wet weather, including snow and frost, was certain to follow. Midwinter seedings were unsuccessful possibly due to the death of seeds when metabolic activity was stimulated by warm sunny conditions followed by severely cold conditions without a protective snow cover. Early seeding was associated with the removal of seeds by rodents, birds, and erosion.

Frost seeding has served to emphasize that wheat, oats, barley, triticale, peas, and flax can be seeded as early in the spring as the soil can be cultivated.

Variations of frost seeding have been employed on a commercial basis with moderate success. Although early seeding may be achieved with seed placed on the frozen soil, the sloppy technique and the inability to cover the seed leave a lot to be desired. In prairie regions, surface seedings may result in loss of seedlings from the drying out of surface soils. Careful placement of seeds and fertilizer is recommended for high-yielding conditions.

Advantages of early seeding become apparent when basic photosynthetic concepts are considered. Early seeding promotes an early date of anthesis (see Table 10–4) so that reproductive development can coincide more closely with high-sunlight energy conditions in late June. Early seeding results in good moisture conditions, freedom from disease, well-differentiated plants, and lodging resistance.

CONCLUSIONS

The examples in this chapter serve to demonstrate applications of the basic photosynthetic concepts that have application to crop production. Understanding basic concepts may be of advantage not only in understanding crop production but to turn a particular situation to advantage.

QUESTIONS

1. Are there any practices that crop producers might use to increase the effective sunlight received by a crop on a practical basis (small- or large-scale)?

2. Suggest possible explanations as to why corn, defoliated at the five leaf stage, caused a yield increase in some cases and a reduction in others. Review carefully circumstances that might lead to a yield increase with frosted corn under field conditions.

3. Which is potentially more dangerous—a severe spring or a fall frost? Explain. What effect might an early fall frost have on grain yield (before physiological maturity), whereby only the upper two or three leaves are damaged? (Review Chapter Two.)

4. Consider the merits of single-eared versus prolific corn hybrids? Under what circumstances might prolific corn hybrids be advantageous?

5. Approximately 25% of the incident solar radiation reaches the ground surface in corn at recommended plant populations when planted in 102-cm wide rows (Tanner, Peterson, and Love 1960; Denmead, Fritschen, and Shaw 1962) and when leaf area has reached its maximum. What problems exist with efforts to try and intercept more light through an increase in plant population? How might plant morphology be used to intercept this light energy?

REFERENCES FOR CHAPTER TEN

Anonymous. 1969. "Row Direction Doesn't Influence Cotton Yields," *Crops and Soils* 22(3):21.

Andrew, R. H., D. A. Schlough, and G. H. Tenpas. 1976. "Some Relationships of a Plastic Mulch to Sweet Corn Maturity," *Agronomy Journal.* 68:422–425.

Arny, D. C., and C. D. Upper. 1973. "Example of the Effects of Early Season Frost Damage on Yield of Corn," *Crop Science.* 13:760–761.

Austenson, H. M., and E. N. Larter. 1970. "Cereal Yields Increase in North-South Rows," *Crops and Soils* 22(7):18–19.

Beevers, L., and J. P. Cooper. 1964a. "Influence of Temperature on Growth and Metabolism of Ryegrass Seedlings I. Seedling Growth and Yield Components," *Crop Science.* 4:139–143.

———. 1964b. "Influence of Temperature on Growth Metabolism of Ryegrass Seedlings II. Variation in Metabolites," *Crop Science.* 4:143–146.

Bidwell, R. G. S. 1974. *Plant Physiology*, pp. 167–168. New York: Macmillan Publishing Co., Inc.

Chin Choy, E. W., J. F. Stone, and J. E. Garton. 1977. "Row Spacing and Direction Effects on Water Uptake Characteristics of Peanut," *Science Society of American Journal.* 41:428–432.

Christie, B. R., and R. J. McLaughlin. 1979. "Modifying the Alfalfa Plant," *Highlights in Agr. Res. in Ontario* 2(1):7–9.

Cloninger, F. D., M.S. Zuber, and R. D. Horrocks. 1974. "Synchronization of Flowering in Corn (*Zea mays* L.) by Clipping Young Plants," *Agronomy Journal.* 66:270–272.

Crookston, R. K., and D. R. Hicks. 1978. "Early Defoliation Affects Corn Grain Yields," *Crop Science.* 18:485–489.

Crookston, R. K., D. R. Hicks, and G. R. Miller. 1976. "Hail. Help in Deciding Whether to Replant Corn and Soybeans," *Crops and Soils* 28(8):7–11.

Denmead, O. T., L. J. Fritschen, and R. H. Shaw. 1962. "Spatial Distribution of Net Radiation in a Corn Field," *Agronomy Journal.* 54:505–510.

Fay, B. 1973. "New Enthusiasm for Prolifics," *Crops and Soils* 26(3):12–13.

Hallauer, A. R. 1969. "Corn with Built-in Insurance," *Crops and Soils* 21(5):16–17.

Harris, R. E. 1965. "Polyethylene Covers and Mulches for Corn and Bean Production in Northern Regions." Proceedings of the American Society for Horticultural Science 87:288–294.

Hicks, D. R., W. W. Nelson, and J. H. Ford. 1977. "Defoliation Effects on Corn Hybrids Adapted to the Northern Corn Belt," *Agronomy Journal.* 69:387–390.

Hughes, H. D., and D. S. Metcalfe. 1972. *Crop Production*, 3rd ed. New York: Macmillan Publishing Co., Inc. 627 pp.

Johnson, R. R. 1978. "Growth and Yield of Maize as Affected by Early Season Defoliation," *Agronomy Journal.* 70:995–998.

MacLeod, J. A., and L. B. MacLeod. 1975. "Effects of Spring N Application on Yield and N Content of Four Winter Wheat Cultivars," *Canadian Journal Plant Science.* 55:359–362.

Pendleton, J. W., C. M. Brown, and R. O. Weibel. 1965. "Effect of Reflected Light on Small Grain Yields," *Crop Science*. 5:373.

Pendleton, J. W., D. B. Peters, and J. W. Peek. 1966. "Role of Reflected Light in the Corn Ecosystem, *Agronomy Journal*. 58:73-74.

Pendleton, J. W., D. B. Egli, and D. B. Peters. 1967. "Response of *Zea mays* L. to a "Light Rich" Field Environment," *Agronomy Journal*. 59:395-397.

Pittman, U. J. 1962. "Growth Reaction and Magnetotropism in Roots of Winter Wheat (Kharkor 22 M.C.)," *Canadian Journal Plant Science*. 42:430-436.

———. 1964. "Magnetism and Plant Growth II. Effect on Root Growth of Cereals," *Canadian Journal Plant Science*. 44:283-287.

———. 1970. "Magnetotropic Responses in Roots of Wild Oats," *Canadian Journal Plant Science*. 50:350-351.

Robinson, R. G. 1975. "Effect of Row Direction on Sunflowers," *Agronomy Journal*. 67:93-94.

Skidmore, E. L., N. L. Nossaman, and N. P. Woodruff. 1966. "Wind Erosion as Influenced by Row Spacing, Row Direction and Grain Sorghum Population," *Soil Sci. Soc. Amer. Proc.* 30:505-509.

Stoskopf, N. C., E. Reinbergs, and G. E. Jones. 1967. "Frost Seeding," *Crops and Soils* 19(5):12-13.

Stoskopf, N. C., et al. 1968. "Increased Yields from Cereals Sown on Frozen Soil," *Canadian Journal Plant Science*. 48:428-430.

Tanner, C. B., A. E. Peterson, and J. R. Love. 1960. "Radiant Energy Exchange in a Corn Field," *Agronomy Journal*. 52:373-379.

Waggoner, P. E., D. N. Moss, and J. D. Hesketh. 1963. "Radiation in the Plant Environment and Photosynthesis," *Agronomy Journal*. 55:36-39.

Weiss, M. G., 1949. "*Soybeans*," *Advanced Agronomy*. 1:78-157.

Willard, C. J. 1947. "Controlling Weeds in Soybeans," *Soybean Digest* 7(11):32-33, 48.

Wilsie, C. P. 1962. *Crop Adaptation and Distribution*. San Francisco: W. H. Freeman & Company, Publishers. 448 pp.

chapter eleven

Crop Mixtures, Monocultures, and Yield

Crop producers have commonly observed plants in the field that are stunted and retarded as a result of double-seeding in the turnaround areas of headlands or from weeds where the sprayer missed. **Competition** for moisture, for fertilizer nutrients, or for sunlight, or any combination of factors is immediately pinpointed as the cause. Competition is an everyday phenomenon, and a simple illustration was provided in the parable of the sower and the seed: "And some (seed) fell among thorns, and the thorns sprang up and choked them" (Matthew, 13:7).

Farmers have come to view competition as undesirable, as something to be eliminated or avoided; yet without realizing it, they subject crop plants to severe competitive stresses in an effort to obtain top yields. Perhaps confusion about the use of the term *competition* is derived from its adoption into science with varied shades of meaning associated with its usage in other areas such as sports, games, and economics. In the area of agriculture and ecology, although the term *competition* may have various meanings, most agronomists and ecologists feel they know what it means. To avoid controversy of what is or is not competition, Harper (1961) used the word *interference* to describe the hardships that are caused by the proximity of neighboring plants.

The understanding of competitive phenomena in plant populations is of interest to people at all levels of crop production. It is a widely researched subject, yet a considerable amount remains unknown. As an introduction to the scope and complexity of the subject, the reader is requested to briefly review the first twenty questions at the end of this chapter, and after reading the chapter, to reconsider the questions. The object of the questions is to stimulate discussion and consideration, rather than to seek definitive answers. The work of Donald (1963) is acknowledged, and it is his work that forms the basis of much of this chapter.

WHAT IS PLANT COMPETITION?

Competition exists when each plant in a population is modified through its response to physical factors around it, such as light, CO_2, moisture, and fertilizer nutrients. No matter how close two plants may be, they do not compete with each other as long as the physical supplies are in excess of the needs of both, but when the immediate supply of a single necessary factor falls below the combined demands of the two plants, competition begins.

Donald (1963) listed five factors for which competition may occur among plants:

1. Water.
2. Fertilizer nutrients.
3. Light.
4. Oxygen.
5. Carbon dioxide.

All of these factors are essential to photosynthesis, and any one or more may be in short supply at any time in the development of the plant; hence competition, with its close association with photosynthesis, is recognized as a key factor in crop production. Note that temperature and humidity, which influence photosynthesis and growth, are not included in the list because they are not commodities in finite supply and hence are not subject to competition. Generally, space is also not included in the list because sufficient space usually exists for more plants, more branches, more leaves, or more roots. However, space limitations can occur under crowded conditions and high yields in potatoes, turnips, carrots, or other root crops (Doney, Plaisted, and Peterson 1965).

In most crops, plant populations and row widths are pushed to the upper limit of plant tolerance to achieve highest yields, and the plants show morphological changes under such intense competition. Donald (1963) noted:

It is a surprising thought that, in the production of a successful and "healthy" field crop, such intense competition is developed that in quantitative terms, the individual plants are markedly subnormal. It is the community of suppressed plants which gives the greatest yield. [p. 5]

PLANT RESPONSE TO COMPETITION

Ford (1975) noted three characteristic responses of plants to competitive stresses:

1. Density-dependent mortality.
2. **Plastic responses.**
3. Hierarchy of exploitation.

These three phenomena can be illustrated from studies of plant competition. Donald (1963) demonstrated "density-dependent mortality" using the wheat cultivar Insignia grown at five seeding rates at Adelaide, Australia (Table 11–1). In the first 119 days of plant development, plant competition reduced the plants per m² by 7.6% and 35.6% at the 62 and 366 kg/ha seeding rates, respectively. For the next 63-day period, competition further reduced plants per m² by a further 9.4% and 35.6% at the two highest seeding rates. Density-dependent mortality did not reduce plant population to a level giving maximum grain yield per unit area (see Table 11–2). Thus, seeding rate remains the responsibility of the crop producer. It can also be concluded from Table 11–1 that competition was expressed in plant mortality for most of the life cycle of the plant and not just for the seedling stage.

TABLE 11–1.   Competitive effects expressed as "density-dependent mortality" on wheat plants per m² at five seeding rates at three stages of crop development grown at Adelaide, Australia

Density of plants per m² (ft²)	Seed Rate in kg/ha (lb/ac)				
	0.4 (0.4)	2.4 (2.1)	12 (10.7)	62 (55.4)	366 (326.8)
At emergence	1.4 (0.13)	7 (0.63)	35 (3.2)	184 (16.6)	1078 (97.0)
At 119 days	1.4 (0.13)	7 (0.63)	35 (3.2)	170 (15.3)	694 (62.5)
At 182 days	1.4 (0.13)	7 (0.63)	35 (3.2)	154 (13.9)	447 (40.2)
Estimated seeds planted per m²	1.4	8.4	42	217	1281
per ft²	0.13	0.76	3.78	19.53	115.3
Percent of seeding rate at emergence	100	83.4	83.4	84.8	84.2

Source: Data from Donald 1963

Results of this same test by Donald (1963) also dramatically illustrate 'plastic responses' of plants (Table 11–3). From the lowest to the highest seeding rate, dry weight per plant decreased 42 times; tillers per plant decreased 34 times; weight per tiller decreased 1.17 times; height increased 1.44 times; and leaf area per plant decreased 51 times. Wheat plants are clearly very plastic in their response to plant competition during vegetative development.

At maturity, plastic responses to competition also were expressed clearly in all agronomic traits including grain yield (Table 11–2). From the onset of reproductive development to maturity, plant plasticity continues to be expressed (compare Tables 11–2 and 11–3). An average of 33.0% of the tillers formed by the time reproductive development began (Table 11–2) did not

TABLE 11–2. Expressions of competition and plant plasticity in agronomic yield components and final grain yield at maturity of wheat plants grown at five densities
grown at Adelaide, Australia

Agronomic Component	Seeding rate in kg/ha (lb/ac)				
	0.4 (0.4)	2.4 (2.1)	12 (10.7)	62 (55.4)	366 (326.8)
Grains per ear	32.9	37.8	29.9	21.5	18.8
Ears per plant	29.4	18.6	7.2	2.2	0.7
Tillers filled as % at anthesis	72.6	63.1	67.9	73.3	58.3
Weight per grain (mg)	34.2	35.0	32.7	33.2	33.1
Seeds per plant	970	705	215	56	12
Grain yield per plant (g)	33.2	24.7	7.1	1.5	0.4
(oz)	1.17	0.87	0.25	0.05	0.01
Grain yield (kg/ha)	460	1730	2470	2340	1850
(bu/ac)	6.8	25.7	36.8	34.8	27.5

Source: Data from Donald 1963.

develop to maturity (Table 11-3). Plastic responses were expressed in all
components except grain weight. Apparently, sufficient food energy was
available for grain development once sink size was established (in this case
grains per ear and ears per plant). Grain yield per plant and seeds per plant
varied by a multiple of 83.0 and 80.8, respectively, between the lowest and the
highest seeding rate; highest yields were obtained at the intermediate seeding

TABLE 11–3. Expressions of competition and plant plasticity in wheat at five plant densities at the
time of reproductive development grown at Adelaide, Australia

Agronomic features	Seeding Rate in kg/ha (lb/ac)				
	0.4 (0.4)	2.4 (2.1)	12 (10.7)	62 (55.4)	366 (826.8)
Dry matter per plant (g)	45.9	47.6	17.2	4.2	1.1
(oz)	1.6	1.7	0.61	0.15	0.04
Tillers per plant	40.5	29.5	10.6	3.0	1.2
Weight per tiller (g)	1.12	1.62	1.73	1.39	0.96
(oz)	0.04	0.05	0.06	0.05	0.03
Height (cm)	75	83	101	108	lodged
(in)	29.5	32.7	39.8	42.5	
Area of leaves per plant (cm²)	2550	2660	861	199	50
(in²)	408	426	138	32	8

Source: Data from Donald 1963.

rates of 12 and 62 kg/ha (10.7 and 55.4 lb/ac). At these seeding rates, individual plants in quantitative terms were markedly subnormal compared to those at lower seeding rates, but as stated, it is the community of suppressed plants that produces the greatest yield.

Plastic responses occur in other crop plants as well as in wheat and cereals (Table 11-4). Plant plasticity is one reason why recommended seeding rates within selected crops vary as much as 100%. Perhaps wheat (and other cereals) that show a tillering capacity ranging from 1.2 to 40.5 per plant may be an example of a crop with a high degree of plasticity. In contrast to corn, a common observation is that above a maximum seeding rate, corn may lose its plasticity from excessive ear shading and nubbin development. Plastic responses and the degree of plant plasticity make seeding rates, row widths, and plant spatial arrangements important considerations in crop production.

When two species or cultivars of the same species are mixed, one may yield less and one may yield more than their performance in pure cultures. One is the aggressor, being able to exploit more than its "share" of the environmental factors, whereas the other is suppressed because it is able to secure only part of the light, water, and nutrients. This was referred to originally by Harper (1967) as the "hierarchy of exploitation" and the concept can be illustrated in corn blends in Ontario (Kannenberg and Hunter 1972). Seven short-season commercial corn hybrids were grown in pure stands and all possible equal-proportion mixtures (alternated within each row) at 76,500 and 103,500 pph (30,100 and 41,900 ppa) in 1969. When grown in pure stands the seven hybrids did not yield equally (Table 11-5), but no significant differences in grain yield were observed between plant densities, and no differences were noted between observed yields of the mixtures and midcomponent means.

TABLE 11-4.   Competition and the reduction in yield per plant at densities giving the maximum yield per unit area at Adelaide, Australia

Crop	At Wide Spacing				At a Spacing Producing Maximum Yield per Unit Area			
	Density		Yield per Plant (g)	(oz)	Density		Yield per Plant (g)	(oz)
Buckwheat (total D.M. of tops)	25/m²	(2.25/ft²)	19.6	(0.69)	400/m²	(36/ft²)	1.6	(0.06)
Subterranean clover	6/m²	(0.54/ft²)	34.0	(1.20)	1500/m²	(135/ft²)	0.6	(0.02)
Wheat (grain)	1.4/m²	(0.13/ft²)	33.2	(1.17)	34/m²	(3.1/ft²)	7.1	(0.25)
Corn (silage) D.M.	1.157 cm of row	(62 in)	748	(26.38)	16/157 cm of row	(62/in)	233	(8.22)
Broad beans	11/m²	(1.0/ft²)	29.7	(1.01)	66/m²	(5.9/ft²)	9.3	(0.33)

Source: Data from Donald 1963.

TABLE 11–5. Yielding ability (t/ha) over all populations at Guelph, 1969

Hybrid	Designation	Yield	(bu/ac)
P.A.G. SX 47	1	7.98	128.2
United 106	2	7.94	127.6
United 108	3	7.30	117.3
Warwick SL 209	4	6.84	109.9
Pride 116	5	6.48	104.1
Pride R100	6	6.22	99.9
Warwick 263	7	6.12	98.3

Source: Data from Kannenberg and Hunter 1972.

When these hybrid cultivars were grown in mixtures the yield contributions of the components were disproportionate: hybrids that yielded highest in pure stands contributed more to a mixture than would be expected, based on their pure stand yield, and the lower yielding hybrids contributed less than expected. Hybrids designated as 1, 4, and 7 (from Table 11–5) were selected to illustrate the magnitude of these changes (Table 11–6).

TABLE 11–6. Contributions of each hybrid in a mixture to grain yield expressed as a percent of performance in pure stands, 1969 at Guelph, Ontario

Mixture	Relative Yield of Each Component (%)	Mean
PAGSX47 + United 106	100.8 + 100.3	100.5
PAGSX47 + United 108	102.5 + 95.9	99.2
PAGSX47 + Warwick SL209	104.8 + 105.3	105.1
PAGSX47 + Pride 116	111.5 + 96.3	103.9
PAGSX47 + Pride R100	109.7 + 94.1	101.9
PAGSX47 + Warwick 263	108.5 + 92.9	100.7
Mean	106.3 + 97.5	101.9
Warwick SL209 + United 106	107.9 + 102.0	105.0
Warwick SL209 + United 108	101.8 + 95.9	98.9
Warwick SL209 + Pride 116	107.9 + 95.1	101.5
Warwick SL209 + Pride R100	110.8 + 94.9	102.8
Warwick SL209 + Warwick 263	114.6 + 92.8	103.7
Mean	108.6 + 96.1	102.4
Warwick 263 + United 106	96.1 + 105.8	101.0
Warwick 263 + United 108	99.3 + 103.8	101.6
Warwick 263 + Pride 116	95.8 + 105.6	100.7
Warwick 263 + Pride R100	106.5 + 101.6	104.1
Mean	99.4 + 104.2	101.8

Source: Data from Kannenberg and Hunter 1972.

NOTE: In no case did the mean of the mixture outyield the highest yielding component of the mixture.

COMPETITION IN MIXTURES OF DIFFERENT SPECIES (BLENDS)

The custom of seeding a mixture of two or more species is probably centuries old. It was reviewed and recommended by Warburton (1915). Forage crops commonly contain a legume and a grass or even several grass and legume species mixed together. Oat and barley mixtures are accepted as sound practice and often contain a third component of spring wheat or peas. Forages may be undersown with a companion crop to establish a stand for the following year; and without adequate weed control, an intended pure crop may become a mixture containing weeds. Milthorpe (1961) suggested that in a mixture of two or more species, the same mechanisms of competition exist as in a pure stand, but they may be more obvious or more severe where large differences in size, habits of growth, and responses to weather exist.

Possibly, the general use of mixtures developed out of the belief that cooperation or mutual benefit might occur. Competition is dependent on individuals having a common need, with total requirements of the population exceeding the supply of whatever commodity is needed. In a grass-legume mixture, nitrogen will be a requirement for growth of the grass; but under conditions of nodulation, nitrogen requirements of the legume may be negligible. In such a case, cooperation for nitrogen may exist. Both plant species will require water, but if the legume is able to utilize a deep water source by a top root, and if the grass can utilize water at a shallow source, cooperation may exist. Likewise, both species will require light, but competition may be minimal if the upright leaves of the grass species allow light to penetrate to the shorter legume with horizontal leaves in the stand. Mather (1961) suggested that situations could exist that are wholly cooperative, wholly competitive, neutral, or competitive in some aspects and cooperative in others. Cooperation among plants could be said to exist when *the yield of a mixture exceeds the yield of the highest yielding component of that mixture.* Competition is more common than cooperation, and definite examples of cooperation among plants are hard to find.

In forage tests at the Kemptville College of Agricultural Technology in Ontario, Moore (1974) demonstrated cooperation in a simple alfalfa-brome mixture and competition in a complex mixture of eight species (Table 11–7). The apparently cooperative alfalfa-brome situation may involve some competition or may change from being cooperative to being neutral, and finally to being actively competitive as environmental conditions change.

In mixtures consisting of species having similarities in their needs, competition rather than cooperation can almost invariably be expected. Donald (1963) developed four general principles of competition that have wide application and add considerably to our understanding of competition. The four principles are:

1. Mixtures will usually yield less than the higher yielding pure culture.
2. Mixtures will usually yield more than the lower yielding pure culture.

TABLE 11–7. Cooperation and competition expressed in forage dry-matter yields (kg/ha) and (lb/ac) under a three-cut system at Kemptville, Ontario

Components	Cut 1	Cut 2	Cut 3	Total
Alfalfa	4,287	3,525	2,508	10,321
(lb/ac)	3,828	3,147	2,239	9,215
Alfalfa + brome	5,660	3,571	2,421	11,652
(lb/ac)	5,054	3,188	2,162	10,404
Alfalfa + alsike + red clover + ladino + bromegrass + perennial ryegrass + timothy + reed canary grass				8,984
(lb/ac)				8,021

Source: Data from Moore 1974.

3. Mixtures may yield more or less than the average of the two pure cultures, but generally less.
4. Although plant cooperation exists, there is limited evidence to show that mixtures can exploit the environment better than a pure stand.

These four principles can be illustrated by results obtained from cereal mixtures grown in Ontario from test plots and field data. To reduce the concern that small test plots do not adequately sample the variations in topography, drainage, and soil types in a large variable field, data from 51 counties and districts in Ontario were prepared by the Ontario Ministry of Agriculture and Food and are presented here in Table 11-8.

TABLE 11–8. Yield comparisons (t/ha) and (lb/ac) and percentage variations for the highest yielding components of pure oats and barley and mixtures of the two in Ontario for the fourteen-year period 1966–1979 presented for three different periods

Component	Yield 1966–1967 (t/ha)	Percent Variation	Yield 1971–1975 (t/ha)	Percent Variation	Yield 1976–1979 (t/ha)	Percent Variation
Oat	1.94	72.0	1.83	68.0	1.77	66.0
(lb/ac)	1,732		1,634		1,580	
Barley	2.69	100.0	2.69	100.0	2.69	100.0
(lb/ac)	2,402		2,402		2,402	
Mixture	2.55	95.0	2.55	95.0	2.50	92.9
(lb/ac)	2,277		2,277		2,232	
Oat and barley mean	2.31	86.0	2.56	84.0	2.23	83.0
(lb/ac)	2,063		2,286		1,991	

Source: Data from Ontario Ministry of Agriculture and Food 1979

The results support all four of the competition principles. These are not firm rules and exceptions can be found, and perhaps it is because of these exceptions that mixed grain production persists in much of the cereal-producing areas of the world. Testing agencies often continue to test oat and barley mixtures, even though no recommendations of specific mixtures are given. The only directive that farmers are given is that, generally, the highest yielding cultivars of oats and barley in pure stands also perform best in mixtures. It should be realized by all growers and testing agencies that cooperation of plants in mixtures comprised of components having similar needs is not going to be found. If crop producers are seeking yield stability, then the production of mixtures may be justified. For an inexperienced grower with variable land, a mixture may be advantageous because it will provide stability of production although not maximum yield. In contrast to cereal producers, corn growers have shifted from four-way cross hybrid cultivars, which have a broad gene base and offer greater yield stability, to single crosses with a narrow gene base.

CONCLUSIONS ON CEREAL MIXTURES

Crop producers may wish to continue the practice of producing **mixed grain** to secure greater yield stability, but they should be under no illusion that mixtures will produce a greater yield even if all conditions are favorable. The idea that yield can be increased stems from a failure to recognize competition principles with regard to published reports. For example, Zavitz (1927) reported the results of six years of testing of barley, oats, spring wheat, and peas grown separately and in various combinations at Guelph. In ten of the eleven comparisons, the yield of the mixture exceeded that of the average yield when each was grown separately. An impression favorable to mixtures is left, but such a situation clearly illustrates all four of the competition principles, generally because peas are a low-yielding component of the mixture. Crop producers should be interested in mixtures that outyield the highest yielding component, and except under unique circumstances, this is not likely to occur.

COMPETITION AMONG CULTIVARS OF THE SAME SPECIES

Crop producers and plant breeders recognize that no single cultivar has all the desirable "built-in" features needed to obtain maximum production, and they reason that if the good features found in each of several cultivars were blended and grown together, a yield advantage might be obtained. Plant breeders and seed producers seek with great determination and take pride in the uniformity of all plants in a cultivar, so much so that concern has been expressed that genetically uniform seeds may lack the ability to adjust to environmental variability. The noted Australian cereal breeder Frankel wrote: "It seems to me that the 'purity concept' has not only been carried to unnecessary lengths, but that it may be altogether inimical (harmful) to the attainment of highest production" (1950, p. 90). He implied that genetic uniformity was sought for reasons other than high yield, such as technical, commercial, historical, psychological, and aesthetic reasons. Simmonds (1962), in reviewing the literature on blends, which dates back many years and covers a wide range of crops in diverse areas of the world, concluded:

The use of deliberately compounded mixtures—this would have only occasional applications—but the evidence suggests that deliberately produced heterogeneity of crop populations has much to offer in terms of improvement in and stability of performance, especially in regard to disease reaction.

The evidence strongly supports the four competition principles in many crops tested.

Cereals

In mixtures of varying proportions of two oat cultivars, Pfahler (1965) found no grain yield advantage for mixtures over the higher yielding cultivar in Gainesville, Florida. In rye, no superiority of a mixture over the highest yielding parent was obtained, but a grain yield equal to the highest yielding component was found.

Yields of mixtures and monocultures of eight oat cultivars and twenty advanced breeding lines were tested at two locations in Iowa for two years (Shorter and Frey 1979). For both grain and straw yield, the highest yielding mixtures were not superior to either their higher yielding component or to the line with the highest monoculture yield, and the authors concluded that blends of oat cultivars or lines to obtain a yield advantage could not be justified. In all cases in oats, however, Jensen (1952) found that the yield of the blend exceeded the average of the pure cultivars.

Soybeans

Composites of soybeans exceeded the mean yields of their pure lines, but not that of the best component (Mumaw and Weber 1957).

Probst (1957) found that soybean blends in general showed no yield superiority over the highest yielding cultivar in any one year or over the average of a four-year period. He suggested that blending had a stabilizing effect on yield.

Four morphologically distinguishable soybean cultivars were used in mixtures so that the performance of each cultivar in the mixture could be observed (Hinson and Hanson 1962). No superiority of the mixtures over the pure stands was observed, but within the mixtures, certain cultivars tended to show an increase in yield at the expense of the other cultivars. Greater plant-to-plant variation of a cultivar was reported within mixtures than in pure stands.

Sorghum

Comparisons of eight parental lines, parental blends, hybrid cultivars and hybrid cultivar blends of grain sorghum were made (Reich and Atkins 1970). Tests were conducted in nine environments in Iowa over a two-year period, 1966 and 1967. Twenty-two of the 32 blends yielded an average of 2% more than the pure stand mean of their components, but the highest yielding individual entry in each environment, except one, was a hybrid. Six of the blends exceeded the mean pure stand of their most productive single component.

Ross (1965) grew five grain sorghum single-cross hybrid cultivars of varying genotypes alone and in 1:1 blends for five years. Average yield of all blends was not meaningfully different from the average of all hybrids alone. No blend yield exceeded that of the hybrid cultivars.

In Ross's study, blends did outyield hybrid cultivars only in one extremely favorable year. Presumably all factors were in adequate supply so that competition among plants for any one of the facts was minimal or nonexistent. Roy suggested that the experiment does not support the theory that blends or mixtures of pairs of uniform (homogenous) grain sorghum hybrids perform advantageously under varied and often stressed growth conditions, the very reason why they were proposed.

Corn

Blends of corn hybrid cultivars have not differed in grain yield from the midcomponent mean. Blending of two or more corn hybrid cultivars either in the same hill, or in alternate hills, or in alternate rows, did not increase grain yields over the mean of the component hybrid cultivars grown separately (Funk and Anderson 1964). Hybrid cultivars showed considerable differences in competitive ability, and the actual contribution of the individual components of mixture differed widely from what would be expected on the basis of their yields in pure stands. Blending corn hybrid cultivars was found to increase yield stability.

The results of Funk and Anderson (1964) appear to be consistent with the finding of others (Stringfield 1959; Thompson 1977; Eberhart, Penny, and Sprague 1964). It is also generally agreed that the performance of mixtures can be predicted with reasonable accuracy from the relative performance of component hybrid cultivars in pure stands (Eberhart, Penny, and Sprague 1964; Funk and Anderson 1964; Kannenberg and Hunter 1972; Jurado-Tovar and Compton 1974; Pendleton and Seif 1962; Stringfield 1959; Thompson 1977).

Forages

To determine if blends of alfalfa cultivars might improve yield, early and medium maturing alfalfa cultivars were grown in pure stands and were blended in all simple combinations (see Table 11–9). No significant differences were obtained among the cultivars or blends in any of the three harvests or in the total seasonal yield of dry matter in the first production year, but the alfalfa blends demonstrated the competition principles.

TABLE 11–9.   Total season yield (t/ha) and (t/ac) from three harvests in the first production year, 1979

Cultivar	Yield in Pure Stand		Mean of Pure Stands		Yield of Mixture	
	(t/ha)	(t/ac)	(t/ha)	(t/ac)	(t/ha)	(t/ac)
Saranac	14.61	6.53	14.64	6.59	14.17	6.38
Thor	14.66	6.60				
Saranac	14.61	6.58	13.93	6.27	14.00	6.30
Vernal	13.25	5.96				
Saranac	14.61	6.57	14.24	6.41	14.05	6.32
Iroquois	13.86	6.24				
Thor	14.66	6.60	13.96	6.28	14.06	6.33
Vernal	13.25	5.96				
Thor	14.66	6.60	14.26	6.42	14.16	6.37
Iroquois	13.86	6.24				
Vernal	13.25	5.96	13.56	6.10	13.95	6.28
Iroquois	13.86	6.24				
Mean	14.10	6.35	—		14.02	6.31

Source: Data from unpublished material, Department of Crop Science, University of Guelph.

MULTILINES

The commercial possibilities for **multilines** of cereals have received much attention in recent years. A multiline is a blend or mixture of lines that are genetically very similar because they are derived from a series of crosses with disease-resistant lines and are backcrossed to a common parent. Components of a multiline are intended to closely resemble each other in height, maturity, yield performance, and leaf characteristics, so that competition between two plants in the multiline is about the same as in a pure cultivar. The various lines comprising the multiline blend differ in gene(s) for disease-resistance. The explicit objective of the multiline is to provide greater yield stability through more stable resistance to disease. When a multiline is grown and a new race of rust appears, only a small percentage of plants in the field are likely to be susceptible. The typical epidemic that spreads from plant to plant is thus checked.

The concept of multilines was proposed by Jensen (1952) and promoted by Borlaug (1959) as an urgent step toward yield stability. In 1977 a winter wheat multiline was released in the Netherlands (Groenewegen 1977), and in 1978 the first multiline wheat in India was released, comprised of nine components (Marko, Nasr, and Uwar 1979). A multiline is supposed to provide various forms of stability: (1) stability against a virulent outbreak of a disease in any one year; (2) stability against the short life of a variety of about five or six years before the fungus disease develops new strains of the organism; (3) stability against the rapid geographical spread of virulent strain so common in systems of monoculture; and, finally, (4) stability of world food production (Allaby 1977). Each time a new strain of the disease is developed, a new resistant line must be found, and there is danger of depleting the readily available sources of genetic resistance. The success of multilines to give these forms of stability is yet to be determined. Evidence obtained in 1975 in the International Multiline Nursery, consisting of 215 entries grown at 30 locations in the world, has revealed widespread disease-resistance (Anonymous 1976). Maximum yield performance, under disease-free conditions, may not reflect the value of multilines. From the standpoint of competition, yield performance is of interest. Twenty-two components of the multiline nursery were used to make twelve multiline composites, each consisting of three-to-eleven components, and these were tested under disease-free conditions (Table 11-10).

Only one composite outyielded the check cultivar, and although the composites consist of genetically similar strains, competition principles appeared to exist. The real value of multilines was demonstrated in 1976 at Toluca, Mexico, under highly diseased conditions, when the Siete Cerros check was destroyed by disease and several multilines withstood the attack and remained completely resistant.

TABLE 11–10. Grain yields of twelve multiline composites of wheat in comparison to a check variety, Siete Cerros, and to the average yield of their components (Yaqui Valley, Mexico, 1975–1976)

Composite Number	Number of Components	Yield of Composite (t/ha)	Yield of Composite (bu/ac)	Percent of Check	Average Yield of Components (t/ha)	Average Yield of Components (bu/ac)	Percent of Check
No. 1	5	7.53	112.0	90	8.16	121.4	97
No. 2	4	7.93	118.0	95	7.60	113.1	91
No. 3	6	8.52	126.8	102	7.87	117.1	94
No. 4	9	8.09	120.4	96	7.73	115.0	92
No. 5	3	7.94	118.2	95	7.68	114.3	92
No. 6	5	7.41	110.3	88	7.39	110.0	88
No. 7	5	6.67	99.3	79	7.29	108.5	87
No. 8	4	7.61	113.2	91	7.45	110.9	89
No. 9	5	7.23	107.6	86	7.48	111.3	89
No. 10	3	7.36	109.5	87	7.78	115.8	93
No. 11	8	7.29	108.5	87	7.30	108.6	87
No. 12	11	7.56	112.5	90	7.77	115.6	93
Siete Cerros check	—	8.38	124.7				

Source: CIMMYT report on wheat improvement 1976.

GENERAL OBSERVATIONS ON COMPETITION IN CROPS

Cereals

Recognition of the intense competition exerted on individual cereal plants, under conditions of high yield performance, led Donald and Hamblin (1976) to propose three ideotypes: (1) the "isolation ideotype" was considered as one able, under spaced conditions, to utilize the environment because of its high tillering ability, large, lax leaves, and relatively tall stature; (2) the "competition ideotype" was proposed as a plant capable of being the aggressor because it is tall, leafy, and capable of a high tillering capacity even under crowded conditions, as well as being able to shade its neighbors and gain a large share of nutrients and water; and (3) the "crop ideotype" plant had an erect stem, upright small leaves, and sparse tillering. If tillering is suppressed under normal crop conditions, why should plant breeders incorporate a high tillering capacity into plants? Efforts to develop a uniculm cereal plant have not met with outstanding success to date. However the dilemma of the crop producer is whether the "competition ideotype" is more desirable than the "crop ideotype." Can a short-statured cultivar associated with a crop ideotype perform well in dense stands? Will a plant classed as a "crop ideotype" under spaced conditions perform competitively?

The other consideration in the cereal crop is yield stability. Two barley cultivars deserve attention. Brock barley was licensed for sale in Canada in 1969 (Reinbergs and Loiselle 1970). The 150 breeder lines of Brock were classified into three distinguishable morphological types, based mainly on leaf width and erectness of the spikes. When tested separately, the three types gave inferior yields in comparison with the composite of all the 150 breeder lines. Consequently, all lines were bulked to produce seed of Brock. The cultivar Brock was recognized as having a greater degree of heterogeneity than other cultivars, yet Brock did not prove to perform satisfactorily under farm conditions and was a short-lived cultivar.

In contrast, the barley cultivar Mingo was licensed in 1979 by Ciba-Geigy Seeds Ltd., the first cultivar to be developed using the doubled haploid method whereby half the normal chromosome complement is obtained in a seedling and is doubled by chemical means (Kasha 1974). Doubling of the chromosome means that the resulting plant is **homozygous** for all genetic traits, and Mingo may be genetically the purest cultivar ever released. In Ontario tests in 1979, Mingo was the highest yielding of fourteen barley cultivars under test.

Forage Establishment

In an effort to obtain a grain yield in the seeding year when establishing a forage stand, forages may be seeded with a **companion crop** of cereals. Use of a companion crop assumes that some of the light is unexploited by the cereal crop, that water and nutrients will not be limiting, and that a weak growth of forages can develop with little effect on the cereal crop. Companion crops develop more rapidly than forage seedlings, which invariably become the aggressor, and competition between the two species can be so severe as to cause forage seedling failures. Failures or weak stands can be expected in dry years, under conditions where forage stands are difficult to obtain, or in the case of species that have a weak competitive ability. If a companion crop is not used, weed control with direct seeded forages is essential because a crop of weeds more competitive than cereals may occur.

Direct-seeded forages, with adequate weed control, suffer no adverse competitive effects from a companion crop, and if seeded in Ontario by early May, are ready for a mid-July harvest with a possible second cut under favorable environmental conditions. In Minnesota, a first-year alfalfa yield of 4030 kg/ha (3598 lb/ac) was reported (Schmid and Behrens 1972). Yields of more than 6700 kg/ha (5982 lb/ac) of alfalfa dry-matter herbage were obtained in Missouri and Connecticut (Peters 1964; Peters 1961). In Wisconsin, yields ranging between 3620 and 6640 kg/ha (3232 and 5929 lb/ac) were reported (Fawcett and Harvey 1977) with yields of 3620 to 6640 kg/ha (3232 and 5929 lb/ac) obtained in the first cutting.

Without competition in the seeding year, higher yields can be measured with direct-seeded forages in the second year (Table 11–11).

TABLE 11–11. Comparison of forage crop yields (kg/ha) and (lb/ac) with various companion crops and direct seeding of an alfalfa-bromegrass mixture at Guelph over a three-year period 1975 to 1977

	Yield			
	Seeding Year (Grain or Forage)		*Second Year Forage*	
Crop	*(kg/ha)*	*(lb/ac)*	*(kg/ha)*	*(lb/ac)*
Oats	3125	2790	5174	4620
Barley	2800	2500	5062	4520
Oats + barley	3158	2820	4906	4380
Direct-seeded	3606 (hay)	3220	5667	5060

Source: From unpublished data provided by R.S. Fulkerson, OAC. (Ontario Agricultural College)

The photograph (Fig. 11–1) illustrates the vigor of forage seedlings when seeded without a companion crop.

Forage Seeding Rate

From 10 to 17 kilograms (9 to 15 lb/ac) of seed per hectare is generally recommended for most hay meadows, but this varies from region to region and within regions, according to the condition of the site to be seeded. Specific

Figure 11-1.   Forages established without a cereal companion crop on the left are strong and vigorous. Forages seeded with the crop of rye on the right are subject to intense competition and are likely to fail. (Courtesy Ontario Ministry of Agriculture and Food.)

seeding rates in Ontario for selected forages are shown in Table 11-12, to illustrate the number of seeds per square metre or per square foot using the reported number of seeds per unit of weight. Such high rates are based on empirical observations, but they appear to be essential for satisfactory seedings. It is competition that is mainly responsible for such seemingly high rates. In humid regions, 70% emergence is considered excellent, but only 40 to 50% of the seed sown is expected to survive the first year (Tesar and Jackobs 1972). Competitive effects were expressed after the seeding year. In October of the seeding year and in October of the fourth year, the average number of plants per square metre of an alfalfa-brome mixture was 100 and 20 (9 and 1.8 ft^2), respectively, at Guelph. Tesar and Jackobs (1972) suggested that, for the northeastern United States, an average seeding rate of 11.2 kg/ha (10 lb/ac) was used, which provided 538 seeds/m^2 (48/ft^2) of which 215 to 269 plants per metre survived (19 to 24/ft^2) the first year, and that 60 to 78 plants per square metre (5.4 to 7.0/ft^2) were adequate for maximum yield in subsequent years.

Because density-dependent mortality plays such a dominant role throughout the life of the forage crop, it raises the question of whether forage seeding rates could be reduced. Because density-dependent competition is a strong force and denser high-yielding stands have been achieved with high seeding rates (see Table 11-13) a reduction in seeding rate may not be possible with conventional seeding methods.

TABLE 11–12. Recommended seeding rates in Ontario for selected forages and mixtures 1980

Component	Seeding Rate (kg/ha)	(lb/ac)	Seeds per Square Metre	(/ft²)
I Alfalfa-based mixtures				
Alfalfa alone	13	11.6	572	51.5
Alfalfa	11	9.8	754	67.9
Bromegrass	9	8.0		
Alfalfa	11	9.8	2108	189.7
Timothy	6	5.0		
Alfalfa	11	9.8	1746	157.1
Timothy	4	3.6		
Bromegrass	6	5.0		
Alfalfa	9	8.0	2100	189.0
Ladino	2	1.8		
Timothy	4	3.6		
Bromegrass	9	8.0		
II Birdsfoot Trefoil-based mixtures				
Trefoil alone	11	9.8	1029	92.6
Trefoil	9	8.0	1383	124.5
Timothy	2	1.8		
III Red Clover-based mixtures				
Red Clover	7	6.3	2047	184.2
Timothy	6	5.0		
Red Clover	7	6.3	5127	461.4
Alsike	2	1.8		
Timothy	6	5.0		
IV Pure Grass				
Reed Canary	9	8.0	573	51.6
Brome	11	9.8	330	29.7
Timothy	9	8.0	2435	219.2
Orchard	9	8.0	1295	116.6

Source: Ontario Ministry of Agriculture and Food 1980.

A similar study in Montana (Table 11-14) but with alfalfa seeded in rows 30.5 cm (12 in) apart showed that a seeding rate of 15.7 kg/ha (14 lb/ac) was required to produce the highest yield in the seedling year (Cooper, Ditterline, and Welty 1979). No advantage of size of seed was found. Seedling density was as good or better with small as with large seeds at the same seeding rate. Donald (1963) explained seeding rate in forages as having to reach equilibria in mixtures. If full production is desired in the first year of a perennial forage crop, then a heavy seeding rate must be used, but this will lead to plant losses

TABLE 11–13.   Influence of seeding rate on dry-matter yield and stand density of alfalfa, three-year mean 1973–75

Seeding Rate		Seeding Year Yield		Second Year Yield		Fourth Year Plants		Fourth Year Total Yield	
(kg/ha)	(lb/ac)	(kg/ha)	(lb/ac)	(kg/ha)	(lb/ac)	per m²	per ft²	(kg/ha)	(lb/ac)
6.7	6.0	2768	2471	4646	4148	54.9	4.9	9050	8080
13.4	12.0	3044	2718	4927	4399	58.5	5.3	8781	7840
20.2	18.0	3108	2775	4909	4383	63.0	5.7	8893	7940
26.9	24.0	3222	2877	4927	4399	—		—	
40.3	36.0	3233	2887	4938	4409	—		—	

Source: Unpublished data from J. E. Winch, OAC.

through competition in subsequent years. If a low seeding rate is used, which will result in a full stand in the second and subsequent years, then first-year production will be low.

Seedling vigor of birdsfoot trefoil was related consistently to seed size (Carleton and Cooper 1972; Hensen and Tayman 1961; Stickler and Wasson 1963; Twamley 1967), but the impact of seed size on density-dependent mortality is not known.

TABLE 11–14.   Effect of seeding rate on plant numbers, percentage emergence, and first- and second-year yield of alfalfa under irrigation in Montana in 1976 and 1977

Seeding Rate		Seedling Number per metre of row	per foot of row	Relative Emergence* (%)	Yield 1976 (t/ha)	(t/ha)	1977 (t/ha)	(t/ac)
(kg/ha)	(lb/ac)							
0.6	0.5	12	3.6	100	4.82	2.17	11.34	5.10
1.1	1.0	20	6.0	81	5.40	2.43	12.24	5.51
2.2	2.0	35	10.5	71	6.36	2.86	13.14	5.91
4.5	4.0	50	15.0	50	6.46	2.91	12.94	5.82
6.7	6.0	73	21.9	48	6.43	2.89	12.10	5.45
7.8**	7.0	92	27.6	52	7.35	3.31	13.42	6.04
9.0	8.0	88	26.4	44	7.78	3.50	14.30	6.44
11.2	10.0	97	29.1	39	7.60	3.42	13.74	6.18
13.4	12.0	123	36.9	42	7.31	3.29	13.25	5.96
15.7	14.0	137	41.1	39	8.31	3.74	13.94	6.27
17.9	16.0	154	46.2	39	7.71	3.47	14.10	6.35
20.2	18.0	154	46.2	35	7.69	3.46	14.39	6.48
22.4	20.0	172	51.6	35	7.40	3.33	13.18	5.93

Source: Data from Cooper, Ditterline, and Welty 1979.

*Based on the assumption that all seeds planted were viable.

**Recommended seeding rate in Montana for irrigated land.

Legume species differ in their abilities to compete, primarily as a result of differences in growth rate. Cooper (1977) suggested that under the stress of competition from a mixture of legumes, the most aggressive will have the greatest survival rate.

Birdsfoot trefoil has received considerable attention in breeding programs for seedling vigor, mostly because of an apparent lack of competitive ability in the seedling stages. By selecting plants capable of producing large seeds and then selecting for the most vigorous seedlings from those plants, Twamley (1974) increased seedling vigor of birdsfoot trefoil (cultivar Leo) 35 to 40%.

Management of Forage Stands

In forage mixtures, it is possible to shift the hierarchy of exploitation by management practices that favor one of the species in the mixture. In a grass-legume mixture, the legume may fix atmospheric nitrogen to the mutual benefit of both. If nitrogen fertilizer is applied to help stimulate the growth of grass, an otherwise nonaggressive grass may produce abundant leaf growth that shades the legume to the point where competition for light is so severe that the legume stand is reduced.

Reid (1970, 1972) studied the long-term effect of nitrogen fertilizer applications on the hierarchy of exploitation of a grass-legume mixture in Great Britain. A seed mixture containing 34 k/g (30 lb) of S23 perennial ryegrass and 3.4 k/g (3.0 lb) of S100 white clover per hectare (30 lb and 3.0 lb/acre) was seeded directly in April 1963. Twenty-one nitrogen fertilizer rates were applied in the spring of each year, and the mean percent clover content of the mixed herbage was determined (Table 11–15). Without nitrogen fertilizer, white clover made up 30.7% of the mixture over a six-year period. At rates of 28 and 56 kg/ha (25 and 50 lb/ac) of nitrogen, clover content dropped to 27.6 and 24.1%, respectively, indicating the extent to which the increased growth of the grass competed with the legume. As nitrogen rates increased, clover content decreased rapidly, in the first year as well as in subsequent years.

Recognizing the management problems associated with a mixed sward, producers seeking to maximize fertilizer and management inputs have questioned the merits of including a legume in a mixture if yields comparable to a mixed sward could be obtained with a pure grass stand under high nitrogen fertility conditions. The work of Reid (1970, 1972), which compared a white clover perennial ryegrass mixture with a pure grass stand given 21 nitrogen fertility levels (Table 11–16), revealed the high levels of nitrogen required to push up pure grass yields. The pure-grass stand responded remarkably consistently in each of the six years with a steady increase in dry-matter **herbage** between 0 to 336 kgN/ha, (0 to 300 lb N/ac) followed by a leveling in the response to additional nitrogen. In no case did a pure grass stand yield as well as a grass-clover mixture, which indicates the practical importance of including clover in the sward.

TABLE 11-15. Mean clover content of a grass-legume mixture subjected to 21 nitrogen fertilizer rates for a six-year period

N Applied		Clover Content (%)					
(kg/ha)	(lb/ac)	1964	1965	1966	1967	1968	1969
0	0	48.3	34.0	36.6	29.3	31.1	5.0
28	25	41.2	27.6	26.0	30.0	26.8	13.9
56	50	36.4	22.4	21.5	30.3	26.4	7.6
84	75	32.8	20.0	13.0	19.2	10.9	0.9
112	100	24.2	18.2	10.0	13.7	5.0	3.2
140	125	20.7	14.5	6.6	10.2	11.6	2.6
168	150	17.7	12.6	4.8	4.5	4.2	0.5
196	175	13.0	5.5	2.1	1.6	0.9	—
224	200	16.0	4.8	1.4	2.8	4.8	1.0
252	225	9.3	2.4	0.9	0.7	0.1	0.6
280	250	9.9	3.2	0.8	0.9	0.8	—
308	275	6.1	1.8	0.2	0.6	0.4	—
336	300	6.5	1.7	0.3	0.4	—	—
392	350	5.9	1.7	0.3	0.2	0.5	0.4
448	400	5.2	0.8	—	0.2	0.2	—
504	450	2.7	0.4	—	0.1	—	—
560	500	3.5	0.7	0.1	—	0.2	—
616	550	3.1	0.4	0.1	—	—	—
672	600	2.5	0.2	—	—	—	—
784	700	1.2	0.1	—	—	—	—
896	800	1.1	0.2	—	—	—	—

Source: Data from Reid 1970, 1972.

Intercropping Competition

Intercropping is the production of two or more crops simultaneously on the same field, in an attempt to increase land equivalent ratios. Fruit-tree growers may plant a second crop between the rows of the developing trees; occasional temporary windbreaks of two corn rows have been grown with soybeans; or alternate rows of two crops may be grown whereby one can be removed at an early stage of development, allowing the second crop to develop. Compared with single cropping, land usage can be improved under conditions of successful intercropping. The competitive effects of intercropping must be recognized and will be a key consideration in its success.

A study was conducted in Minnesota using corn hybrids intercropped with soybeans at conventional row spacings in various combinations: single alternate rows, three rows of each crop alternated, six rows of each crop alternated, and twelve rows of each crop alternated (Crookston and Hill 1979). Both crops were planted on the same day. Although corn yields increased in some combinations, accompanying soybean yields were always reduced to the extent that none of the combinations showed intercropping to be superior over single cropping (Table 11-17).

TABLE 11-16. Total yield comparisons of a grass-legume sward and pure grass at 21 nitrogen fertility levels over the six-year period 1964–1969

N Applied		Herbage Dry Matter (t/ha) (tons/ac)			
(kg/ha)	(lb/ac)	(Grass plus clover)	(tons/ac)	(Grass alone)	(tons/ac)
0	0	7.46	3.36	2.59	1.17
28	25	8.01	3.60	4.14	1.86
56	50	8.46	3.81	4.92	2.21
84	75	8.82	3.97	6.03	2.71
112	100	9.35	4.21	6.37	2.87
140	125	9.58	4.31	7.73	3.48
168	150	10.38	4.67	8.41	7.57
196	175	10.80	4.86	9.61	4.32
224	200	11.01	4.95	10.55	4.75
252	225	11.66	5.25	10.81	4.86
280	250	12.30	5.54	11.58	5.21
308	275	12.47	5.61	11.97	5.39
336	300	13.28	5.98	12.48	5.62
392	350	13.50	6.08	13.12	5.90
448	400	14.01	6.31	13.58	6.11
504	450	14.33	6.45	13.99	6.30
560	500	14.44	6.50	14.20	6.39
616	550	14.34	6.45	13.97	6.29
672	600	14.40	6.48	14.33	6.45
784	700	14.64	6.59	14.25	6.41
896	800	15.00	6.75	14.28	6.43

Source: Data from Reid 1970, 1972.

TABLE 11-17. Corn and soybean yields when the two crops were grown in patterns of alternating 76-cm (30-in) spaced rows averaged over a three-year period in Minnesota 1975 to 1977

Planting Pattern	Yield			
	Corn		Soybeans	
	(t/ha)	(bu/ac)	(t/ha)	(bu/ac)
Control	7.8	125.3	3.0	48.2
12:12	8.2	131.7	2.8	45.0
6:6	8.1	130.1	2.7	43.4
3:3	8.8	141.4	2.6	41.8
1:1	10.3	165.5	2.1	33.7

Source: Data from Crookston and Hill 1979.

Competition from Weeds

Crop losses due to weed competition are severe, consistent, and predictable; weed control is regarded as an essential aspect of crop production. Without weed control, direct-seeded forages are doomed to failure because forage seedlings lack competitive ability. In larger, more competitive crops, losses due to weed competition can be measured. A survey in western Canada indicated that 266 weeds per square metre (24/ft²) in cereal grain crops caused a 15% yield loss (Friessen 1963). In spring wheat, 83 and 226 wild oat plants per square metre (7 and 24 ft²) reduced yield by 30 and 53%, respectively (Bowden and Friessen 1967), and Shebeski (1955) found that 119 wild mustard plants per square metre (10.7/ft²) reduced yield in wheat by 30%. The effects of weed competition are evident from Figure 11-2.

Figure 11-2. Crop losses due to weed competition may be severe. The corn on the right is unsatisfactory due to competition. The corn on the left has received the same management inputs plus herbicidal weed control. (Courtesy Ontario Ministry of Agriculture and Food.)

Studies in Ontario (Anonymous 1973) showed that 3.3, 6.6, and 10 lamb's-quarter plants per metre of row in corn (1.0, 2.0 and 3.0/foot of row) reduced yields 10, 18, and 30%, respectively. In Illinois, grain corn losses ranged between 28 and 100% with an average of 81% in unweeded plots (National Academy of Sciences 1975); and in Minnesota, losses in corn averaged 51% and ranged from 16 to 93% in unweeded corn plots over a three-year period (Behrens and Lee 1966).

In soybeans, extensive tests in Minnesota and Nebraska showed a 45% yield reduction in unweeded plots (National Academy of Sciences 1975).

The competitive nature of weeds may be related to a variety of reasons. Grummer (1961) proposed that some weed species produce toxic root substances that hamper the growth of adjacent plants. No toxic effect whatsoever could be verified, however (de Wit et al. 1960), and the significance of toxic substances may be of little importance in field crop production.

CONCLUSIONS

Despite considerable research in the area of plant competition, it remains an obscure and difficult topic with which crop producers must deal. Until a fuller understanding of competition is obtained, progress in blending mixtures and plant populations may have to continue to be based on empirical results.

QUESTIONS

1. Is competition in field crops desirable or undesirable? Can competition be used to increase plant yield, or is it always associated with reducing yield? Can plants derive benefit from each other's presence and, in this sense, cooperate rather than compete? (See Mather 1961; and Gustafsson 1957.)

2. Are recommendations concerning seeding rates, plant population, row widths, and spatial arrangements designed to reduce or maximize plant competition?

3. When does competition occur—in the seedling stage, during vegetative growth, or during reproductive development? Are competitive effects cyclical or seasonal, or do they occur steadily with increasing or decreasing intensity as the crop develops? (See Jensen 1978.)

4. Do competitive effects exist only between two plants or can competitive effects occur within a single isolated plant? Why do fruit growers reduce the fruit set on a tree? Expressed otherwise, does an isolated plant compete for external environmental factors such as CO_2, O_2, light, water, or fertilizer nutrients among the plants, or does it compete only for internal factors?

5. Can the effects of competition be observed and measured? How are competitive effects expressed?

6. Are plants of a uniform cultivar more or less competitive with adjacent plants than a mixture of plants from the same species or among plants of different species?

7. Does root competition occur?

8. If the effects of competition are reflected in grain yield, can competition account for differences in yield between an early or a late seeded crop, between a spring or a winter type, between one cultivar and another?

9. Is competitive ability in plants a genetic trait? (See Sakai 1961.) If

competition is a heritable trait, should plant breeders attempt to develop strains that are strongly or weakly competitive?

10. How might a strongly competitive line differ phenotypically from a weakly competitive line?

11. Is yield stability or wide adaptation an expression of competitive ability?

12. Does such a thing as a neutral balanced or harmonious situation among plants exist?

13. Can a mixture of lines or species result in a synergistic effect; i.e., can the yield of a mixture be greater than the yield of the components of the mixture?

14. Can competitive effects be avoided by mixing selected types within a species?

15. Should farmers use blends of seed, i.e., a mixture of crop cultivars, to obtain greater yields?

16. Is "competitive ability" a desirable feature to incorporate into a crop cultivar?

17. How does competitive ability of wild plants compare with that of domesticated plants? (See Sakai 1961, p. 260.)

18. How would you attempt to measure differences in competitive ability?

19. List factors that might influence competition among plants.

20. What role might competitive ability have played in the evolution of plants?

21. Compare seeds planted per m² with plants required to maximize yield. Suggest reasons why such a large seeding rate is recommended. How might the seeding rate be effectively reduced?

22. Discuss Frankel's argument (1950) that the purity concept of crop cultivars is harmful to highest production. Does this concept apply to cross-pollinated and self-pollinated crops equally?

23. Discuss the reasons why plant breeders seek with such determination, uniformity in a cultivar. Explain each of the five reasons suggested by Frankel (1950).

24. What is the most suitable nitrogen fertilizer rate for a grass-legume mixture if the clover in the stand is to be favored?

25. How long a time was required for nitrogen fertility rates to cause a shift in the hierarchy of exploitation in the grass-clover mixture?

26. Using current values for nitrogen and protein, calculate the economic practicality of using 112 kg N/ha (100 lb N/ac) to replace the legume in a mixture. (See Table 11-17.)

27. In an effort to increase the effective LAI in soybeans, a rippled canopy has been proposed by alternating rows of tall and short cultivars in four-row sets, or by using a skip-row planting pattern in combination with tall and

short cultivars and high and low seeding rates (Ryder and Beuerlein 1979). Another proposal was to alternate rows of two cultivars of different maturities. Comment on these proposals from a competitive viewpoint.

28. Eighty-three and 226 wild oat plants per m² (7 and 24/ft²) reduced spring wheat yields 30 and 53%, respectively. Establish the density of wild oats in relation to a stand that might be established at a recommended seeding rate of 130 kg/ha (115 lb/ac) if there are 35 (15,875) seeds per gram (pound).

29. Why are yield losses resulting from weed competition so severe? Stated otherwise, why are weeds so competitive?

REFERENCES FOR CHAPTER ELEVEN

Allaby, M. 1977. *World Food Resources: Actual and Potential*, pp. 251–252. London: Applied Science Publishers Ltd.

Anonymous. 1973. *A Statement of the Canada Weed Committee's Position on Herbicide Use*. Canada Weed Committee, Agriculture Canada.

Anonymous. 1976. *CIMMYT Report on Wheat Improvement*, pp. 5–6. Mexico City: International Improvement Centre for Wheat and Corn.

Behrens, R., and D. C. Lee. 1966. *Weed Control. Advances in Corn Production*, pp. 331–352. Ames, Iowa: Iowa State University Press.

Borlaug, N. E. 1959. "The Use of Multilineal or Composite Varieties to Control Airborne Epidemic Diseases of Self-Pollinated Crop Plants," *Proceedings International Wheat Genetics Symposium*, pp. 12–16.

Bowden, B. A., and G. Friesen. 1967. "Competition of Wild Oats (*Avena fatua L.*) in Wheat and Flax," *Weed Research* 7:349–359.

Carleton, A. E., and C. S. Cooper. 1972. "Seed Size Effects upon Seedling Vigor of Three Forage Legumes," *Crop Science* 12:183–186.

Cooper, C. S. 1977. "Growth of the Legume Seedling," *Advances in Agronomy* 29:119–139.

Cooper, C. S., R. L. Ditterline, and L. E. Welty. 1979. "Seed Size and Seeding Rate Effects upon Stand Density and Yield in Alfalfa," *Agronomy Journal* 71: 83–85.

Crookston, R. K., and D. S. Hill. 1979. "Grain Yields and Land Equivalent Ratios from Intercropping Corn and Soybeans in Minnesota," *Agronomy Journal* 71: 41–44.

de Wit, C. T., et al. 1960. "Competition and Non Persistency as Factors Affecting the Composition of Mixed Crops and Swards," *Proceedings Eighth International Grassland Congress* Paper 14/B3, pp. 736–741.

Donald, C. M. 1963. "Competition among Crop and Pasture Plants," *Advances in Agronomy* 15:1–18.

Donald, C. M., and J. Hamblin. 1976. "The Biological Yield and Harvest Index of Cereals as Agronomic and Plant Breeding Criteria," *Advances in Agronomy* 28:361–405.

Doney, D. L., R. L. Plaisted, and L. C. Peterson. 1965. "Genotypic Competition in Progeny Performance Evaluation of Potatoes," *Crop Science* 5:433-435.

Eberhart, S. A., L. H. Penny, and G. F. Sprague. 1964. "Intraplot Competition among Maize Single Crosses," *Crop Science* 4:467-471.

Fawcett, R. S., and R. G. Harvey. 1977. "Field Comparison of Seven Dinitroaniline Herbicides for Alfalfa Seedling Establishment," *Weed Science* 26:123-127.

Ford, E. D. 1975. "Competition and Stand Structure in Some Even-Aged Plant Monocultures," *Journal of Ecology* 63:311-333.

Frankel, O. H. 1950. "The Development and Maintenance of Superior Genetic Stocks," *Heredity* 4:89-102.

Friessen, H. A. 1973. "Herbicides Popular in Western Canada," *Weeds Today* 4(3): 20-22.

Funk, C. R., and J. C. Anderson. 1964. "Performance of Mixtures of Field Corn (*Zea mays* L.) hybrids," *Crop Science* 4:353-356.

Groenewegen, L. J. M. 1977. "Multilines as a Tool in Breeding for Reliable Yields," *Cereal Research Communications* 5(2):125-132.

Grummer, G. 1961. "The Role of Toxic Substances in the Interrelationships between Higher Plants," in *Mechanisms in Biological Competition, Symposia for the Society Experimental Biology* 5:219-228.

Gustafsson, A. 1957. "The Cooperation of Genotypes in Barley," *Hereditas* 39:1-18.

Harper, J. L. 1961. "Approaches to the Study of Plant Competition," in *Mechanisms in Biological Competition, Symposia for the Society Experimental Biology* 15: 1-39.

———. 1967. "A Darwinian Approach to Plant Ecology, *Journal of Ecology* 55: 247-270.

Hensen, P. R., and L. A. Tayman. 1961. "Seed Weights of Varieties of Birdsfoot Trefoil as Affecting Seedling Growth," *Crop Science* 1:306.

Hinson, K., and W. D. Hanson. 1962. "Competition Studies in Soybeans, *Crop Science* 2:117-123.

Jensen, N. F. 1952. "Intra-Varietal Diversification in Oat Breeding," *Agronomy Journal* 44:30-31.

———. 1978. "Seasonal Competition in Spring and Winter Wheat Mixtures," *Crop Science* 18:1055-1057.

Jurado-Tovar, A., and W. A. Compton. 1974. "Intergenotypic Competition Studies in Corn (*Zea mays* L.) I. Among Experimental Hybrids, *Theoretical and Applied Genetics* 45:205-210.

Kannenberg, L. W., and R. B. Hunter. 1972. "Yielding Ability and Competitive Influence in Hybrid Mixtures of Maize," *Crop Science* 12:271-277.

Kasha, K. J., ed. 1974. "Haploids from Somatic Cells," in *Haploids in Higher Plants. Advances and Potential.* Proceed. First International Symposium, Guelph, Ontario.

Marko, P., H. Nasr, and H. Uwar, eds. 1979. *Bithoor—First Multiline Released in India* Wheat Team Field Notes No. 8. CIMMYT.

Mather, K. 1961. "Competition and Cooperation," in *Mechanisms in Biological Competition. Symposia for the Society Experimental Biology* 15:264-281.

Milthorpe, F. L. 1961. "The nature and analysis of competition between plants of different species," in *Mechanisms in Biological Competition. Symposia for the Society of Experimental Biology.* 15:330-355.

Moore, C. E. 1974. *Production and Handling of Forages.* Ontario Ministry of Agriculture and Food. Pub. 369. (20 pp.)

Mumaw, C. R., and C. R. Weber. 1957. "Competition and Natural Selection in Soybean Varietal Composites," *Agronomy Journal* 49:154-162.

National Academy of Sciences. 1975. "Pest Control: An Assessment of Present and Alternative Technologies," Vol. II. *Corn—Soybean Pest Control.* Washington, D.C.: National Academy of Sciences.

Pendleton, J. W., and R. D. Seif. 1962. "Role of Height in Corn Competition," *Crop Science* 2:154-156.

Peters, E. J. 1964. "Pre-emergence, Preplanting and Post Emergence Herbicides for Alfalfa and Birdsfoot Trefoil, *Agronomy Journal* 56:415-419.

Peters, R. A. 1961. "Legumes Establishment as Related to the Presence or Absence of an Oat Companion Crop," *Agronomy Journal* 53:195-198.

Pfahler, P. L. 1965. "Environmental Variability and Genetic Diversity within Populations of Oats (Cultivated Species of Avena) and Rye (*Secale Cereale* L.)," *Crop Science* 5:271-275.

Probst, A. H. 1957. "Performance of Variety Blends in Soybeans," *Agronomy Journal* 49:148-150.

Reich, V. H., and R. E. Atkins. 1970. "Yield Stability of Four Population Types of Grain Sorghum, *Sorghum bicolor* (L.), Moench, in Different Environments," *Crop Science* 10:511-516.

Reid, D. 1970. "The Effects of a Wide Range of Nitrogen Application Rates on the Yields from a Perennial Ryegrass Sward with and without White Clover," *Journal of Agricultural Science* (Cambridge). 74:227-240.

Reid, D. 1972. "The Effects of the Long-Term Application of a Wide Range of Nitrogen Rates on the Yields from Perennial Ryegrass Swards with and without White Clover," *Journal of Agricultural Science* (Cambridge). 79:291-301.

Reinbergs, E., and R. Loiselle. 1970. "Brock Barley," *Canadian Journal of Plant Science* 50:205-206.

Ross, W. M. 1965. "Yield of Grain Sorghum (*Sorghum vulgare* Pers.) Hybrids Alone and in Blends," *Crop Science.* 5:593-594.

Ryder, G. J., and J. E. Beuerlein. 1979. "Soybean Production: A Systems Approach," *Crops and Soils* 31(6):7-9.

Sakai, K. J. 1961. "Competitive Ability in Plants. Its Inheritance and Some Related Problems," in *Mechanisms in Biological Competition. Symposia Society of Experimental Biology.* 15:245-263.

Schmid, A. P., and R. Behrens. 1972. "Herbicide versus Oat Companion Crop for Alfalfa Establishment," *Agronomy Journal* 64:151-159.

Shebeski, L. H. 1955. "Weed Competition as Affected by Time of Spraying, *Proceedings of 8th Western Canadian Weed Control Conference*, p. 40.

Shorter, R., and K. J. Frey. 1979. "Relative Yields of Mixtures and Monocultures of Oat Genotypes," *Crop Science* 19:548-553.

Simmonds, N. W. 1962. "Variability in Crop Plants, Its Use and Conservation," *Camb. Philosophical Society Biological Review* 37:422–465.

Stickler, F. C., and C. E. Wasson. 1963. "Emergence and Seedling Vigor of Birdsfoot Trefoil as Affected by Planting Depth, Seed Size, and Variety, *Agronomy Journal* 55:781.

Stringfield, G. H. 1959. "Performance of Corn Hybrids in Mixtures," *Agronomy Journal* 51:472–473.

Tesar, M. B., and J. A. Jackobs. 1972. (In) *Alfalfa Science and Technology*, pp. 415–433. C. H. Hanson, ed., Agronomy Monograph, Vol. 15. *American Society of Agronomy* Publication.

Thompson, D. L. 1977. "Corn Hybrid Mixtures in a Southern Environment," *Crop Science* 17:645–646.

Twamley, B. E. 1967. "Seed Size and Seedling Vigor in Birdsfoot Trefoil, *Canadian Journal of Plant Science* 47:603–609.

———. 1974. "Recurrent Selection for Seedling Vigor in Birdsfoot Trefoil," *Crop Science* 14:87–90.

Warburton, C. W. 1915. "Grain Crop Mixtures," *Journal of the American Society of Agronomy* 7:20–29.

Zavitz, C. A. 1927. *Forty Years' Experiments With Grain Crops*. Guelph, Canada: Ontario Agricultural College Bulletin No. 332.

chapter twelve

Biological Nitrogen Fixation

Possibly ranking second to photosynthesis in biological importance is nitrogen fixation in legume plants. Biological nitrogen fixation is the process whereby bacteria form a symbiotic relationship with legume plants such as alfalfa, soybean, clover, peanut, beans, and peas, to convert the nearly inert atmospheric nitrogen, N_2, into compounds useful to living organisms. The importance of biological nitrogen fixation lies in the fact that legume plants are at least partially self-sufficient in nitrogen, an essential constituent of protein, which is required in large amounts by all forms of life.

Nitrogen is probably the most common limitation to growth for both plants and animals. Although the earth's atmosphere is comprised of nearly 80% nitrogen, the nitrogen is so inert that it is biologically useless to the vast majority of organisms. Atmospheric nitrogen must be "fixed" or combined with hydrogen or oxygen to be useful to biological systems. Because nitrogen always has been a limited agricultural resource, and because nitrogen in adequate supply for agriculture is a critical factor contributing to human food supply, the importance of biological nitrogen fixation is clearly demonstrated by the fact that the annual contribution of nitrogen fixation to agricultural productivity worldwide probably exceeds all the nitrogen fertilizer manufactured in any one year by four to five times (Brill 1977). As the exploding world population demands more food each year, the value of biological nitrogen fixation becomes even more important.

Several biological systems fix nitrogen, but in agriculture, **rhizobium** bacteria are most important. Rhizobia cause nodules to form on legume roots; without the legume, rhizobia cannot fix nitrogen; without rhizobia, legume plants cannot fix nitrogen. Nitrogen is fixed in nodules only through this symbiotic relationship. (See Figure 12-1.)

Nitrogen fixation can be achieved industrially through a process whereby N_2 gas and hydrogen are combined under conditions of high temperature and pressure, normally with natural gas as the energy source. Energy equivalent to 82 litres of gasoline is required to make 100 kg of inorganic nitrogen fertilizer (20 gallons per 100 lb of N). As supplies of petroleum products become more

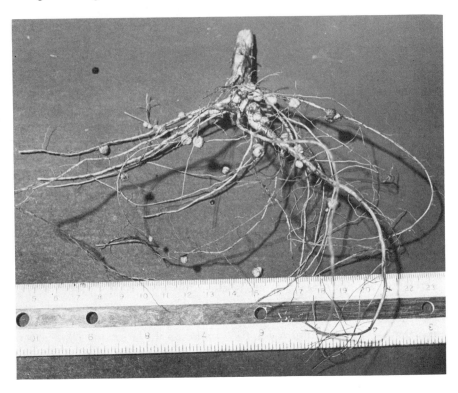

Figure 12–1. Nitrogen-fixing nodules on the root of a soybean. Nitrogen fixation ranks second to photosynthesis in biological importance. (Courtesy Ontario Ministry of Agriculture and Food.)

expensive and perhaps limited, the huge potential of biological nitrogen fixation assumes considerable importance. Not only can legumes supply part of their own nitrogen needs and produce quality protein, but they also provide available nitrogen for subsequent crops following the legume. Rotation of crops is a method of making nitrogen fixed by legumes available to subsequent crops, and although industrial fertilizer must be added for maximum yield, the amount needed is reduced. Plant proteins are formed by combining nitrogen produced in the nodule with the products of photosynthesis. When legume plants are plowed under, the protein in their tissues and the nitrogen in the nodules are broken down to ammonia (NH_3) or nitrate nitrogen (NO_3), which can be used by other plants. Some nitrogen is returned to the atmosphere as inert N_2 gas by bacteria causing **denitrification.** As a result of denitrifiers, the reservoir of fixed nitrogen in the soil must be continually replenished. Each time a crop is removed from a cropping area (marketed), the nitrogen it contained is lost to the soil.

In biological nitrogen fixation and in the industrial process, the immediate product is ammonia, an effective fertilizer itself, which can be converted into other useful and common compounds such as urea and nitrates. Whether nitrogen is fixed biologically (organic) or industrially (inorganic), the product is identical, and plants cannot distinguish one source from the other.

AMOUNTS OF NITROGEN FIXED

Varying estimates of the amount of nitrogen fixed by different legumes have been reported and in 1975 were summarized for the United States (Table 12-1) by researchers and reported by Hume (1978).

TABLE 12-1.   Amounts of nitrogen fixed by various legume crops in the United States, 1975

Crop	Kilograms Nitrogen Fixed per Hectare per Year	Pounds per Acre per Year
Pure alfalfa	128–300	114–268
Pure clovers	104–220	93–196
Mixed legumes in pastures	118–125	105–112
Soybeans	57– 94	51– 84
Peas	50– 85	45– 76
Peanuts	47– 91	42– 81

Source: Data as reported by Hume 1978.

These U.S. values conform with measurements from research studies in Ontario in 1977 (Table 12-2), except for soybeans. In Ontario a higher value was obtained (Hume 1978), and this value coincides more closely with measurements obtained by Hardy et al. (1968).

TABLE 12-2.   Amounts of nitrogen fixed by various legume crops in Ontario, 1977

Crop	Kilograms Nitrogen Fixed per Hectare per Year	Pounds per Acre per Year
Alfalfa	150	134
Birdsfoot trefoil	125	112
Soybeans	100	89
White Beans	20	18
Other legumes	50	45

Source: Data as reported by Hume 1978.

FACTORS AFFECTING NITROGEN FIXATION

As nitrogen fertilizer costs rise, crop producers are keenly interested in assuring nitrogen fixation in their legume crops by using management practices that will promote maximum levels of fixation. Crop producers should identify the degree of successful nodulation in their crop in the seedlings. Roots can be inspected to determine if a spring-seeded crop is nodulated by mid-to-late July. The size of individual nodules is of little consequence; total volume or weight of nodules per plant is of greater consequence. Soybean plants are considered to be well-nodulated if they have ten to twenty nodules per plant a month after planting. Active nodules appear pink to the naked eye, indicating active **haemoglobin.** Producers should recognize the following factors in relation to nitrogen fixation.

1. Proper Inoculation

Each legume is associated with a distinct species of Rhizobium; the bacteria that will form nodules on alfalfa roots will not infect soybean roots, but they can infect sweet clover roots. Groups of rhizobia that will cause nodulation on specific crops are called **cross-inoculation** groups, and each cross-inoculation group requires its own inoculant (Table 12–3).

Inoculation means the application of the appropriate cross-inoculation group to the field at seeding time. Various types of commercial formulations of inoculant can be purchased as follows:

(a) *Peat-based inoculant.* This is a common form that can be applied either directly to the dry seed in the seed box or to seed moistened with a sticky syrup such as honey or molasses.
(b) *Liquid formulations.* These can be applied to the soil close to the seed in the seedbed.
(c) *Granular inoculants.* These have the potential of greatly increasing rhizobial numbers in the seed vicinity and have particular value on new fields or on fields known to have a nodulation problem.

TABLE 12–3.   Cross-inoculation groups for major legumes in North America

Group	Legumes Included in Group
Alfalfa	Alfalfa, sweet clover (yellow and white)
Clover	Red clover, alsike, white clover, crimson clover
Pea	Field, garden and sweet pea, common and hairy vetch, broad bean
Bean	Garden and field beans
Lupine	Blue and yellow lupine
Soybean	Soybeans
Cowpea	Cowpeas, lima bean, peanut, kudzu
Lotus	Birdsfoot trefoil and other members of the trefoil species

Granular inoculants are applied to the soil in the seedbed along with the seed.

(d) *Pre-inoculated seed.* Encapsulated seed may contain additional compounds to promote seedling growth.

2. Fertility Status

The results of studies with a number of legume crops all show that when the applied nitrogen exceeds that necessary for growth increase, it tends to replace the fixation process; i.e., as fertilizer nitrogen was applied to legumes, biological nitrogen fixation diminished (Allos and Bartholomew 1955, 1959; Norman and Krampitz 1945). In a grass-legume mixture, farmers face the conundrum of adding nitrogen fertility to promote forage grass production, but also of reducing nitrogen fixation and increasing competition from the grass (see Chapter Eleven). Some nitrogen fertilizer may be applied with advantage to promote seedling growth and in moderate amounts in later stages. Allos and Bartholomew (1959) observed that fixation processes never supplied sufficient nitrogen for maximum growth under greenhouse conditions. Soybeans, alfalfa, sweet clover, ladino clover, and birdsfoot trefoil all exhibited an apparent capacity to supply by fixation only about one-half to three-fourths of the total nitrogen that could be used by the plant. The remainder of the crop's nitrogen requirements should be applied as commercial fertilizer. Beard and Hoover (1971) found that as nitrogen fertility was applied to the soil, nodulation in soybeans declined. Fewer nodules were produced when 56 kg N/ha (50 lb N/ac) or more was applied, but up to 112 kg N/ha (100 lb N/ac) did not affect nodule numbers if applied at flowering, but neither did such a treatment increase soybean yields. If nodules have not formed on soybeans a month after planting, top-dress with 75 to 100 kg/ha (67 to 89 lb N/ac) of nitrogen in the form of urea, unless the reason for poor nodulation is high levels of nitrogen already in the soil.

Cobalt, a minor soil element, is required by rhizobia in the formation of vitamin B_{12}, which in turn is essential to the formation of haemoglobin needed for nitrogen fixation. There are numerous references to research carried out under controlled conditions with legumes adequately inoculated that failed to develop nodules in the absence of cobalt (Tisdale and Nelson 1966).

3. Dry Soil Conditions

Serious losses in the number of bacteria are likely to occur if inoculated seed is left suspended in dry soil. For best results, sow inoculated seed in a moist seedbed (Vincent, Thompson, and Donovan 1962; Alexander and Chamblee 1965).

As plants develop, soil moisture deficiencies reduce fixation. At Guelph, Hume (1978) found that irrigation increased fixation in peanuts and soybeans, although there was no increase in soybean yield. Inherently, droughty, sandy soils may have low levels of viable rhizobia present, which can reduce nodulation unless the soils are properly inoculated.

4. Soil Acidity

As soil pH declines, survival of rhizobia is reduced, and the availability of **molybdenum** decreases. Molybdenum is important for nitrogen fixation because it is needed in the biochemical pathway of fixation. Ideally, **acidic soil** conditions should be corrected by the application of lime, but where lime is not applied, inoculation should be practiced to ensure the presence of abundant rhizobia, preferably with the addition of molybdenum.

5. Frequency of Inoculation

Studies have shown that yield increases were not obtained from inoculation when soybeans had been grown on an area during the preceding few years (Caldwell and Vest 1970; Elkins et al. 1976; Ham, Cardwell, and Johnson 1971; Kapusta and Rouwenhorst 1973). The lack of response to inoculation has been explained by the fact that only about 5% of the nodules found on plants grown on soil previously cropped to soybeans originated from the inoculum added (Johnson, Means, and Weber 1965). Because of competition among rhizobial strains, little benefit from inoculation would be expected even if added rhizobia were more efficient in nitrogen fixation than "native" strains. Fields that have grown well-nodulated soybeans or other crops in the previous year or two generally produce the best nodulation; on fields that have not produced soybeans or other legumes for three to four years previously, inoculation is regarded as cheap insurance; and seeds of soybean and other legumes should always be inoculated for planting on fields where legumes have not been produced for five or more years.

In northern areas, in particular, it is difficult to obtain adequate nodulation the first time soybeans are grown in a field. Under such situations seedbox treatments seldom give as good results as with a sticker, which helps many more bacteria to adhere to the seed. Although more expensive, farmers introducing soybeans to their farm for the first time should consider granular inoculants. In a three-year study in Ontario, a 15 to 20% yield advantage was obtained over a conventional powdered peat inoculant (Anonymous 1977). A possible reason is that 20 to 40 times more rhizobia are applied with granular than with powdered peat inoculants. Granular applications may be superior because the soybean roots grow right through the needed bacteria, whereas bacteria adhering to the seed coat may be removed from the root zone if the soybean skin adheres to the cotyledons as they emerge above the soil.

Granules that provide a protective peat coat for the rhizobia have a better chance of survival in hot, dry soil. The results of using granular inoculant for two years at two sites are shown in Table 12–4 (Hume and Beversdorf 1978).

TABLE 12–4. Yields of soybeans grown with granular or peat inoculates at two Ontario sites in 1976 and 1977

| | Two-Year Average Yields | | | |
| | Woodstock (New Soybean Land) | | Elora (Former Soybean Land) | |
Treatment	(kg/ha)	(bu/ac)	(kg/ha)	(bu/ac)
Granular	3158	47.0	2621	39.0
Powdered peat Inoculant	2621	39.0	2621	39.0
Control (no inoculant	2083	31.0	2554	38.0

Source: Data from Hume and Beversdorf 1978.

Soils previously cropped to soybeans were found to contain about 64,000 rhizobia per teaspoon of soil (Johnson 1977). In areas with a high natural population, bacteria may be spread with farm machinery and wind. Under conditions with high natural populations of rhizobia, granular formulations represent the most likely means of altering rhizobia populations in the soil, but this was not found to be true in Indiana (Nelson, Swearingin, and Beckham 1978; Johnson 1977).

RESEARCH ON SOYBEANS

Commercial soybean yield increases of about 1% per year have been less than the advances obtained in most other crops. It was reasoned that nitrogen supply may limit soybean yields when one considers the amount needed to produce a grain yield of 3000 to 4000 kg/ha (45 to 60 bu/ac) with a protein content of 40 to 44%. Although nodules provide a prime source of nitrogen, studies at the University of Minnesota revealed that soybean nodules began to senesce during the early and critical pod-filling period (Lawn and Brun 1974a). The decline may present a "yield barrier" to soybean production because nitrogen may limit yield. Using a simulation model, Sinclair and de Wit (1976) showed that seed yield was clearly limited by the duration of seed fill and by the nitrogen supply rate; both nitrogen supply and seed fill period were restricted by a decline in available nitrogen. The loss of activity by nodules during early pod-filling, may be explained by a limited supply of food energy to the roots, resulting from internal competition from the developing pods and seed. Increasing plant photosynthesis to allow a greater proportion of food energy to be used to support nitrogen assimilation would be expected to increase seed yields. Soybean producers should carefully note this concept and implement management practices such as applying irrigation water, CO_2 enrichment, or nitrogen applications at this critical stage. Field-grown soybean plants were given 1200 ppm of CO_2 for five-week periods corresponding to vegetative flowering and pod-filling stages of development. As a result CO_2 enrichment during pod-filling increased seed yield; and CO_2 enrichment during flowering caused increased node numbers, leaf and stem dry weights, and pod number; but because seed size decreased, seed yield was unaffected. Perhaps a greater sink size was developed than could be filled from the given photosynthetic source. CO_2 enrichment during the vegetative stage had no effect on the parameters measured (Hardman and Brun 1971).

Nitrogen fixation was increased in soybeans from 90 to 112 kg/ha (80–100 lb/acre) to more than 450 kg/ha (400 lb/ac) by increasing CO_2 in an open-topped enclosure to 800 to 1200 ppm (Hardy and Havelka 1974). CO_2 did not escape because it is heavier than the air. Perhaps such results demonstrate that it is experimentally possible to break the nitrogen input barrier in soybeans. However, with present technology, CO_2 enrichment under field conditions is impractical.

Another attempt to increase soybean yields involved the application of 0, 224, and 448 kg N/ha (0, 200 and 400 lb N/ac) as ammonium nitrate at the end of flowering to the cultivars Chippewa 64 and Clay (Lawn and Brun 1974*b*; Lawn, Fischer, and Brun 1974). A seed yield increase of 7.3 and 8.8% at 224 and 448 kg N/ha (200 and 400 lb N/ac), respectively, occurred in the cultivar Clay. Seed protein was increased in Chippewa 64 by 1.8% at the high fertility level (Table 12–5).

RESEARCH ON ALFALFA

The symbiotic relationship between legume plants and rhizobia was discovered for alfalfa by two German scientists, Hellriegel and Wilfarth in 1888. They established the concept of inoculation in 1890 by transferring soil, presumably with rhizobia strains with bacteria to a new soil area, and demonstrated

TABLE 12–5. Effect of supplemental nitrogen applications at the end of flowering on various yield characters for two soybean varieties in Minnesota in 1972

Cultivar	Nitrogen Applied (kg/ha)	(lb/ac)	Seed Yield (kg/ha)	(lb/ac)	Seed Protein (%)	Protein Yield (kg/ha)	(lb/ac)
Chippewa 64	0	0	3452	3082	43.5	1504	1343
	224	100	3376	3014	43.7	1476	1318
	448	200	3540	3161	44.3	1586	1416
Clay	0	0	3848	3436	41.1	1580	1411
	224	100	4128	3686	41.9	1732	1546
	448	200	4188	3739	42.2	1786	1595

Source: Lawn and Brun 1974

improved growth and nodulation. In the past 90 years, alfalfa has become widely grown; successful inoculants have been developed; and biological nitrogen fixation in alfalfa is highly successful.

Alfalfa grown for forage production is cut three to five times annually, which has a depressing effect on nodule activity but in essence eliminates the decline in haemoglobin activity observed in soybeans at flowering time.

Alfalfa is believed to be able to fix 128 to 300 kg N/ha (114 to 268 lb/ac) (see Table 12-1), and although it is the highest producer of fixed nitrogen, producers are asking if double the amount could be fixed. Perhaps more fibrous root systems, rather than a large top root, might be more effective. Could alfalfa be bred for its ability to fix more nitrogen?

Could a superstrain of Rhizobium be developed? More efficient strains are being sought, but it is unlikely that a new superstrain will be more effective because it must compete with native populations in the soil, and it may not be adapted over a wide geographic area. It is also unlikely to be effective on all cultivars (Caldwell and Vest 1970; Brill 1977; Anonymous 1978).

ENERGY RELATIONSHIPS IN NITROGEN FIXATION

The industrial manufacture of nitrogen is an energy-intensive process; biological nitrogen fixation is also an energy-intensive process, and legume plants do not fix nitrogen without an energy cost. It is not possible to get something for nothing.

In the plant, food energy diverted to nitrogen fixation is not available for growth, carbohydrate production, or general plant vigor. Hume (1977) suggested that 10 to 20% of the potential yield of legumes is lost to nitrogen fixation. It is incorrect to conclude that a 7400 kg/ha (6600 lb/ac) crop could be increased by 10 to 20%, however, if it didn't fix nitrogen. Legumes obtain nitrogen from two sources—nitrate nitrogen in the soil from fertilizers and fixed nitrogen—and it apparently takes as much energy to utilize one form as the other.

CONCLUSIONS

The range of plants that possess the ability to biologically fix nitrogen is wide, but not included are the world's major food crops—wheat and rice—or the major forage crop—grass. The possibility of wheat, rice, corn, and grass fixing nitrogen does exist, and it is an attractive thought for all countries of the world, especially where the high price of nitrogen fertilizer is a limiting factor, and in the tropics where nitrate nitrogen is quickly leached out. The problems and prospects for nitrogen fixation in grasses have been extensively reviewed by Neyra and Dobereiner (1977) and Hardy (1975).

QUESTIONS

1. Note the protein content of soybean seed and the amount of seed protein produced per hectare (Table 12-5). Compare the amounts produced in soybean with other crops including legumes and nonlegumes.

2. On the basis of the current value for nitrogen, calculate the value of the nitrogen fertility added by the crops tested in Table 12-2.

3. Speculate on the impact of biological nitrogen fixation in grasses on world food production.

4. What impact may high nitrogen fertilizer prices have on systems of crop production such as forage mixtures or rotations?

5. Review the reasons why CO_2 enrichment during pod-filling increased seed yield in soybeans.

6. Will the statement that biological nitrogen fixation may be second in importance only to photosynthesis be true in the future?

7. Review the scope and extent of biological nitrogen fixation in your region. Could it be improved?

8. Do you believe that crops may someday be grown entirely for their nitrogen-fixing ability? Explain.

9. List reasons why alfalfa is said to be the "queen of the legumes."

10. What circumstances led early agriculturists to develop a crop rotation?

REFERENCES FOR CHAPTER TWELVE

Alexander, C. W., and D. S. Chamblee. 1965. "Effect of Sunlight and Drying on the Inoculation of Legumes with Rhizobium Species," *Agronomy Journal.* 57: 550-553.

Allos, H. F., and W. V. Bartholomew. 1955. "Effect of Available Nitrogen on Symbiotic Fixation," *Science Society America Proceedings.* 19:182-184.

Allos, H. F., and W. V. Bartholomew. 1959. "Replacement of Symbiotic Fixation by Available Nitrogen," *Soil Science.* 87:61-66.

Anonymous 1977. "Invest Your Money in Granular Inoculants," *Cash Crop Farming* 40(5):28.

Anonymous 1978. "Nitrogen Fixation in Alfalfa. Is It an Inherited Trait? Can Breeding Increase It?" *Crops and Soils* 30(7):7-8.

Beard, B. H., and R. M. Hoover. 1971. "Effect of Nitrogen on Nodulation and Yield of Irrigated Soybeans," *Agronomy Journal.* 63:815-816.

Brill, W. J. 1977. "Biological Nitrogen Fixation," *Scientific America.* 236(3):68-81.

Caldwell, B. E., and G. Vest. 1970. "Effects of *Rhizobium japonicum* on Soybean Yields," *Crop Science.* 10:19-21.

Elkins, D. M., et al. 1976. "Effect of Cropping History on Soybean Growth and Nodulation and Soil Rhizobia," *Agronomy Journal.* 68:513-517.

Ham, G. E., V. B. Cardwell, and H. W. Johnson. 1971. Evaluation of *Rhizobium japonicum* Inoculants in Soils Containing Naturalized Populations of Rhizobia," *Agronomy Journal.* 63:301-303.

Hardman, L. L., and W. A. Brun. 1971. "Effect of Atmospheric Carbon Dioxide Enrichment at Different Developmental Stages on Growth and Yield Components of Soybeans," *Crop Science.* 11:886-888.

Hardy, R. W. F., and U. D. Havelka. 1974. "The Nitrogen Barrier," *Crops and Soils* 26(5):10-13.

———. 1975. "Nitrogen Fixation Research: A Key to World Food?" *Science* 188: 633-643.

Hardy, R. W. F., et al. 1968. "The Acetylene-ethylene Assay for N$_2$ Fixation: Laboratory and Field Evaluation," *Plant Physiology.* 43:1185-1207.

Hume, D. J. 1977. *Nitrogen Fixation: Free Fertilizer?* Information for Industry Personnel. Ontario Ministry of Agriculture and Food. 141/544.

———. 1978. "Nitrogen Fixation—Nature's Fuel Saver," *Highlights of Agricultural Research in Ontario* Vol. 1(2):6-7.

Hume, D. J., and W. D. Beversdorf. 1978. *Use of Granular Soybean Inoculant.* Information for Industry Personnel. Ontario Ministry of Agriculture and Food. No. 141/23.

Johnson, H. W., U. M. Means, and C. R. Weber. 1965. "Competition for Nodule Sites between Strains of Rhizobium japonicum Applied as Inoculum and Strains in the Soil," *Agronomy Journal.* 57:179-185.

Johnson, R. R. 1977. "Soybean Inoculation: Is It Still Needed?" *Crops and Soils* 29(5):11-12.

Kapusta, G., and D. L. Rouwenhorst. 1973. "Influence of Inoculum Size on Rhizobium japonicum sero Group Distribution Frequency on Soybean Nodules," *Agronomy Journal.* 65:916-918.

Lawn, R. J., and W. A. Brun. 1974a. "Symbiotic Nitrogen Fixation in Soybeans I. Effects of Photosynthetic Source—Sink Manipulation," *Crop Science.* 14:11-16.

———. 1974b. "Symbiotic Nitrogen Fixation in Soybeans. III. Effect of Supplemental Nitrogen and Intervarietal Grafting," *Crop Science.* 14:22-25.

Lawn, R. J., K. S. Fischer, and W. A. Brun. 1974. "Symbiotic Nitrogen Fixation in Soybeans. II. Interrelationship between Carbon and Nitrogen Assimilation," *Crop Science.* 14:17-22.

Nelson, D. W., M. L. Swearingin, and L. S. Beckham. 1978. "Response of Soybeans to Commercial Soil Applied Inoculants," *Agronomy Journal.* 70:517-518.

Neyra, C. A., and J. Dobereiner. 1977. "Nitrogen Fixation in Grasses," *Advances in Agronomy.* 29:1-38.

Norman, A. G., and L. O. Krampitz. 1945. "The Nitrogen Nutrition of Soybeans: II." *Soil Science Society of America Proceedings.* 10:191-196.

Sinclair, T. R., and C. T. de Wit. 1976. "Analysis of the Carbon and Nitrogen Limitations to Soybean Yield," *Agronomy Journal.* 68:319-324.

Tisdale, S. L., and W. L. Nelson. 1966. *Soil Fertility and Fertilizers,* 2nd ed. New York: Macmillan Publishing Co., Inc. 694 pp.

Vincent, J. M., J. A. Thompson, and K. A. Donovan. 1962. Death of Root-Nodule Bacteria on Drying," *Australian Journal of Agricultural Research.* 13:258-270.

chapter thirteen

Cropping Programs and Soil Productivity

THE SITUATION

Following World War II, agriculture became highly specialized, partly by choice and partly by pressure. Crop producers chose specialization because it freed them from the rigorous routines of mixed farming and allowed them to concentrate their managerial ability. Except for a concentrated period at seeding and harvest, specialization offered a freer life style; monocultures allowed the choice of growing only the highest yielding crop that could be produced and marketed in their area; mechanization reduced the necessity for diversification to spread the workload; genetic resistance, precise maturity ratings, and a selection of effective pesticides gave crop producers a feeling of security against crop failure; and high levels of inexpensive commercial fertilizers were believed to be a suitable substitute for manure or legume crops in a **rotation.**

Crop producers were pressured into specialization by a shortage of qualified farm labor, which had to be replaced by very expensive machinery; and to keep machinery costs down, specialization was essential. In turn, expensive machinery was associated with large farms that benefited from economies of scale. Production became subject to a specialized market-orientation involving horizontal and vertical integration, enlargement of scales, and specialty products. Every attempt was made in the system to dominate or even eliminate the constraints of nature. Many farmers sold out, and their land was incorporated into larger units with fewer and fewer farmers, so that agriculture became a "minority occupation" with the majority of agricultural production coming from fully integrated farm enterprises. Farming became a sector of "agribusiness," which manipulated land, water, crops, and animals as production factors. Commuter trains and affluence encouraged the urbanization of rural agriculture, and country estates helped force up land values. This forced many farmers into a renting system and short-term land leases, which discouraged long-term rotation plans. High land values and rentals along with high machinery costs and high operating expenses, forced farmers to put short-term profits ahead of conservation practices, not because of greed but because of the growing indebtedness for sharply rising costs. Consumers, faced with inflation and accustomed to inexpensive food, pressured governments for a continuation of a cheap food policy to help reduce inflation and to be assured that everyone in the nation could afford to buy his daily bread.

For many years, specialization continued to produce ever larger harvests (Chapter Three), the system flourished, and all of society benefited (Barrons 1971). In England, Stanhill (1976) reported wheat yields had increased more than sixfold during the last 750 years, from 0.5 t/ha (0.23 tons/ac) in the first half of the thirteenth century to 3.43 (1.54 tons/acre) t/ha for the period of 1946 to 1973, with 1973 to 1976 yields at nearly 4.5 t/ha (2.03 tons/ac) per year. Since the end of World War II, the national wheat yield in England increased 0.079 t/ha/year (0.036 tons/ac/yr), twice as fast as any previous period. The yield increases are associated with the extensive use of nitrogenous fertilizers; new lodge-resistant cultivars; mechanized cultivation and harvesting; and chemical weed control, all of which can be optimally timed.

In the United States, under a system of taking land out of production to reduce surpluses, crop output in the United States increased. From 1949 to 1969, 23.5 million hectares (58 million acres) or 15% of U.S. farmland was annually withheld from production, but total crop output during the period increased 50% (Pimentel et al. 1976).

Specialization flourished, but at a cost. The cost included a reduction in the level of soil **organic matter** (Rennie 1977), which led to soil compaction and, in turn, to reduced water infiltration and increased soil **erosion** losses (Rennie 1979). Large fields with fence rows removed allowed soil erosion to occur uninhibited over large areas. The soil filled in watershed systems, and **eutrophication** occurred when eroding soil carried fertilizer elements into streams and lakes. **Monoculture** production favored specific pest species build-up, more pesticides were needed, and resistant strains developed. This led to strong criticisms and group protests, and the concern and indignation of the public was aroused by publications such as *The Rape of the Earth* by Jacks and Whyte (1956) and *Silent Spring* by Rachel Carson (1962). Critics allege that present-day agriculture contributes to the deterioration of the environment. Indeed, a study of the runoff from eleven watersheds in Ontario showed that 35 to 40% of the phosphorus reaching Lake Erie had agricultural origins (animal and soil), with about 70% of that coming from cropland (Robinson 1978; Robinson and Miller 1978), mainly from fine-textured soils planted to row crops such as corn and soybeans.

An accurate assessment of the situation was difficult, for in all of history a delicate balance has always existed between agriculture and the natural environment; cropland has always been more vulnerable to erosion, and some deterioration of the environment was considered as unavoidable. But a stronger negative impact of agriculture on nature does appear to have developed under intensive agricultural systems. The adverse impact of environmental deterioration no longer was of purely local concern with the loss of soil and crop productivity on a single field, but it now affected all people in society (Nair 1969). The total impact of intense specialization was starting to be recognized for its effect on environmental problems on a mass scale (Alexander 1974). The need for change was becoming increasingly evident, and this need was accelerated by the fossil fuel crisis in 1973, which caused farm costs (for nitrogen, fertilizer, fuels, machinery) to dramatically rise and raised the threat of shortages. Jansen (1974) described many countries of Western Europe, caught up in a modern market economy, as contributing to the deterioration of the natural environment in their efforts to keep going in the "race without a finish." Following the 1973–1974 crisis, high costs and high prices led to rapid expansion of certain crops and even to excessive cultivation of marginal lands without a long-term view or knowledge of the impact of intensification on such soils.

The need for technical change was evident. A social change was needed as well. As early as 1955, Bromfield, an outspoken writer and agricultural critic, deplored the dehumanizing impact of specialization on the individual and wrote: "The greatest creative and intellectual vice of our times and a factor which causes increasing distress and even tragedy, is overspecialization" (Bromfield 1955, p. 12). He reasoned that over time, the superspecialist loses an understanding and sense of the universe, or even of the small world that surrounds him. Jansen (1974) observed how different interest groups in society viewed and defined production agriculture. The large producer considers agriculture to be a profession and a source of income; small farmers, who own their operation, consider it a way of life and the basis of their subsistence; agriculture is viewed by industry as a furnisher of raw materials and a market for its products; bankers look at agriculture for investment possibilities; environmentalists consider agriculture from the viewpoint of stewards of the land and water resources and as caretakers of the natural environment; many urbanites see rural areas as a place for recreation and relaxation; wildlife people view rural areas as sanctuaries for game and as a place to hunt; and others see a communal, self-sufficiency, a return to the land as a new life style. In a European workshop seminar, agriculture was seen as a place of encounter and reconciliation between man and nature (Jansen 1974), with the precise statement formulated as follows: "Agriculture should facilitate man's progression towards the stabilized mental state necessary for his happiness and the joy which contact with nature and life can bring" (p. 78).

AN INTEGRATED APPROACH

Farmers generally agree that modifications in their cropping programs are needed, but their dilemma is the lack of a precise, profitable, and sound system to follow, based on long-term and proven results. No "hard" scientific results are available that incorporate current problems and situations, or that take an integrated approach to the problem. The risk is that suggestions may reflect a critical, biased, emotional, or polarized view, rather than a sound and proven course of action. If erosion control and/or maintenance of a soil with high fertility and a high rate of water absorption and water-holding capacity are the objectives, then reforestation or establishment of a grass and legume crop may well be recommended. However, a farming system must produce food and attempt an economic profit. Some areas may be best reforested or seeded to forages, but it may be equally true for other areas that a monoculture of corn or a simple soybean-corn rotation is best. The precise system for each farm enterprise must be developed in accordance with clearly defined goals compatible with the goals of other segments of society. The system that is best will evolve from observation, experimentation, research, and perhaps some sacrifice, and will be worked out by a team approach. Research directed toward the application of science to one component or subcomponent (rotations, for example) will be of value only if integrated into the entire system. It was with this view that Harper (1974) became editor of a new journal entitled *Agro-Ecosystems* devoted to an integrated approach.

INTEGRATION THROUGH AGRO-ECOSYSTEMS

It is a fact that each component within an agricultural system may affect all other components in varying, obscure, and unpredictable ways. Farmers know that crop rotation, pest and weed control, animal stocking rates, soil underdrainage, manuring, soil fertilization, the return or removal of residues, and frequency and depth of tillage all influence soil productivity, erosion, compaction, and the profit-loss sheet. In Canada, Shaw, Lavkulich, and Kitts (1977) described the principal characteristics of **ecosystems** and how farming transforms the natural ecosystem into an **agro-ecosystem.** The point that Shaw and his coworkers stressed is the identification and integration of the many factors of an agro-ecosystem. Included are the returns of all crop residues grown in sequence, the inputs over a time period, the impact on favorable microorganisms, the chemical environment (pH, fertility levels, pesticide residues), the physical environment (**clay,** sand, soil structure, compaction, water-holding capacity, soil temperature), the long-term effects of these factors, economic considerations, and others. Computerization may be needed to integrate all the factors, but it is only when this is done and all the interactions are identified and understood that an agro-ecosystem approach develops.

To understand an agro-ecosystem, it is best to begin by examining a **natural ecosystem.**

The Natural Ecosystem

When the first European settlers moved into North America, except for a minimum of cropped land by the native Indians, they found a natural ecosystem. Large areas were either forested or in grasslands, as determined by environmental factors such as rainfall and temperature. The principles on which a natural ecosystem operates are as follows:

1. Solar energy is the force driving the system. As tree leaves or grasslands photosynthesize, energy is captured and stored to eventually be returned to the soil in the immediate area. Virtually nothing is removed, even by wild animals, and the system is self-sustaining; and although the efficiency of sunlight energy conversion is considerably below the 1% conversion level often found in agriculture, the amount of food produced determines the wildlife carrying capacity. As human populations expanded, the capacity of a natural ecosystem fell short of their food needs, and controlled food production (agriculture) was adopted in an attempt to increase food output by increasing the level of conversion of solar energy.

2. Nutrients are recycled through the system by microorganisms decomposing the organic matter produced by photosynthesis. Environmental fluctuations may cause wide variations in the number of organisms in the system, but always the system is resilient because time is of minor consequence.

3. Generally, natural ecosystems are highly diverse, containing large numbers of different plant and animal species, each having a specific role or function in the system. Although some highly developed natural ecosystems are dominated by one or a few species, they are still characterized by resilience. Diversity and resilience provide a high level of stability, inhibiting the spread of a particular disease, and even benefiting from change.

The Agro-Ecosystem

Agriculture imposed major changes on the natural ecosystem, but the principles on which the new agro-ecosystem operates are the same. Management of the agro-ecosystem must guard against deviations that are so severe as to threaten the collapse of the system. Consider:

1. Agro-ecosystems are not self-sustaining. The products of photosynthesis are harvested and are removed from the system. Productivity cannot be sustained unless energy is injected into the system. Energy inputs include the return of organic matter, fertilizer nutrients, rhizobia bacteria, machinery inputs, and others. Early agriculturists did not appreciate the need to return energy forms, and initially production was high until the energy reserves from the natural ecosystems were depleted. Either new land was developed, energy inputs made, or society collapsed. Numerous examples of the rise and fall of civilizations can be found in history.

2. Agro-ecosystems lack diversity and therefore resilience. According to

Harlan (1976), agriculture has selected only seven major food crops in the world, and these in order of total world output can be listed as follows:

- Wheat
- Rice
- Corn
- Potato
- Barley
- Sweet Potato
- Cassava

Successful cultivars of each of these crops may be grown to the exclusion of others, and should a new virulent strain of a pathogen develop, yields in the entire region may be reduced. As new strains of the pathogen develop, breeding programs attempt to incorporate new sources of genetic resistance, with the threat that natural genetic resources existing in plants will be quickly depleted (Lawrence 1975). Development of a multiline crop may offer a greater degree of resilience in an agro-ecosystem.

3. In agro-ecosystems, plant and animal populations are not self-regulating. The corn or wheat plant would not reproduce itself, let alone maintain current levels of production, if left up to its own devices. Animal stocking rates are determined by the manager, and overstocking can result in reduced pasture productivity or reduced winter hardiness if the land is grazed during the critical fall harvest dates. Management in an agro-ecosystem is an essential input.

4. In contrast to the natural ecosystem, where time offered a major advantage, in an agro-ecosystem, production practices are often influenced by the marketplace, quotas, or other legislation or restrictions that usually influence the short-term rather than the long-term productivity. Sound practices that make good sense in the long term must be encouraged, but these often are not or cannot be considered because they are not economically sound in the short term.

APPROACHES TO AN AGRO-ECOSYSTEM

Three approaches to an agro-ecosystem are considered, but without a conclusion or recommendation because they all either lack wide application or are in the formative stages and need to be tried and tested.

1. MacKinnon (1975) adopted an agro-ecosystem approach; he emphasized the long-term implications and treated a dairy farm in terms of

regulated energy flows rather than as a financial system with regulated money flows. He used a computerized program, and energy was the criterion used to assess the interaction of inputs including machinery, buildings, and plant and animal components, with the objectives of maximizing energy output as food per unit area, minimizing food output fluctuations under environmental variations, and optimizing energy-material transformation. Constraints included limitation of energy flows from an ecological and cost standpoint. MacKinnon's work is an excellent guide to the further development of a sound agricultural ecosystem.

2. **Organic farming,** once associated with romantic dreams of a "return to the land," or considered applicable only to small gardens, has been advocated as a possible solution to soil depletion for large-scale farmers (Tucker 1979). Organic farming methods advocate (1) the return of all organic matter byproducts to the land, including garbage and human waste, (2) the reduction of soil chemical changes by the excessive use of fertilizers and pesticides, and (3) the use of rotations. The scientific principles of organic farmers have not been clearly and totally spelled out and are still being fiercely debated. Aldrich (1977) suggested that 30% more land would be required to maintain 1977 output levels with organic farming methods. Tucker (1979), in a convincing and moderate article, wrote:

Organic farming is one aspect of the vast concerns about ecology over the last ten years that may be about to bear fruit. Many university scientists are now admitting that some aspects of organic agriculture are workable. [P. 39]

Perhaps organic agriculture is really an ecological approach, another version of an agro-ecosystem. The establishment of resource centers such as the Ecological Agriculture Project on the Macdonald Campus of McGill University is an indication of the increasing interest in this area, and represents a recognition of the situation and a willingness to take a fresh look at how the food system operates (Hill 1979). Hill suggests that the problem with the agro-ecosystem approach is that it promotes the examination of all interrelated factors, with the result that it delays practical solutions, and those that are proposed tend to be too complex to be adopted. Such has been the case, but perhaps the problem is related to the lack of research data to feed into the system.

The organic position is one of feeding the soil rather than feeding the plant, as has been the common custom, and it argues that if the soil is kept in good condition, plant response will be excellent. The soil is regarded as more than just a medium to physically support the plant while the farmer supplies it with nutrients; the soil is recognized as a living, vibrant entity. Organic farming methods may cause a yield reduction and require more labor, but Tucker (1979) suggests that there is

little sense in continuing our current large-scale overproduction when it may mean that our soils will be less able to produce crops at some future point. If American agriculture were made more efficient—by lowering energy input and reducing surplus problems—it would free up resources that would make the rest of the economy more efficient. [P. 49]

Recognition of organic materials as a resource rather than a waste could reduce environmental problems.

Intelligent consideration of an agro-ecosystem may find common ground between the hard-core environmentalists who want to ban chemical use for agriculture and those who insist on its copious use. Organic farming should also not be associated with a vegetarian life style because livestock are an integral and sound part of a successful agro-ecosystem. In a rapidly changing world organic farming may be "legitimate" (Carter 1980).

3. A third approach was the development in Europe of a research project entitled "Farming in Society Towards the Year 2000." Workshops were held throughout Western Europe, and the first inventory of tentative goals for the future development and role of agriculture in society was made (see Jansen 1974). The long-term objectives for agricultural development were considered as an indispensable guide to an integrated approach.

SOIL PRODUCTIVITY PROBLEMS—A COMPREHENSIVE VIEW

Soil Degradation

Clearing the forest or breaking virgin sod exposes the soil to sunlight, increases aeration, and weathering is speeded up. Organic matter is reduced by the action of microbial populations, which use the organic matter as an energy source. Under a cropping program, the harvesting (and removal) of organic material means that commercial fertilizers must be added to maintain fertility. In Japan and China, large human populations have been fed by farming land that has been under continuous cultivation for 4000 years. It was the observation of oriental agricultural methods that led to the formation of the organic agricultural movement (Tucker 1979). In Canada, many fields have been cropped only 100 times, but already soil problems are sufficiently apparent to raise concerns about productivity in the next 100 years. Less than one-half of the original organic matter in prairie soils is believed to remain (Shaw, Lavkulich, and Kitts 1977; Rennie 1977). Organic matter is the major natural source of the elements needed for plant growth, but perhaps these can be replaced by commercial fertilizers. Organic matter losses raise concern about soil structure because as organic matter levels decline, compaction increases and water infiltration decreases, resulting in water runoff and erosion. Reduced organic matter levels reduce microorganism activity, and the overall impact is reduced yields.

A common practice on the Great Plains is a system of leaving the land idle for one season to allow for water accumulation. The fields are kept bare by repeated cultivation to prevent transpiration from weeds, a practice that Rennie (1977) regarded as excessive and the key factor accounting for the 50% organic matter reduction. Summer fallowing wastes soil nitrogen by leaching, causes deterioration in surface soil structure through weathering, is a highly inefficient use of available water, and leads to **soil salinity.**

Changes in soil quality in the prairie region of North America, perhaps, have been more profound than in any other part of the continent. The monoculture wheat-fallow system in Canada has directly resulted in extensive saline seepage into prime agricultural land (Rennie 1979), reducing crop yields by an estimated average of 50% over about a million hectares (2½ million acres). The problem is also of concern in the United States (Anonymous 1974) and is considered the current number-one soil problem throughout the northern Great Plains region. In Saskatchewan, it is estimated that 10% of the soils under cultivation in the late seventies have been adversely affected by higher than desirable amounts of soluble salts. The action of water is shown in Figure 13-1.

Soil erosion is a growing problem. In a survey of five watersheds in sloping land areas in south-central Wisconsin, where corn is the dominant annual crop, Brink, Densmore, and Hill, (1977) found soil losses of up to 19.3 metric tons per hectare per year (8.7 tons/acre/year) (Table 13-1). In 70% of the 93 quarter-sections sampled, estimated soil losses, on the average, were more than twice the amounts considered compatible with permanent agriculture.

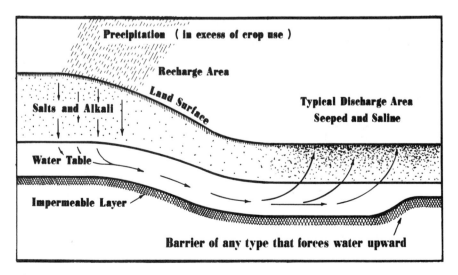

Figure 13-1.   Illustration of the formation of a saline seep. (Courtesy American Society of Agronomy, Crops and Soils Magazine.)

According to expert opinion, 7 tonnes per hectare (3.12 tons/ac) per year is the maximum rate for the maintenance of crop yields (Ketcheson and Moore 1978). The authors suggest that from isolated findings of others, the results shown in Table 13-1 are meaningful for a large area in the United States where row cropping is prevalent on sloping land.

Figure 13-2 dramatically illustrates the decline in soil productivity from erosion and soil mismanagement.

Pimentel et al. (1976) estimated that in the United States during the previous 200 years, 81 million hectares (200 million acres) were lost to crop production by soil erosion or have been so severely eroded that the land is now only marginally suited for crop production. This amounts to 42% of what the United States cropped in 1976. The levels or erosion cited, 22.4 t/ha/yr (10 tons/ac/yr) for continuous corn or cotton culture, is of the order of several hundred to several thousandfold that of natural ecosystems, regardless of slope.

To offset oil imports, the United States exports agricultural commodities or produces **gasohol,** which caused Van Bavel (1977) to question the logic of

Figure 13-2. Soil erosion has become a major agricultural problem. Low levels of organic matter resulting from monoculture production is a primary cause. The unharvested corn crop in this field represents the decline in productivity related to poor crop and soil management. (Courtesy Ontario Ministry of Agriculture and Food.)

TABLE 13–1. Distribution of cropland and estimated soil losses from five watersheds in Wisconsin in 1975

Watershed Area	Total Farmland Cropped	Corn	Hay and Rotation Pasture	Oats	Other Crops	Average (tons/ac/yr)	(Highest Quarter Section)	(tons/ac/yr)
	Cropland (%)					Estimated Soil Losses (t/ha/yr)		
1	80	76	12	8	1	13.3	6.0	30.8	13.9
2	66	70	12	4	4	19.3	17.2	55.8	25.1
3	76	61	27	5	4	16.9	15.1	46.3	20.8
4	79	50	35	12	2	14.6	13.0	27.0	12.2
5	57	34	57	7	0	11.7	10.4	36.0	16.2

Source: Data from Brink, Densmore, and Hill 1977.

intensive agricultural production. Since 1971, when it was estimated that 67% of all cropland in the north central United States needed conservation treatment, highly erosive and sloping soils have been placed in production, resulting in a soil loss of 20 tonnes h/yr (9 tons/ac/yr).

Potential sheet-erosion losses from thirteen agricultural watersheds were investigated to assess the effects of agricultural land use on soil erosion levels (Van Vliet, Wall, and Dickinson 1976). The sizes of agricultural watersheds, ranging in size from 19 to 54 km² (48.7 to 138 sq mi), were representative of the predominant livestock and cropping management systems of southern Ontario. Row crops yielded the highest annual sheet-erosion losses (Table 13-2); and beans and continuous corn have a potential to cause more than twice as much soil loss as crops in rotation.

Results of this investigation show erosion losses considerably below those reported in the United States (Van Bavel 1977; Pimentel et al. 1976; Brink, Densmore, and Hill, 1977). Shrader (1974) reported sheet-erosion losses from two Iowa watersheds, averaged 24.6 to 33.6 t/ha/year over a seven-year period (11 to 15 tons/ac/yr). However, Ontario results are seasonal and ignore the effects of off-season freezing, thawing effects, or snow melt; so the Ontario figures may be an underestimation of the true soil erosion losses.

Erosion levels for woodlands and permanent pasture are useful in helping determine if soil erosion is increasing. Replacing forage pastures and small grains with row crops can result in at least a doubling or tripling of soil losses. Also larger fields can result in soil losses four to ten times the soil losses on smaller fields (Dickinson 1979).

A comprehensive soil erosion control program was initiated in the United States in the 1930s and is regarded as one of the most important conservation developments in U.S. history. Carter (1977) has reviewed reports that suggest that soil erosion problems on a third of all U.S. cropland are too great to be sustained without an ultimately disastrous decline in crop productivity, despite the conservation service and large expenditures.

In the tropics, **laterite** soils are commonly found, and because they form a bricklike structure when exposed to oxygen, when the protective leaf canopy is removed and when organic matter declines, they are regarded as disastrous

TABLE 13–2. Magnitude of potential sheet-erosion losses from agricultural cropping systems in southern Ontario during the 1974 growing season

Crop	Mean Soil Erosion Loss (t/ha/yr)	(tons/ac/yr)	Range of Average Soil Erosion Losses Between Watersheds (t/ha/yr)	(tons/ac/yr)	Number of Watersheds in Which Crops Occur
Horticultural crops (potatoes, tomatoes)	9.1	4.1	6.6 –12.2	3.0 –5.5	4
Beans (soybean and white)	7.6	3.4	5.5 – 9.8	2.5 –4.4	6
Continuous corn	6.7	3.0	2.9 –11.7	1.3 –5.3	9
Corn in rotation	3.7	1.7	0.9 – 6.9	0.4 –3.1	13
Tobacco	3.5	1.6	2.1 – 4.9	0.9 –2.2	3
Small grains	3.4	1.5	1.5 – 6.9	0.6 –3.1	13
Meadow in rotation	2.6	1.2	0.9 – 5.0	0.4 –2.3	11
Permanent pasture	0.4	0.2	0.1 – 0.8	0.05–0.4	13
Woodlands	0.2	0.1	0.05– 0.4	0.02–0.2	13

Source: Data from Van Vliet, Wall, and Dickinson 1976.

handicaps to agriculture (McNeil 1964). The process occurs when lateritic soils are dried out, and it is irreversible. Much controversy among geologists and soil scientists exists, and a suitable management system has not yet been resolved.

Lateritic soils are capable of producing lush vegetative growth, which provides a protective cover and prevents brick formation (laterization). Once the protective vegetation is cleared, such areas can develop into a near desert even under high rainfall conditions. In the Amazon basin, areas have been cleared from the forest in the belief that the underlying soil was fertile and productive, but when cleared, organic matter levels quickly declined, and under equatorial conditions the iron-rich soil baked into brick, and in less than five years the cleared fields became pavements of rock.

In West Africa, the small country of Dahomey was transformed into agriculture on a wholesale scale, and within 60 years large areas of lateritic soils were converted into brick-like soil structure.

In history, ancient civilizations cleared an area in the forest and when the soil was exhausted, cleared a new area. Many civilizations flourished and then perished primarily because of lateritic soil deterioration.

Laterization offers a challenge to workers in the tropics and may hold a key to the management of all soils low in organic matter.

Pest Problems with Crop Specialization

Cultivation of one or a few crops species on the same land for several years or decades shifts the population balance of soil microorganisms, plant pathogens, insect pests, and weeds to those strains favored by the crops grown. For example, a strain of the weed lamb's-quarters was found to be resistant to a herbicide—triazine (Bandeen and McLaren 1976; Bandeen and Souza Mach-ado 1978). Intensive agriculture that promotes crop growth also encourages pest development, thereby developing an imperative need for **pesticides** to prevent pests from threatening crop productivity.

For insect pest species with long life cycles or for insects with a restricted migratory capacity, a monoculture system provides an optimum food supply that frequently allows a previously minor pest to become a major problem. The availability of a cheap and effective **insecticide** in the 1950s appeared to make rotations obsolete as a means of controlling such pests. However, insect resistance to insecticides and increased costs of such insecticides are causing renewed interest in rotations. For example, northern corn rootworm is an insect problem that can develop to severe proportions under continuous corn production (Barnes 1980). Since the corn rootworm feeds only on corn roots, a crop rotation is the easiest and most highly recommended method of control. One season of forage production would be sufficient to reduce the rootworm population to where corn production would not be affected the following year.

According to U.S. survey data collected from 1942 to 1978, insect losses increased nearly twofold, from 7 to 13% inspite of a tenfold increase in insecticide use (Pimental et al. 1978). Weed losses declined from 13.8 to 9.0% and losses from plant pathogens, including nematodes, increased slightly from 10.5 to 12.0%. The reasons for an increase in insect problems were listed by Pimentel et al. (1978) as follows: (1) the use of insect pest-susceptible crop varieties, (2) destruction of natural enemies, (3) reduced crop rotations, (4) development of pesticide-resistant strains, (5) reduced crop diversity, (6) monoculture production, (7) increased "cosmetic" standards by processors and retailers and consumers, (8) reduced tillage, which left unsanitary residues, and (9) the spread of crops outside their zone of adaptation.

The use of pesticides has caused more unfavorable public reaction, even hysteria, than any other agricultural practice. However, Borlaug (1972) warned that crop losses could reach 50% of current production, and food prices would increase four to five times if pesticides were banned.

Although pesticides are a relatively new addition to agriculture, three stages can be identified: exploratory, exploitive, and reflective (McEwen 1978). Agriculture in relation to insecticide use is now in the reflective stage whereby the concept of pesticides as the dominant input in pest control is not valid. A new rationale is being developed that recognizes pesticides as indispensable tools in pest control, but which must be integrated into the system. Such an approach could become a part of an agro-ecosystem.

CONCLUSIONS

The farmers of the world will decide ultimately on how they are going to farm, but unhappily their decisions may continue to be based primarily on short-run economic factors, and the overriding question will be the return on investment and labor. Perhaps change should be directed, therefore, through the economists and through the consumer.

QUESTIONS

1. Consider an agro-ecosystem approach by identifying all the factors to be considered and how they might interact for a selected farming operation, soil type, and topography.
2. What are the objectives of an agricultural society? Who should decide the objectives of the agricultural society? Are present-day agricultural objectives compatible with long-term crop productivity?
3. How many crops make up 75% of the production of your area? How many unrelated cultivars are there available and grown of the major crops? To what extent are monocultures produced?
4. What percentage of the farms in your area have livestock and market the majority of their produce through livestock and return the manure to the land or include a forage legume in the rotation?
5. Cite evidence of environmental changes on your farm or region that have occurred over the past 25 years. To what do you attribute these changes?
6. What caused the Irish potato famine? Trace the events leading up to the failure. Could the famine have been avoided?
7. What action has been taken in your area to combat soil erosion? Is the action adequate? What further action would you suggest?
8. Trace the use of insecticides through the exploratory, exploitive, and reflective stages (see McEwen 1978).
9. Trace the history and extent of summer fallowing. Comment on the practice in dry areas of the prairies and in higher rainfall areas.
10. Comment on the "cosmetic" standards demanded by processors and consumers. Would a lessening of standards be acceptable? Would a reduction in standards have a favorable or unfavorable input on the environment? Is 100% pest control possible or desirable?

REFERENCES FOR CHAPTER THIRTEEN

Aldrich, S. R. 1977. "Organic Farming Can't Feed the World," *Farm Chemicals* 141:17,20,22.

Alexander, M. 1974. "Environmental Consequences of Rapidly Rising Food Output," Agro-Ecosystems 1:249–264.

Anonymous 1974. "Saline Seeps. Their Cause and Prevention," *Crop and Soils* 26:12–13.

Bandeen, J. D., and R. D. McLaren. 1976. "Resistance to *Chenopodium Album* L. to Triazine," *Canadian Journal Plant Science* 56:411–412.

Bandeen, J. D., and V. Souza Machado. 1978. "Weeds Resist Triazine," *Highlights of Agricultural Research in Ontario* 1(2):1-2.

Barnes, G. 1980. "Insect Control: Crop Rotation vs. Monoculture," *Crops and Soils* 32(4):15-17.

Barrons, K. C. 1971. "Environmental Benefits of Intensive Crop Production," *Agricultural Science Review* 9(2):33-39.

Borlaug, N. E. 1972. "Mankind and Civilization at Another Crossroads in Balance with Nature—A Biological Myth," *BioScience* 1:41-43.

Brink, R. A., J. W. Densmore, and G. A. Hill. 1977. "Soil Deterioration and the Growing World Demand for Food," *Science* 197 (4304):625-630.

Bromfield, L. 1955. *From My Experience*. New York: Harper & Row, Publishers, Inc. (355 pp.)

Carson, R. 1962. *Silent Spring* New York: Fawcett World Library. 368 pp.

Carter, L. J. 1977. "Soil Erosion: The Problem Persists Despite the Billions Spent on It," *Science* 196:409-411.

Carter, L. J. 1980. "Organic Farming Becomes "Legitimate," *Science* 209 (4453): 254-256.

Dickinson, W. T. 1979. "Soil Erosion: Is It Getting Worse?" *Crops and Soils* 31(6):24.

Harlan, J. L. 1976. "The Plants and Animals That Nourish Man," *Scientific American* 235(3):88-97.

Harper, J. L. 1974. "Agricultural Ecosystems," *Agro-Ecosystems* 1:1-6.

Hill, S. B. 1979. "Eco-Agriculture. The Way Ahead?" *Agrologist* 8(4):8-11.

Jacks, G. V., and R. O. Whyte. 1956. *The Rape of the Earth: A World Survey of Soil Erosion* London: Faber & Faber Ltd. (312 pp.)

Jansen, A. J. 1974. "Agro-Ecosystems in Future Society," *Agro-Ecosystems* 1:69-80.

Ketcheson, J. W., and N. Moore. 1978. "Management for Water and Wind Erosion Control," *Notes on Agriculture* 14(3):12-14. Published by the University of Guelph.

Lawrence, E. 1975. "Future in the Past," *Nature* 258:278-279.

MacKinnon, J. C. 1975. "Design and Management of Farms as Agricultural Ecosystems," *Agro-Ecosystems* 2:277-291.

McEwen, F. L. 1978. "Food Production—The Challenge for Pesticides," *BioScience* 28(12):773-777.

McNeil, M. 1964. "Lateritic Soils," *Scientific American* 211(5):97-102.

Nair, K. 1969. *The Lonely Furrow: Farming in the United States, Japan and India* University of Michigan Press, Ann Arbor, Michigan. (314 pp.)

Pimentel, D., et al. 1976. "Land Degradation: Effects on Food and Energy Resources," *Science* 194 (4261):149-155.

Pimentel, D., et al. 1978. "Benefits and Costs of Pesticide Use in U.S. Food Production," *BioScience* 28(12):772, 778-784.

Rennie, D. A. February 1977. "Research Data Documents Drop in Soil Productivity: Who Will Pay the Price?" Mimeograph paper presented to Arable Land Seminar, Centre for Continuing Education, University of British Columbia.

———. 1979. "Intensive Cultivation: The Long-Term Effects," *Agrologist* 8(4):20-22.

Robinson, J. B. 1978. "PLUARG Findings: Implications for Agriculture." Department of Environmental Biology OAC. Mimeograph paper presented to the Ontario Soil and Crop Improvement Association, Oshawa, Jan. 31, 1979.

Robinson, J. B., and M. H. Miller. 1978. "Erosion and Water Quality," *Notes on Agriculture* 14(3):14-17. Published by the University of Guelph.

Shaw, M., L. M. Lavkulich, and W. D. Kitts. 1977. "Agro-Ecosystems," In *Proceed. Symp. on Canada and World Food* Royal Soc. Canada and Agricultural Institute, Canada. Ottawa, Aug. 22-27, 1977. pp. B4-1 to B4-18.

Shrader, W. D. 1974. "What Is Happening to Our Soil Resources: Losses to Erosion," in *Land Use: Persuasion or Regulation?* pp. 18-21. Proceedings 29th Annual Meeting Soil Conservation Society of America.

Stanhill, G. 1976. "Trends and Deviations in the Yield of the English Wheat Crop during the Last 750 Years," *Agro-Ecosystems* 3:1-10.

Tucker, W. 1979. "The Next American Dust Bowl—and How to Avert It," *Atlantic Monthly* 244 (1):38-49.

Van Bavel, C. H. M. 1977. "Soil and Oil," *Science* 197(4300):213.

Van Vliet, L. J. P., G. J. Wall, and W. T. Dickinson. 1976. "Effects of Agricultural Land Use on Potential Sheet Erosion Losses in Southern Ontario," *Canadian Journal of Soil Science* 56:443-451.

chapter fourteen

Cropping Practices in an Agro-Ecosystem

Soil productivity and soil erosion are determined by many factors including soil type and depth, angle and length of the slope of the land, cultivation practices, amount of organic matter present, and duration and intensity of winds and rainfall. These parameters vary from one agro-ecosystem to another.

Soil productivity can and must be maintained. Well-tested engineering and biological methods can be used to assure continued soil productivity and include contour plowing, terracing, alternating strips of a protective cover crop with crops that subject the soil to erosion, rotating crops, use of harvest residues as protective litter, renovating erosion-prone land into pastures, planting shelterbelts or windbreaks, and practicing reduced or conservation tillage. Minimum tillage methods used with a soil-holding legume cover crop and chemically suppressing the cover crop each spring to reduce competition is an example of how two or more of the above inputs may be combined (Elkins et al. 1979), but such approaches need to be fully evaluated. **Relay cropping,** or seeding a crop into another crop prior to harvest, such as seeding winter wheat into a standing crop of soybeans, is a method that can increase flexibility of cropping programs (Sanford, Myhre, and Merwine 1973).

Some seemingly sound practices may have little appeal in an intensive agriculture system because they may not result in maximum yields in any single year, or because they have not been adequately or widely tested. Considerable research has been directed toward row crops perhaps at the expense of others. The inclusion of forages in a rotation, especially legumes, is widely accepted as being desirable, but because research in the production, harvesting, utilization, and marketing of traditionally bulky forages has lagged, the handling, marketing, and market value have raised doubts about the practicability of including forage legumes in a cash crop rotation. Plowing under a legume crop is recognized as beneficial, but it raises the question of profitability in the short run. Careful economic and energy flow considerations will be required to determine the extent and frequency of including specific management inputs into a cropping program such as forage legumes in a cash crop enterprise.

Livestock can be a valuable input into an agro-ecosystem because of their ability to utilize cellulose—a plant byproduct—and inorganic nitrogen sources for the production of protein and valuable manure. Animal efficiency averages about 10%; i.e., on the average, animals convert about 10% of their food intake to growth, and this has resulted in a school of thought that livestock production cannot be tolerated in future agriculture. This is a highly unlikely prospect, certainly in areas where soil topography prevents intensive cropping and on **unstable soils.** When livestock are viewed as converters, gleaners, and harvesters of marginal land areas and of products and byproducts that might go to waste, they form a valuable asset in an agro-ecosystem. Rennie (1979) noted that soil quality and productivity "improved" under a forage crop-livestock system in contrast to a natural ecosystem. However, livestock are not essential to a successful agro-ecosystem. A study in the United Kingdom (Agricultural Advisory Council 1970) found that the disappearance of livestock from farming areas in England and the replacement of grasslands and farmyard manure by chemical fertilizers has not led to any loss of inherent fertility. But the reduction in organic matter levels on unstable soils (those with inherent high instability of soil structure, poor natural drainage, and low organic matter content) was expressed as a serious concern. Soils containing high amounts of fine sand or silt are considered as having an inherent high instability. The use of inorganic fertilizers does not compensate for losses in productive capacity of these soils due to their impaired physical condition caused by the loss of farmyard manure and organic matter.

SOIL ORGANIC MATTER

Three kinds of organic matter can be found in soils: (1) the visible root system of plants, (2) the partly decomposed remains of plants still retaining much of their original plant structure, and (3) the well-decomposed brown or black organic matter, commonly called **humus.**

A dense, well-developed visible root system is an indication of a well-drained aerated soil. The restorative value of sod is largely due to the development of a dense root system, which provides fresh organic matter to stabilize the structures the roots have created. Fresh organic matter in the soil is more beneficial in improving structural stability of soils than is the older more stable organic matter. The amount of partly decomposed organic matter is an indicator of the resources available for maintaining soil structure or for restoring it after it has been impaired. For these resources to be effective, soil conditions must avoid anaerobic conditions because under such conditions deterioration (or no improvement) of soil structure occurs.

Partly decomposed organic matter provides a means of holding nitrogen in soils. The nitrogen in the organic matter is consumed by microorganisms in the soil and is used for their own multiplication to decompose the organic material, thus making the nitrogen unavailable for crop growth. Fresh or partly decomposed organic matter contributes little to plant nutrition until decomposition is complete; and crops may show nitrogen-deficiency symptoms unless adequate supplemental nitrogen is provided. Favorable decomposition occurs in the presence of oxygen and suitable temperature conditions.

Nitrogen can be expected to be tied up by residues having a carbon to nitrogen ratio (C/N) of 33:1 (40% carbon, 1.2% nitrogen) (Bartholomew 1965). The carbon-nitrogen ratio of corn is almost always greater than 33:1, depending on the nitrogen fertility level under which the crop is grown. In one test (Ketcheson and Beauchamp 1978) the carbon-nitrogen ratio in the stover was reduced from a high of 88:1 where no fertilizer was applied to the growing crop, to a low of 34.5:1 when 168 kg N/ha was applied. At a C/N ratio of 88:1, a depression in corn yields in a subsequent corn crop without nitrogen fertilizer was measured at Guelph (Table 14-1). Where the stover was removed, yields were less than where manure or stover was returned. Organic matter levels were reduced when stover was removed, but nitrogen did not conserve organic matter.

TABLE 14-1.   Yield of corn grain (t/ha) and (tons/ac) with residue and nitrogen treatments, 1963–1966 average, at Guelph

| | Treatment | | | | | | | |
| | Stover Removed and Manure Returned | | | Stover Removed | | | Stover Returned | | |
	Yield (t/ha)	(tons/ac)	Percent OM*	Yield (t/ha)	(tons/ac)	Percent OM	Yield (t/ha)	(tons/ac)	Percent OM
No nitrogen	5.67	(2.56)	3.9	5.01	(2.25)	3.7	4.94	(2.22)	3.9
112 kg N/ha	5.43	(2.44)	3.7	5.17	(2.32)	3.4	5.41	(2.43)	3.7

Source: Data from Ketcheson and Beauchamp 1978.

*OM = Organic Matter.

In a twelve-year study, Barber (1979) found that highest yields were obtained at a 200 kg/ha (179 lb/ac) of supplemental nitrogen when all residues were returned. Nitrogen applications at both 67 and 200 kg/ha (60 and 179 lb/ac) increased organic matter levels (Table 14-2).

Humus is usually well-mixed with the soil and adds structural stability. However, it is less able than fresh or partly decomposed organic matter to continue to produce products of decomposition that help to stabilize the soil. The value of humus in producing structural stability and ease of cultivation becomes apparent when it declines to low levels and instability becomes a soil feature.

TABLE 14–2. Effect of nitrogen fertilization rate on corn grain yield and soil organic matter with all corn residues returned. Tests were conducted on the Purdue University farm, Lafayette, Indiana

Nitrogen Applied Annually		Corn Grain Yield Twelve-Year Average		Soil Organic Matter After Twelve Years
(kg/ha)	*(lb/ac)*	*(kg/ha)*	*(bu/ac)*	*(%)*
0	0	2450	39.1	2.70
67	60	5520	82.1	2.85
200	179	8840	131.5	2.90

Source: Data from Barber 1979.

The levels of organic matter at which structural stability and ease of cultivation are reduced are not the same for all soils. Soil texture has a marked influence. In soils containing high amounts of sand, very fine sand, or silt, instability of structure can develop when soil organic matter falls below 3% (Agricultural Advisory Council 1970). Under cropping systems in which the only returns of fresh organic matter are from the root systems of crops and the residues of their tops, levels of 2 to 3% organic matter are commonly found. Organic matter levels of less than 2% may be found in very light loamy sands.

Soil organic matter levels appear to equilibrate under Canadian conditions at approximately 75% in the 4 to 5% level typical of a cereal grain-hay rotation (Rennie 1979). In Indiana, Barber (1979) found that organic matter decreased to 90% of the original level after ten years; in Iowa Larson et al. (1972) found that organic matter decreased to 87% of the initial level after twelve years when all surface residues were removed. The decline in organic matter is slow, but it is also a slow process to increase it.

In Quebec, Ketcheson, Martel, and MacLean (1979) noted that ten years of continuous row cropping lowered soil carbon to 2.4% from an initial 3.7% developed under a previous rotational system associated with a dairy farm. On another soil with an organic carbon value near 2.4%, developed under a forage-based rotation, no further decline resulted from continuous row cropping.

Organic matter levels of 3% may be satisfactory for cereal production on stable soils, and this level can be maintained if all residues are returned. On unstable soils or under intensive cultivation, 3% organic matter levels are inadequate.

Rarely are organic matter levels in agricultural soils that are subjected to cultivation and fertilization higher than 7 to 8%. An organic matter level of 8.1% was reported (Bolton 1977) in the top 10 cm (4 in) of soil in 1951 following continuous bluegrass sod grown on Brookston clay. The level in the top 30 cm (12 in) averaged 4.7%. Levels higher than 5% are difficult and are expensive to achieve and maintain because of the tremendous volume of organic material involved. Consider that topsoil to a 17-cm (7 in) depth weighs about 2,260,000 kg/ha (2,017,860 lb/ac). If 5% of the weight of this soil was organic material, it would weigh about 113,000 kilograms (100,900 lb), eight to ten times the amount of roots and residue produced in an average corn crop.

Not all crop residues are converted to organic matter or humus. Organic matter levels in soil result from the balance between the rate of loss from decomposition of the soil organic matter and the rate of gain from formation of organic matter during decomposition. Cultivation and summer fallowing greatly influence the rate of loss from decomposition. The rate of conversion of crop residues into stable organic matter or humus was estimated by Larson et al. (1972) to be 18% and by Barber (1979) to be 8 to 11%. To raise the organic matter level reported in Table 14–2 from 2.70 to 5.0%, a 2.3% increase, and considering a 13% conversion rate (mean of 8 and 18%), would require the addition of 4522 t/ha (2035 tons/ac) of organic matter or humus. At a harvest index value of 50%, a 6200 kg/ha (100 bu/ac) grain corn yield would produce 6200 kg/ha (5600 lb/ac) of leaves and stocks plus roots. Warncke and Barber (1974) estimated that in the corn crop, roots accounted for 17% of the dry weight of the shoots, and total residues returned from continuous corn production could equal 7250 kg/ha (6,473 lb/ac) of residue. At this rate of return and at 13% conversion to organic matter, 624 years would be required to achieve a 5% organic matter level. Seventy-five years would be required to raise the organic matter level by 1% if manure was applied at 22.4 t/ha (10 tons/acre) each year (Anonymous 1977*b*).

The inevitable conclusion is that every effort should be made to guard the precious level of organic matter that exists. The purchase of feedstuffs and the return of manure can help increase farm fertility. Organic matter build-up may be more rapid when manure is applied to bluegrass sod if nothing is removed. Although desirable, this procedure is too expensive to be practical. In any event, a soil having an organic matter level of 5.0% represents a formation many, many years in the making.

A study in Iowa (Larson et al. 1972) with continuous corn produced on a silty clay loam over a twelve-year period indicated that 6 t/ha (2.7 tons/ac) of residue (dry weight basis) was required to maintain the organic matter level in the soil. Under any cropping system, organic matter levels reach a state of equilibrium, but whether or not a state of equilibrium was achieved in the above experiment would be determined only if the experiment was conducted for a very long time.

In addition to the need for crop residues as a source of organic matter and nutrients, or as a protection against wind and water erosion, they also are desired as a source of feed (Mowat 1976, 1979), for bedding, or as a biomass source for energy production. In these cases, residues may eventually be returned to the soil. If crop residues are used for industrial uses (Fenster, Follett, and Williamson 1978); if they are regarded as a nuisance and burned; or if they are otherwise permanently removed from the soil, serious implications exist for soil productivity. The use of crop residues should be determined not just for their immediate cash value but also for their long-range value in an agro-ecosystem.

In grain corn production, residues may be returned; in silage corn production, they are removed. A study was conducted in Indiana to determine the extent that residue removal would reduce soil organic matter content and decrease long-term productivity (Barber 1979). The soil was a silt loam, and one experiment had continuous corn production for eleven years—with residues removed, annual residues returned, and double residues returned. A continuous fallow treatment for six years, followed by five years of residues returned, was part of the study. No yield differences were measured due to treatment during the years of this study (Table 14-3), but soil samples after six years showed organic matter levels were 0.2% lower where residues were removed and 0.4% higher where double residues were applied over the situation with annual residues returned. After six years of fallow, soil organic matter was 0.4% lower. Samples taken in the eleventh year had 0.3% lower soil organic matter content where residues were removed, and 0.3% higher soil organic matter content where double residues were returned. In terms of the top 17 cm (7 in) of soil, a 0.3% change in organic matter would amount to 6780 kg/ha (6054 lb/ac), or an average annual change of 678 kg/ha (605 lb/ac).

At Ames, Iowa, a nineteen-year study with continuous corn production, with residues returned and removed, resulted in reduced soil organic matter levels. In 1957, the organic matter level was 4.5%, and after continuous removal of only the grain, the soil organic matter declined to 4.1% when grain only was removed and to 3.7% when all top growth was removed (Barnhart, Shrader, and Webb 1978).

When residues are removed, the crop root system must be relied upon to help maintain organic matter levels. After six years of fallow, with no residue return, the level of organic matter was 2.6% compared to 2.8% where a crop was grown and all above-ground residues removed (Barber 1979). Root residues accounted for a 0.2% increase.

TABLE 14–3.   Effect of residue management on corn yield (kg/ha) and soil organic matter (%) as shown by an eleven-year study in Indiana 1962 to 1972

Treatment	Average Corn Grain Yield				Soil Organic Matter	
	0–11 Years	(lb/ac)	6–11 Years	(lb/ac)	6th Year (%)	11th Year (%)
Residues removed	9020	8054	9420	8411	2.80	2.75
Residues returned	9180	8196	9670	8634	3.00	3.05
Double residues returned	8950	7991	9450	8438	3.40	3.36
Fallow for six years, then corn with residues returned for five years			10,035	8960	2.60	2.72

Source: Data from Barber 1979.

Organic matter levels in soil affect the flexibility of cultural practices, the impact of drought, the earliness of getting on the land, and the amount of CO_2 released. Low levels of organic matter in association with compacted soils may influence mechanical impedance of roots. Root growth in compacted low organic-matter level soils is not well understood, but it can be assumed that the extra energy required for root growth is not available for yield.

Farmers today may have to return to the former practice of plowing down organic matter to maintain crop productivity and reduce erosion (Fig. 14-1).

CROP ROTATIONS

Research results with rotations clearly and often dramatically demonstrate the value of rotational systems. Numerous research examples could be quoted to show the yield advantage of a selected crop in a rotational sequence as compared to continuous rotations. The work of Sheard (1977) is typical (Table 14-4). Highest corn yields were obtained immediately following a legume. Forages are unquestionably desirable in a rotation because of: (1) the nitrogen fixation by legumes; (2) the year-round protection offered by a perennial; (3) the favorable effect of grass roots on soil aggregation, reduced erosion, and water infiltration; and (4) the fact that forages effectively break pest cycles associated with continuous cropping. But on cash crop farms, forages are often regarded as a crop of yesterday's agriculture. What value can be placed on forages in a rotation? Could additional nitrogen be economically

Figure 14-1. The return of crop residues, once regarded as a nuisance in crop production, is an essential step to prevent erosion and soil compaction and to maintain soil productivity. (Courtesy Ontario Ministry of Agriculture and Food.)

applied to compensate for a lack of rotation? Experiments in Ontario showed a greater grain corn yield following a legume-grass crop than where corn was grown after cereal grain or corn under three fertility regimes (Table 14-5). On the loams and silt loams, additional nitrogen did not increase yields where the previous crop had been a legume-grass sod. On the finer textured clay loam and clay soils, additional fertilizer resulted in a yield increase, but at the levels applied was insufficient to reach the yield level obtained with corn following the legume-grass sod.

TABLE 14-4. The effect of crop sequence on the five-year average yield of grain corn on Haldimand clay (a heavy clay soil) in Ontario 1970-74

Crop Sequence	Average Yield of Corn	
	(t/ha)	(tons/ac)
Continuous corn	3.14	1.41
Corn-oats (sown to red clover, plowed in fall)-corn	3.95	1.78
Corn-oats-red clover, 2 years timothy hay-corn	3.39	1.52

Source. Data from Sheard 1977.

TABLE 14-5. Average yield of grain corn (t/ha) and (bu/ac) over a seven-year period obtained from experiments on private farms across southern Ontario 1968 to 1974

Soil and Cropping History	No Fertilizer		0-20-20* at 448 kg/ha	at 67 kg/ha (400 lb/ac)	10-10-10 (60 lb/ac)	
Loams, silt loams						
After legume-grass sod	4.33	69.6	5.71	91.7	5.71	91.7
After cereal grain or corn	3.89	62.5	5.21	83.7	5.52	88.7
Clay loam, clay						
After legume-grass sod	4.52	72.6	5.90	94.8	6.21	99.8
After cereal grain or corn	2.82	45.3	3.20	51.4	4.20	67.5

Source: Data from Sheard 1977.

*0-20-20 is a standard method of expressing fertilizer analysis—the figures refer to the percentage of nitrogen, phosphorus, and potassium, respectively.

Rohweder, Shrader, and Templeton (1977) found that corn grown in a rotation that included a legume sod always outyielded continuous corn, regardless of how much nitrogen was applied. First year corn following alfalfa and without any nitrogen fertilizer applied yielded as much as corn fertilized with 168 kg N/ha (150 lb N/ac). In the second year of corn after alfalfa, 84 kg N/ha (75 lb N/ha) was needed for the corn to equal the yield obtained with 168 kg N/ha (150 lb N/ac). By the third year of corn in the rotation, the amount of nitrogen required for equal yield was about the same as that for continuous corn. The total nitrogen equivalent produced by the full stand of alfalfa in these rotations was 252 kg/ha (225 lb N/ac).

What economic worth, energy value, or index rating can be placed on the reduction of soil erosion and water runoff? Soil losses from erosion in a corn field following two years of an alfalfa-brome hay mixture were only 3% of that lost from continuous corn. The loss of water was also substantially reduced by forages in a rotation (Table 14-6).

TABLE 14–6. The effect of crop sequence on soil erosion losses and water runoff from a Guelph loam soil in 1956

| | Losses | | | |
| | Soil | | Water | |
Crop Sequence	(t/ha)	(tons/ac)	(cm)	(in)
Continuous bluegrass	none		none	
Continuous corn	38.70	17.42	7.25	2.85
Hay-hay-corn-oats:				
first-year hay	0.79	0.36	1.25	0.49
first-year corn	1.23	0.55	1.25	0.49
first-year oats	32.30	14.54	6.50	2.56

Source: Data from Sheard 1977.

Inclusion of a forage crop in a rotation enhanced the ease of seedbed preparation and water penetration into the soil whereas continuous corn reduced organic matter and soil aggregation (Table 14-7) over that obtained where an alfalfa-brome forage mixture was included (Sheard 1977). Thus, two years of forages in the rotation will alleviate to a significant degree the deleterious effects of continuous corn on soil structure.

Rotations are of benefit in reducing the build-up of pests because life cycles are broken. During the seventies, continuous or frequent inclusion of barley in rotations resulted in the build-up of the fungus disease Helminthosporium root rot or blight, with accompanying yield reductions. Where a forage crop preceded barley in a rotation, the incidence of the disease was reduced (Table 14-8), and higher yields were produced.

TABLE 14–7. The effect of crop sequence and strip cropping on soil physical properties on a Guelph loam soil in 1961

Crop Sequence	Organic (%)		Aggregation— Percent of Soil Particles 0.25 mm in Diameter
Continuous bluegrass	4.87		72.6
Continuous corn	3.60		22.1
Corn in rotation		(first year	
(*corn*-oats-hay-hay)	4.67	of rotation)	48.1
Oats in rotation		(second year	
(corn-*oats*-hay-hay)	4.18	of rotation)	46.1

Source: Data from Sheard 1977.

TABLE 14–8.   Helminthosporium root rot infestation ratings on barley and yield (kg/ha) and (bu/ac) at two Ontario sites in 1975

County	Previous Crop	Disease Rating on July 10	Barley Yield	
			(kg/ha)	(bu/ac)
Wentworth	Legume sod	Moderate	2849	53.0
	Barley	High	2527	47.0
Oxford	Hay crop	Light	2150	40.0
	Barley	Very high	1935	36.0

Source: Data from Sheard 1977.

The advantages of rotations are clear from the results presented in the preceding tables; these results are not unexpected by producers. The problem is that rotations often do not fit well into an intensive and specialized cropping program. Hog and poultry producers need to maximize high energy feed grain requirements; cash crop producers wish to maximize output by maximizing the production of the highest yielding crop in their region; and all producers are confronted with additional machinery costs associated with rotations.

TILLAGE PLANTING AND CROPPING SYSTEMS

Soil structure and soil organic matter are influenced by the crops grown, the residue or manure returned, and the **tillage-planting system.** The objective of tillage operations is to prepare a fine, firm seedbed capable of rapidly warming under sunshine, and once achieved there is no advantage to additional tillage. The conventional answers to the objectives of tillage may include such reasons as the burial of trash, to kill weeds, loosen soil, reduce lump size, destroy hardpans, increase moisture infiltration, improve appearance, increase soil temperature, incorporate chemicals, and avoid condemnation by peers. Some of these reasons are sound; others are not.

In any consideration of the objectives of tillage, an optimum level is sought, and such considerations must distinguish between a root bed and a seedbed. A fine, firm seedbed assures contact between seed and soil and good capillary moisture movement from below, and reduces the risk of soil crusting. A seedbed may be a narrow strip (5 cm or 2 in wide) in which the seed is planted or may refer to that area cultivated by shallow secondary tillage.

The root bed is much less clearly defined, and unless a compacted layer or problem zone exists, root beds may need little attention. The root bed is a massive area, and when the volume and weight of soil moved are considered, the energy involved is colossal. Freezing and thawing action in the temperate zones, deep root penetration, and tile drainage under certain conditions may be all that is needed for a satisfactory root bed.

Tillage may be viewed as a necessary evil that destroys soil structure, reduces soil organic matter, brings previously buried weed seeds to the surface, and often leaves soil in a state subject to all forms of erosion. A wide choice of machinery and chemical weed control offers enormous scope in the extent of tillage operations, and the task is to find the combination to maximize yields, minimize costs and soil damage, and optimize energy flow.

Richey, Griffith, and Parsons (1977) described high, moderate, and low energy systems ranging from a thorough loosening of the soil 12 to 20 cm (4.7 to 8 in) deep, to minimum strip or slot tillage in conjunction with vegetation control by herbicides, to a "no-till" colter system. There is no one best system because of the large number of variables depending on the extent of tillage. Eight considerations of tillage-planting-cropping systems are reviewed here and include:

- Soil erosion prevention
- Speed and timeliness of planting
- Soil compaction and four causes:
  1. excessive tillage
  2. untimely field operations
  3. cropping programs
  4. farm implement design and operation
- Fertilizer placement
- Weed control
- Insect and disease control
- Energy input
- Soil moisture

Soil Erosion Prevention

Production systems that provide a protective cover of crop residues or leave a rough surface protect soil from erosion but may clog seeding equipment or delay spring warm-up. Fall plowing is traditionally preferred in northern areas to allow freezing and thawing action to break lumps, to prevent spring delays in seeding, and to facilitate secondary tillage. However, fall plowing subjects soils to excessive erosion losses (Table 14–9), as reviewed by Richey, Griffith, and Parsons (1977). Soil erosion losses by fall plowing and with no cover crop are shown in Fig. 14–2.

The use of no-tillage corn production systems minimizes soil moisture and wind erosion losses and has also opened the way for seeding into a cover crop killed by a herbicide. Rye has been used as a mulch cover crop, but Mitchell and Teel (1977) found heavy growth of rye was associated with irregular and low corn populations. Mulch covers are less reflective than the bare soil surface, resulting in mulch surface temperatures up to 10°C (18°F) higher than those with unprotected soil surface in Delaware. Temperatures immediately below the mulch covers, however, were more than 10°C (18°F) lower than

TABLE 14–9.   Total soil erosion loss (t/ha) and (tons/ac) comparisons with three fall tillage treatments in 1974 and 1975 on corn and soybean fields in Illinois with a 5% slope on silt loam

	Corn 1974		Soybeans 1975	
Treatment	(t/ha)	(tons/ac)	(t/ha)	(tons/ac)
Fall moldboard plow	25.3	11.4	54.5	24.5
Disk-chisel	1.93	0.87	15.4	6.9
No fall tillage	2.3	1.04	7.9	3.6

Source: Data from Richey, Griffith, and Parsons 1977.

those at the soil surface, which was a problem in delaying spring growth in northern areas, this was also reported by Mock and Erbach (1977). Mitchell and Teel (1977) studied the use of leguminous cover crops for no-tillage mulches to supply biological nitrogen using hairy vetch and crimson clover mixtures. The legume cover crops were seeded in August or September, and killed in springtime just prior to no-till corn seeding. Corn grain yields were comparable to those obtained by the application of 112 kg N/ha (100 lb N/ac). Approximately one-third of the total nitrogen from mulch covers was released to the corn in a single season.

Figure 14–2. The protective winter wheat cover on the upper part of the slope effectively prevented erosion, but without the protective cover on the lower section, erosion losses were severe. (Courtesy Ontario Ministry of Agriculture and Food.)

A common practice with no-till corn is to kill the existing vegetation so that it will not compete with the corn for nutrients and moisture. In southern Illinois, field studies were conducted over a three-year period of no-till planting of corn into grass sod sprayed with a growth-retardant at sublethal herbicide rates to suppress the sod but allow recovery for erosion control (Elkins et al. 1979). If the grass sod was not adequately suppressed, competition reduced corn yields substantially. In the best treatment plots, corn yields were reduced 9% due to competition from the suppressed grass plants. If a legume sod could be established and suppressed, nitrogen fixation may accompany the reduction in soil erosion associated with such a system. A yield advantage was reported (Anonymous 1977a) when no-till corn was seeded into a winter-annual legume cover crop at the University of Delaware.

Because of the desirability of rotations, farmers must examine every possibility for fitting crops into a sequence and devising methods of making rotations workable. A desirable cash crop rotation may be corn-soybeans-winter wheat, but in short-season areas or with full-season soybean crops in other areas, the crop may not be removed sufficiently early to allow for seedbed preparation and a timely seeding of winter wheat in a conventional manner. In Ontario, farm field seedings of soybeans into standing wheat have been successful if seeded between September 5 to 10 in the Guelph area. The seed is broadcast into the stand before the soybean leaves senesce and drop, and is timed to precede a rainfall so that the soil is moist. Weed control must be adequate in the soybean crop. This system is a method of **double cropping** and makes the desirable three-way rotation workable. Such a rotation is desirable because it includes three high return crops, provides a winter cover crop one out of three years, includes a nitrogen-fixing legume, reduces energy input requirements because a no-tillage system is involved, provides an opportunity to break disease cycles, and reduces erosion. In an experiment conducted in Mississippi from 1973 to 1977, overseeding wheat in standing soybeans on October 1 was found to be an acceptable method of establishing the wheat crop when compared with conventionally seeded wheat after soybean harvest November 1 to 10 (Table 14–10) (Sanford 1979).

TABLE 14–10. Yields (kg/ha) and (bu/ac) of overseeded and conventionally seeded wheat following soybeans, in Mississippi (1973–1977)

Year	Overseeded Oct. 1		Seeded After Soybean Harvest	
	(kg/ha)	(bu/ac)	(kg/ha)	(bu/ac)
1974	3051	45.4	2845	42.3
1976	3823	56.9	3322	49.4
1977	2737	40.7	2283	34.0
Mean	3204	47.7	2817	41.9

Source: Data from Sanford 1979.

A seeding rate of 101 kg/ha (90 lb/ac) was found to be optimum. Overseeded wheat failed in this study in 1975 because the soybean canopy was only 50% closed due to late seeding of the soybeans (July 15). Under such conditions the overseeded wheat only emerged directly beneath the protective soybean canopy. Rows were 100 cm (40 in) wide, and narrower rows probably would assure a closed soybean canopy.

Speed and Timeliness of Planting

An optimum date of seeding can be established for each crop, and delays in seeding usually result in a yield penalty. To assure timeliness of planting dates, farmers with a large hectarage (acreage) must start planting as early as an adequate and healthy stand can be established. Speed of operation is important, and primary and secondary tillage systems must consider the impact on soil moisture, soil temperature, and soil crusting that might hinder seedling emergence.

Studies at Guelph, Ontario clearly reveal the importance of timely seeding. At the northern latitude of Guelph the air and soil temperatures in the spring increase slowly. When seeding time arrives, speed of operation is important.

For corn at Guelph there is a 60 kg/ha/day (1 bu/ac/day) decrease for crops seeded after May 10. This varies from year to year, with the decrease starting later in a late spring and earlier in an early spring. For corn planted after June 15 for grain, there is likely to be a crop failure. Spring grain yields decrease 50 to 70 kg/ha/day (44 to 62 lb/ac/day) if planting is delayed after May 1.

In addition to losses in yield because of late planting, there are also losses in yield when a crop remains in the field after it is ready to harvest. Accumulated data indicate that after corn is ready to harvest, yields drop by 300 kg/ha/week (267 lb/ac/week) (Hume 1976). Farmers must consider the machine and method that provide timeliness of operations.

A soil does not warm until it has dried, and tillage may be necessary in areas with soil slow to warm in the spring. Fall plowing with conventional spring tillage produced the highest soil temperatures (Richey, Griffith, and Parsons 1977) in Indiana. Systems that produce surface residues, such as no-till planting in shredded corn stalks, were slower to warm because of shading. In Indiana, plowing treatments resulted in soil temperatures 2.5°C (4.0°F) higher at a 10-cm (4 in) depth when averaged over an eight-week period following corn planting than with no-till planting. Fall plowing subjects soil to erosion (Figure 14–3) but may be necessary in short-season areas to help the soil warm up and to allow weathering to break soil clods.

Because of cool temperatures associated with heavy clay soils, no-till systems in cool regions have been conducted mainly on light-textured soils. There is also concern about the effect of no-till on the physical effects on clay loam soils. In Minnesota, Gantzer and Blake (1978) found evidence that clay loam soils managed under no-till acquired physical characteristics quite

Figure 14–3. Fall plowing or tillage subjects soil to erosion losses. In short-season areas, fall plowing without any secondary tillage is recommended. Freezing and thawing over winter reduce large clods of soil and promote a fine, firm seedbed. (Courtesy Ontario Ministry of Agriculture and Food.)

different from those found under conventional tillage up to a 30 cm (12 in) depth.

The Minnesota data show that soil under no-till has significantly greater bulk densities both spring and fall in the plow zone. Values ranged from 1.24 to 1.32 g/cm³ (0.72 to 0.76 g/in³) for no-till, compared to 1.05 to 1.12 g/cm³ (0.61 to 0.65 g/in³) for conventional tillage. This increase occurred in spite of the fact that the mean number of channels 1 mm in diameter and larger, developed by earthworms and decomposing rootlets, were greater for no-till. Channel counts for no-till ranged from 666 to 1732/m² (555 to 1443/yard²), compared to 243 to 1475/m² (202 to 1229/yd²) for conventional tillage. Reduced evaporation with no-till resulted in greater water content, which is associated with cold soils; and if gaseous exchange is restricted, germination and seedling development conditions are less than optimum.

Research at the University of Guelph (*Annual Crop Science Report* 1979) showed fall moldboard plowing (Table 14–11) produced the highest yields.

Details of grain yields under nine tillage treatments for a three-year period in Ontario are shown in Table 14–12. Grain corn yields on silt loam soils averaged 4.9 t/ha (78.7 bu/ac), 0.4 t/ha (6.4 bu/ac) lower than on sandy loam. **Zero tillage** over all soils averaged 5.1 t/ha (81.9 bu/ac), 0.7 t/ha (11.2 bu/ac)

TABLE 14–11. Grain corn yields on silt loam soil at the Elora Research Station, University of Guelph, for an eight-year period under four tillage treatments 1971 to 1978

Tillage Treatment	Grain Yield	
	(t/ha)	(tons/ac)
Moldboard plow (fall)	5.7	2.6
Chisel plow (fall)	4.8	2.2
Ridge (fall)	4.9	2.2
Zero tillage	4.5	2.0

Source: *Annual Crop Service Report* 1979.

*Disced and harrowed once prior to planting.

lower than the 5.8 t/ha (93.2 bu/ac) for fall plowing coupled with secondary spring tillage. Spring plowing and secondary tillage produced a 5.6 t/ha (90 bu/ac) grain yield, and although lower than with fall plowing, erosion losses associated with fall plowing should be considered. An offset disc in spring plus harrowing produced a yield (5.6 t/ha) (90 bu/ac) that was comparable to spring plowing (5.6 t/ha) (90 bu/ac) and also comparable to spring plowing plus secondary tillage (5.6 t/ha) (90 bu/ac).

There is no one best tillage-planting system, and each farmer must consider the merits of each system on his own farm. None of the tillage planting systems listed in Table 14–12 can be described as unacceptable on the basis of grain yield alone. A number of variables in addition to yield need to be considered.

Soil Compaction

A compact soil is undesirable because root growth is limited, and this in turn limits nutrient and water uptake and causes a yield reduction (Richey, Griffith, and Parsons 1977) (Table 14–13). Small root-induced yield reductions may be masked if water and nutrients never become limiting factors. This is demonstrated in Table 14–13. At a low nitrogen fertility level, high compaction reduced yield 534 kg (8½ bu/ac) or 11.6%; at the higher nitrogen level of 135 kg/ha (121 lb/ac), nitrogen masked the adverse effect of reduced root size and the reduction was 408 kg/ha (6½ bu/ac) or 5%. Compaction caused by wheel traffic adjacent to rows of soybeans reduced soybean nodulation by 20 to 36% (Voorhees, Carlson, and Senst 1976).

Compaction is expressed by the bulk density of a soil. Compact soils have a high bulk density, and a given volume of soil will weigh more than less compacted soils. Robertson and Erickson (1978) suggested that bulk densities of 1.46 to 1.66 g/cm^3 (0.84 to 0.96 oz/in^3) found in subsoil materials in Michigan are too compact for rapid root growth and high crop yields. Corn and peanut root growth were greatly reduced when bulk density values were

TABLE 14-12. Grain corn yield (t/ha) (bu/ac) comparisons with nine tillage treatments on three soil textures in Ontario for a three-year period, 1976–1978

Tillage Treatment	Sandy loam			Loam			Silt loam		
	1976	1977	1978	1976	1977	1978	1976	1977	1978
Zero tillage	5.8 (93.2)	5.6 (90.0)	4.5 (72.3)	6.0 (96.4)	4.7 (75.5)	4.6 (73.9)	4.9 (78.7)	5.8 (93.2)	4.1 (65.9)
Fall moldboard plow	6.5 (104.4)	6.6 (106.0)	5.2 (83.5)	—	5.8 (93.2)	5.0 (80.3)	5.0 (80.3)	6.8 (109.2)	4.8 (77.1)
Fall plow; spring disc, harrow	6.3 (101.2)	6.3 (101.2)	5.5 (88.4)	6.2 (99.6)	5.3 (85.1)	5.2 (83.5)	5.8 (93.2)	6.8 (109.2)	5.0 (80.3)
Spring moldboard plow	6.1 (98.0)	6.3 (101.2)	5.4 (86.8)	6.3 (101.2)	5.6 (90.0)	5.1 (81.9)	5.0 (80.3)	6.5 (104.4)	4.2 (67.5)
Spring plow disc and harrow	5.8 (93.2)	5.9 (94.8)	5.2 (83.5)	5.9 (94.7)	5.2 (83.5)	5.3 (85.1)	5.5 (88.4)	6.6 (106.0)	4.7 (75.5)
Fall chisel plow	5.8 (93.2)	6.0 (16.4)	—	—	4.6 (73.9)	4.3 (69.1)	5.0 (80.3)	5.9 (94.8)	4.5 (72.3)
Fall chisel; spring disc, harrow	5.9 (94.8)	6.4 (102.8)	—	6.0 (96.4)	5.5 (88.4)	5.3 (85.1)	5.4 (86.8)	6.4 (102.8)	5.1 (81.9)
Offset disc in spring, harrow	6.2 (99.7)	6.2 (99.6)	5.3 (85.1)	5.8 (93.2)	5.6 (90.0)	5.2 (83.5)	5.2 (83.5)	6.1 (98.0)	4.5 (72.3)
Rototill in spring	—	—	5.2 (83.5)	—	—	5.3 (85.1)	—	—	4.4 (70.7)

Source: Data from Annual Crop Science Report 1979.

TABLE 14–13. Compaction effects on corn grain yields (kg/ha) and (bu/ac) at two nitrogen levels in central Illinois (1960 and 1961)

Compaction Level	Grain Corn Yields at Nitrogen Rates of:			
	45 kg/ha	(40 lb/ac)	135 kg/ha	(121 lb/ac)
Medium	4614	73.6	7219	115.1
High	4080	65.1	6811	108.6

Source: Data from Richey, Griffith, and Parsons 1977.

more than 1.3 g/cm³ (0.75 oz/in³) in Iowa and Alabama, respectively. Yields of spring wheat were 24 to 51% (average of 36%) lower in wheel traffic areas on a silt loam soil at Lethbridge, Alberta over a seven-year period on fields receiving chemical weed control instead of tillage between crops (Lindwall and Anderson 1977). Bulk densities of 1.2 g/cm³ (0.69 oz/in³) or greater in the 0- to 5-cm (0–2 in) soil depth appeared responsible. Robertson and Erickson (1978) suggested that compaction problems are a greater and more extensive problem than many scientists and crop producers suspect. A comparison of soil material from the fence row compared with soil in the adjacent field had 18% more total pore space, 15.5% more air (noncapillary) pore space, was less compact, and had a bulk density of 1.11 g/cm³ (0.64 oz/in³) compared to 1.48 g/cm³ (0.86 oz/in³) in the adjacent field. Other workers have suggested that freezing and thawing and wetting and drying cause soils to expand and contract and are natural forces that reduce compaction, but the problem is more closely associated with organic matter levels. Climatically warms areas under intensive cultivation may speed up the loss of organic matter, and compaction in these areas may be more severe. Buckingham (1975) did suggest that compaction problems are greater in southern United States.

Soil compaction is associated with increased soil erosion, lower soil moisture levels, cooler soil temperatures, reduced soil aeration, and increased energy requirements for tillage; and under wet conditions, water remains standing on the surface because of reduced efficiency of underdrainage. The compaction problem is aggravated by surface traffic on wet soil when soil granules are crushed and are rearranged so that the soil contains less pore space for both air and water. This is the type of soil that becomes lumpy when worked.

Causes of Soil Compaction

1. EXCESSIVE TILLAGE

Excessive tillage reduces aggregate size; small aggregates are less stable than larger ones and this leads to soil compaction, soil crusting, increased soil erosion, and reduced yields, as well as increasing the cost of production and wastes energy. At Elora, Ontario, corn yields were reduced when aggregate size was reduced by excessive tillage on a silt loam soil (Table 14-14). On Brookston clay soil at Woodslee, Ontario, excessive fall or spring tillage caused reductions in corn grain yields (Table 14-15). Discing soil in the fall can lead to soil compaction and lower yield. It is widely accepted that farmers carry out more secondary tillage than is needed.

TABLE 14-14.   Excessive tillage effects on grain corn yield in 1978 at Elora, Ontario

Treatment	Grain Yield (t/ha)	(bu/ac)
Spring moldboard plow plus		
-one cultivation	3.3	53.0
-two cultivations	3.0	48.2
-slow speed rototill	3.8	61.0
-high speed rototill	3.7	59.4

Source: Data from *Annual Crop Science Report* 1979.

TABLE 14-15.   Yield of grain corn (t/ha) and (bu/ac) at Woodslee, Ontario, with different amounts of tillage (four-year average)

Primary Tillage in Fall	Secondary Tillage in Spring Using a Tandem Disc-Harrow			
	2 discings		7 discings	
Moldboard plow	7.53	121.0	7.02	112.8
One-way disc	6.77	108.8	6.90	110.8
Tandem disc	6.77	108.8	6.46	103.8

Source: Data from Bolton 1977.

Soil crusting should be avoided to ensure plant emergence (Figure 14-4).

With fine-textured (clay) soils, or soils slow to warm in the spring, there appears to be no alternative to the moldboard plow as a primary tillage implement to secure highest yields. But with coarse-textured soils or in areas where cool temperatures are no problem, many crops can be grown without any tillage at all (Baeumer and Bakermans 1973). Yields may even be higher than with conventional tillage because of improved soil properties. If zero-tillage is not appropriate to your area or conditions, reduced tillage is certain to be of advantage.

Where primary tillage is necessary, there appears to be no advantage in plowing more than 10 to 15 cm (4 to 6 in) deep. This depth is sufficient to facilitate secondary spring tillage even on Brookston clay soil and is inducive to highest grain yields (Table 14-16) (Southwell and Ketcheson 1978). Increasing the depth from 10 to 20 cm (4 to 8 in) more than doubled the draft

Figure 14–4. Excessive tillage may reduce soil aggregate size, which can lead to soil crusting and poor plant emergence. The soybean seedlings in this photo have emerged successfully. (Courtesy Ontario Ministry of Agriculture and Food.)

and energy used. Also deep plowing brings up subsoil, which is not inducive to crop production and brings dormant weed seeds to the surface where they germinate and grow.

2. UNTIMELY FIELD OPERATIONS

Field operations when the soil is wet can result in a deterioration of soil physical condition. Hauling manure, harvesting corn silage, or other traffic when the soils are wet should be avoided. Studies at Guelph showed the adverse effect of soil cultivation under wet conditions (Table 14–17). Spring

TABLE 14–16.   Effect of plowing depth on grain corn yields (kg/ha) and (bu/ac) on Brookston clay loam at Woodslee, Ontario (seven year average) 1970 to 1976

Depth		Grain Yield	
(cm)	*(in)*	*(kg/ha)*	*(bu/ac)*
10	4	8781	140.0
20	8	8154	130.0
30	12	8028	128.0

Source: Data from Southwell and Ketcheson 1978.

plowing may lead to greater compaction, and in Table 14–17 spring plowing produced lower yields. Cash cropping systems, as opposed to livestock programs, involve more tillage operations and lead to greater soil compaction.

TABLE 14–17. Effect on grain corn yield (t/ha) and (bu/ac) of cultivation on a wet loam soil at Elora

	Grain Yield					
Treatment	1976		1977		1978	
Fall moldboard plow plus						
cultivation when wet*	6.1	98.0	6.8	109.2	4.6	73.9
cultivation at ideal						
moisture level**	6.4	102.8	7.1	114.1	4.7	75.5
Spring moldboard plow plus						
cultivation when wet	5.8	93.2	5.8	93.2	4.2	67.5
cultivation at ideal						
moisture level	6.1	98.0	6.3	101.2	4.2	67.5

Source: Data from Annual Crop Science Report 1979.
*Soil above lower plastic limit.
**Soil below lower plastic limit.

3. CROPPING PROGRAMS

Burning or removal of crop residues, or the production of crops with small residues such as beans or sugarbeets, leads to soil structure problems. Large amounts of crop residues are of little value, however, when anaerobic soil conditions associated with compaction cause slow decomposition. The use of green manure crops, especially deep-rooted legumes and **fibrous roots** of grasses, are essential for restoring compacted soils.

4. FARM IMPLEMENT DESIGN AND OPERATION

Large flotation tires or dual wheels spread the machine weight over a greater area and compact soils to a lesser depth, and thus are considered less damaging than tandem wheels. Deep compaction may result if the tractor wheel is operated in the furrow when plowing, rather than on the field. Plowing at the same depth each year can cause a hard plow pan or plow sole to develop. Water and roots may have difficulty penetrating a hard layer, and deep tillage practices such as subsoiling and chiseling may not noticeably improve hardpans. Since the poor physical condition of the soil is the effect and not the cause of the problem, it is difficult to improve the soil's physical condition by mechanical means. The remedy may lie in treating the cause by the use of forages in a rotation.

Fertilizer Placement

Three considerations in regard to fertilizer placement exist: (1) a general overall application, (2) a starter application placed on either side and below the seed, and (3) side-dressing with nitrogen. Plowing and secondary tillage that thoroughly mix the soil may promote deep root formation to avoid dry weather vulnerability, as compared to no-till methods where roots are encouraged to develop near the surface because of soil firmness, high surface moisture under a mulch residue, and cooler soil temperatures at a greater depth. Nitrogen can be injected into the soil with loosened plowed soil.

Band placement of starter fertilizers 5 cm (2 in) to the side and 2 to 3 cm (3/4 to 1 in) below the seed is more effective than broadcasting fertilizers. Large amounts of fertilizers cannot be banded or placed with the seed because of salt injury to seedlings, which can produce uneven stands.

Phosphorus and potassium tend to move very little, and should be placed in the zone of root development. Surface applications in zero-tillage systems will be of little value in the year of application. Large amounts of fertilizer can be applied broadcast in preparation for zero-tillage. Nitrogen can also be injected into the soil.

The lack of incorporating fertilizer materials has been a concern in continuous no-tillage systems. Soil properties and nutrient availability were studied over a five-year period of continuous no-tillage corn with conventional tillage in Kentucky (Blevins, Thomas, and Cornelius 1977). Nitrogen applications of 0, 84, 168, and 336 kg/ha (0, 75, 150 and 300 lb/ac) were made to both. It was found that no tillage with moderate rates of nitrogen most nearly preserved the soil chemical characteristics found under the original 50-year-old bluegrass sod. Nitrogen levels were higher under the no-till system, apparently due to the higher levels of organic matter and to the nitrogen being carried in solution to the root zone.

Weed Control

Numerous weed seeds can be found in the top 20 cm (8 in) of soil, but most remain dormant until brought to the surface by plowing and secondary tillage. Weeds that remain at a depth due to zero or reduced tillage systems are a contained population, and such a system reduces the need for herbicides. However, when weeds germinate under a zero or no-till system, surface trash residue may prevent effective herbicidal coverage of the soil or of the weed seedlings.

Specific tillage operations favor or discourage selected weeds, and those weeds not controlled by the herbicide may dictate the extent and form of tillage. Milkweed for example is not killed by selective herbicides used for weed control in corn; and when milkweed infestations become severe, plowing may be practiced as a control measure. Under a no-till system, however, interrow cultivation may not be possible because of trash or firm soil, and should herbicide action be incomplete, a weed control program is difficult.

Insect and Disease Control

Some insects and diseases overwinter in and on residue from diseased plants of the previous year, and often some control is offered by complete burial of such residues. Northern corn rootworm damage was less, however, when residues were not incorporated because freezing destroyed the insects' eggs. Residue incorporation reduced the incidence of European corn borer but is not complete enough for effective control. In Nebraska, crop rotations and minimum tillage were combined into an **ecofallow** system that effectively decreased stalk rot of grain sorghum and corn. A three-year rotation of winter wheat-grain sorghum-fallow broke the disease cycle that commonly built up in the residues (Doupnik and Boosalis 1980).

Energy Inputs

Most agronomists agree that the so-called intensive or conventional tillage has been applied too extensively at too much expense and energy input. As discussed, no-till is not suited to heavy cold soils, but forms of minimum tillage can be adopted that produce satisfactory yields (Table 14–18) (Baldwin 1979).

TABLE 14–18. Corn yields *(t/ha)* and *(bu/ac)* as affected by tillage practices on sand loam at Ridgetown, Ontario, Canada

		Yield	
Tillage Applied	*Description*	*(t/ha)*	*(bu/ac)*
No till	Corn planted between old rows,	5.56	89.3
Minimum tillage	Plowed, cultipacked and planted	6.42	103.1

Source: Data from Baldwin 1979.

Richey, Griffith, and Parsons (1977) reviewed the energy requirements for corn and soybeans with various tillage-planting systems in soils with draft classifications of low, medium, and high in Nebraska (Table 14–19).

Similar results were shown for soybeans. It is not the intent of Table 14–19 to reveal the best tillage-planting system but to indicate that the maximum fuel savings on high draft soil between the highest energy input (moldboard plow-conventional) and the lowest energy input (rotary strip) is 38.6 litres (8.5 gallons) which does not provide a compelling economic incentive to reduce tillage, particularly in cases where there is a possibility of a yield reduction with reduced tillage. Under conditions of limited fuel supply or with high prices, reduced tillage may be of greater economic value, especially if it is compatible with the entire agro-ecosystem. Thus, factors other than energy inputs may make reduced tillage more attractive.

TABLE 14–19.   Total energy for corn tillage planting and weed control in litres (and gallons per acre) of diesel fuel equivalent per hectare (1974 and 1975)

Tillage-Planting System	Soil Draft Classification					
	Low		Medium		High	
	litres	gallons	litres	gallons	litres	gallons
Moldboard plow—conventional	52.0	(4.6)	63.6	(5.7)	78.2	(7.0)
Moldboard plow—wheel track plant	32.3	(2.9)	43.0	(3.8)	57.1	(5.1)
Chisel plow—conventional	50.9	(4.5)	61.6	(5.5)	72.2	(6.4)
Disc—conventional	44.3	(3.9)	49.0	(4.4)	53.1	(4.7)
Ridge (furrow-mulch)	44.6	(4.0)	49.6	(4.4)	55.2	(4.9)
Till plant	37.3	(3.3)	40.7	(3.6)	43.2	(3.8)
No-till colter	41.6	(3.7)	42.6	(3.8)	44.2	(3.9)
Rotary strip	35.4	(3.1)	37.0	(3.3)	39.6	(3.5)

Source: Data from Richey, Griffith, and Parsons 1977.

Soil Moisture

In water-short areas such as the midwest prairies, water supplies are barely adequate to permit a profitable agriculture to exist on an annual basis. A common practice aimed at increasing water for crop production is a system of summer fallowing. To prevent the loss of water from the soil through plant transpiration, and to allow the soil to soak up rain and melting snow, a field is left uncropped for an entire season. The summer fallow land is frequently cultivated to prevent water loss from transpiration of weeds. The unused moisture in the soil is stored for use in crop production in the year following summer fallow. Saskatchewan is the driest of the three Canadian prairie provinces and in 1978 had 51% of its cropped hectarage (acreage) in summer fallow; Alberta had 35% and Manitoba 25% (Anonymous 1978).

Summer fallowing is not a very effective method of conserving moisture because the soil is fully exposed to the direct rays of the sun, and evaporation is high. About 30 to 50 cm (12 to 20 in) of rainfall occur annually on parts of the Canadian prairies, with only 15.0 to 20.0 cm (6 to 8 in) falling during the growing season, which is sufficient to produce minimum vegetative growth of cereals. Additional moisture is required if grain production is to be satisfactory. Every 2½ cm (in) of water over and above the 15.0- to 20-cm (6 to 8 in) minimum will produce 400 to 470 kg/ha (357 to 420 lb/ac) of grain. A cereal crop in the prairie regions requires about 26 to 27 cm (10 to 11 in) as a minimum amount of rainfall. A 30-cm (12 in) moisture situation could produce a crop of 1600 to 1900 kg/ha (1430 to 1700 lb/ac) under mid-western U.S. and Canadian conditions.

In the dry land area around Regina, the cropping program is based on a crop-fallow system in which the rotation of crops and fallow is managed as follows:

1. Year one—spring wheat is harvested in late summer or fall.
2. After harvest, the soil is cultivated to hasten decomposition of crop stubble.
3. The following spring no crop is planted, and the field is cultivated periodically to control weeds; this is the fallow period during which precipitation is stored in the soil.
4. Spring wheat is seeded and the cycle is repeated.

In the southern Great Plains region of the United States where rainfall effectiveness is reduced due to increased evapotranspiration under warm temperature conditions, alternate winter wheat and fallow are generally considered a practical sequence when the annual precipitation is less than about 55 to 60 cm (22 to 24 in). In the Central Great Plains, the rainfall level at which summer fallowing is practical is 45 to 50 cm, (18 to 20 in) and in the northern Great Plains, 35 to 40 cm (14 to 16 in).

Estimates vary on the efficiency of summer fallowing. Falk (1975) estimated that 15 to 20% of the moisture received from summer rainfall is held in the upper two metres of soil where the roots can reach it. Another estimate suggested 10 to 30%. About 50% of the summer precipitation evaporates, and the remaining 30 to 35% percolates too deeply into the soil for the roots to reach. Consider a water budget as follows: total annual precipitation—30 cm (12 in). If 15 to 20% is held in the soil, it equals 4.5 to 6.0 cm (2 to 2½ in). Rainfall during the crop growing period amounts to 20 cm (8 in). Total seasonal rainfall plus stored moisture = 24.5 to 26.0 cm (9 to 10 in) and is capable of producing a cereal crop.

The summer fallowing system is effective but not efficient, and although it offers some advantages, it also offers some disadvantages. Consider the following:

1. Before the advent of herbicidal weed control, summer fallowing offered an effective method of reducing a weed population because cultivation encouraged weed seeds to germinate, which were then destroyed by further cultivation. The need to summer fallow has been reduced greatly by herbicides.
2. Summer fallowing speeds up the decomposition and mineralization of residues in the soil; and before the widespread use of commercial fertilizers, this was a necessary function to restore soil fertility and productivity.
3. The fallow system has served to stabilize grain yields and reduce the extreme yearly fluctuations in yield due to drought.

4. Under conditions of inexpensive land and horse-drawn implements, a system of summer fallowing allowed for a methodical approach to a cropping program that was effective under such conditions. Under present-day conditions, summer fallowing may be regarded as outdated, ineffective, and too expensive.
5. Fallow fields are subject to serious losses due to wind and water erosion.
6. Summer fallowing may lead to saline seep areas (Anonymous 1974).

A protective mulch cover can reduce water loss by evaporation and runoff but in northern regions mulches prevent soils from warming up properly in the spring (Fig. 14-5).

Figure 14-5. Cool soil temperatures in northern regions are often associated with surface residues. A protective mulch can reduce evaporation and delays soil warm-up in the spring. Zero-tillage in this situation is possible but should be restricted to light soils that warm up readily. (Courtesy Ontario Ministry of Agriculture and Food.)

ALTERNATIVE MEANS OF MOISTURE CONSERVATION

The key to moisture management is to capture a part of the available precipitation that has been wasted in the past. Solutions include: (1) capturing and retaining snowfall on fields, (2) irrigation, and (3) reduced evaporation.

About 20 to 30% of the precipitation in the northern Great Plains falls as snow. Nearly half of this snow blows off the fields and gathers in gullies or fence rows where it causes runoff and erosion and prevents uniform drying of the field. It also means delayed seeding. If this snow could be retained on the fields, it might provide the margin of moisture necessary for a crop each year without the loss of a crop when the field is summer-fallowed. Potentially 7.5 to 15 cm (3 to 6 in) of additional soil moisture could be captured by the successful retention of snowfall.

A barrier to slow the wind as it blows across the fields would result in the snow being deposited on the fields and would prevent the snow from being picked up again and moved elsewhere as drifts. For spring wheat producers there is either too much or too little moisture. The inclusion of snow barriers means a soggy field, a delay of spring seeding, and the dissection of large fields into narrow strips, and for these reasons short barriers are not extensively used. There also may be too little moisture at the critical period of reproductive development for satisfactory grain yields. For winter wheat, additional snow cover may be more attractive because a snow cover serves as insulation; it is also an effective protection against frost-heaving and winter damage, it reduces water erosion, and the need for early spring seeding does not exist.

One of the simplest methods of holding snow is with stubble and crop residues. Therefore, Lindwall and Anderson (1977) suggested that farmers in the brown soil zone of Alberta should consider eliminating preseeding cultivation, thereby leaving stubble and crop residues on the field. Also specialized machinery could be used to provide effective seed and fertilizer placement with minimal soil disturbance in the wide range of soil and residue conditions encountered with no-till seeding.

Parallel strips of tall perennial wheatgrass were found to be an effective windbreak to hold snow (Anonymous 1979). The wheatgrass was planted in double rows, at 9- to 18-metre (300 to 600 feet) intervals, and grew to about a metre (3 feet) in height. The barriers reduced the wind speed by as much as 20% at a distance of up to 18 metres (600 feet), sufficient to cause a 32 km/hr (20 mph) wind to deposit half the snow it is carrying. Also, one-half as much snow will be blown off a field with these barriers as compared to a fallow field. A combination of stubble and wheatgrass barriers retained 136% more moisture in the soil than bare summer fallow (Table 14–20).

A number of plant species and permanent hedgerows adapted to local regions could be satisfactory wind barriers. Wheatgrass and other grass species

TABLE 14–20. Soil moisture (cm) and (in) as influenced by stubble and wheatgrass barriers (data for eight years at Montana, 1966–1973)

	Moisture without Barrier		Moisture with Barrier	
	(cm)	(in)	(cm)	(in)
Fallow soil	7.0	(2.8)	10.5	(4.1)
Standing stubble	11.5	(4.5)	16.5	(6.5)

Source: Data from Falk 1975.

offer greatest ease of operation because they can be quickly removed, or crossed with machinery, and they are not damaged by herbicides used on cereal grains.

Wind barriers may also influence yields during the crop growing season by lessening the effect of hot dessicating summer winds on evapotranspiration; they influence CO_2 levels and relative humidity. Work in Nebraska and Texas (Brown and Rosenberg 1975) found favorable effects from corn windbreaks in fields of sugarbeets, peanuts, soybeans, and horticultural crops. In eastern Canada the removal of wind barriers has caused concern (*Cash Crop Farming* 1979).

QUESTIONS

1. What are the organic matter levels commonly found on soils in your area? Have the levels increased or decreased in recent years? Are the levels satisfactory?

2. Describe the physical properties of soils that are of interest to successful long-term crop production.

3. Determine the economic value of including a legume in a rotation in terms of value of the nitrogen fertilizer, return from the forage crop, cost of pest control (such as northern corn rootworm), and other factors using current values.

4. Describe and evaluate the tillage-planting system commonly used on your farm or area in terms of (a) soil erosion, (b) speed and timeliness of planting, (c) soil compaction, (d) moisture conservation, (e) fertilizer placement, and (f) pest control.

5. Trace the agricultural history in your area, and if possible trace the story of rotations. What changes are most evident in relation to soil management or cropping practices?

6. What types of machinery are currently available for planting without plowing?

7. Is summer fallowing practiced in your area?

8. What rotation system is common in your area? On cash crop farms? On livestock farms?

9. What insect, disease, or weed problems exist in your area as a result of specific cropping practices?

10. Describe a saline seep area.

REFERENCES FOR CHAPTER FOURTEEN

Agricultural Advisory Council. 1970. *Modern Farming and the Soil.* Report of the Agricultural Advisory Council on Soil Structure and Soil Fertility. Ministry of Agriculture, Fisheries and Food. Published by Her Majesty's Stationery Office, London. (119 pp.)

Annual Crop Science Report. 1979. Department of Crop Science, Ontario Agricultural College, University of Guelph, Guelph. (pp. 35–38).

Anonymous. 1974. Saline Seeps. Their Cause and Prevention,'' *Crops and Soils* 26(9):12–13.

Anonymous. 1977*a*. "Legume Cover Crops Boost Corn Yields," *Crops and Soils* 29(9):20–21.

Anonymous. 1977*b*. "On Looking After the Land," (editorial in) "Corn in Canada," *Agri-Book Magazine* 3(2):4.

Anonymous. December 1978. "Open Door," *Canada Grains Council Market Newsletter*, p. 3.

Anonymous. 1979. "Grass Barriers Greatly Reduce Erosion," *Crops and Soils* 31(7):27.

Baeumer, K., and W. A. P. Bakermans. 1973. "Zero-Tillage," *Advances in Agronomy.* 25:77–123.

Baldwin, C. S. 1979. "Tillage Practices and Their Effects on Erosion," *Highlights of Agricultural Research in Ontario.* 2(3):10–11.

Barber, S. A. 1979. "Corn Residue Management and Soil Organic Matter," *Agronomy Journal* 71:625–627.

Barnhart, S. L., W. D. Shrader, and J. R. Webb. 1978. "Comparison of Soil Properties under Continuous Corn Grain and Silage Cropping Systems," *Agronomy Journal* 70:835–837.

Bartholomew, W. V. 1965. "Mineralization and Immobilization of Nitrogen in the Decomposition of Plant and Annual Residues, in W. V. Bartholomew and F. E. Clark, eds., "Soil Nitrogen," *Agronomy* 10:285–306.

Blevins, R. L., G. W. Thomas, and P. L. Cornelius. 1977. "Influence of No-Tillage and Nitrogen Fertilization on Certain Soil Properties after 5 Years of Continuous Corn," *Agronomy Journal* 69:383–386.

Bolton, E. F. 1977. "Tillage and Soil Compaction," *Proceedings of the Annual Meeting of the Ontario Soil and Crop Improvement Association.* pp. 182–183.

Brown, K. W., and N. J. Rosenberg. 1975. "Annual Windbreaks Boost Yields," *Crops and Soils* 28(7):8–11.

Buckingham, F. 1975. "Controlled Traffic Can Stop Compaction of Agricultural Soils," *Crops and Soils* 27(8):13–15.

Cash Crop Farming. February 1979. "Windrow Removal Can Cut Yields," pp. 38–42.

Doupnik, B., Jr., and M. G. Boosalis. 1980. "Ecofallow—A Reduced Tillage System and Plant Diseases," *Plant Disease* 64:31–35.

Elkins, D. M., et al. 1979. "No-Tillage Maize Production in Chemically Suppressed Grass Sod," *Agronomy Journal* 71:101–105.

Falk, D. E. 1975. "Summerfallow. Is There a Better Way to Store Water? *Crop and Soils* 28(3):9–11.

Fenster, C. R., R. H. Follett, and E. J. Williamson. 1978. "How Should Crop Residues Be Best Used? *Crops and Soils* 30(9):19–22.

Gantzer, C. J., and G. R. Blake. 1978. "Physical Characteristics of Le Sueur Clay Loam

Soil Following No-Till and Conventional Tillage," *Agronomy Journal* 70: 853–857.

Hume, D. J. 1976. "Effect of Timeliness on Crop Yields," *Proceedings of the Annual Meeting of the Ontario Soil and Crop Improvement Association*. pp. 208–209.

Ketcheson, J. W., and E. G. Beauchamp. 1978. "Effects of Corn Stover, Manure, and Nitrogen on Soil Properties and Crop Yield," *Agronomy Journal* 70:792–797.

Ketcheson, J. W., Y. A. Martel, and A. MacLean. 1979. "Eastern Canadian Soils. Trends in Productivity," *Agrologist* 8(4):16–17.

Larson, W. E., et al. 1972. "Effects of Increasing Amount of Organic Residues on Continuous Corn. II. Organic Carbon, Nitrogen, Phosphorus, and Sulfur," *Agronomy Journal* 64:204–208.

Lindwall, C. W., and D. T. Anderson. 1977. "Effects of Different Seeding Machines on Spring Wheat Production under Various Conditions of Stubble Residue and Soil Compaction in No-Till Rotations," *Canadian Journal Soil Science* 57:81–91.

Mitchell, W. H., and M. R. Teel. 1977. "Winter—Annual Cover Crops for No-Tillage Corn Production," *Agronomy Journal* 69:569–573.

Mock, J. J., and D. C. Erbach. 1977. "Influence of Conservation—Tillage Environments on Growth and Productivity of Corn," *Agronomy Journal* 69:337–340.

Mowat, D. N. 1976. "Alternate Sources of Animal Feeds," *Notes on Agriculture University of Guelph* 7(4):12–14.

———. 1979. "Crop Residues: Feed Potential for Cattle," *Highlights of Agricultural Research in Ontario* 2(3):12–13.

Rennie, D. A. 1979. "Intensive Cultivation. The Long-Term Effects," *Agrologist* 8(4):20–22.

Richey, C. B., D. R. Griffith, and S. D. Parsons. 1977. "Yields and Cultural Energy Requirements for Corn and Soybeans with Various Tillage-Planting Systems," *Advances in Agronomy* 29:141–182.

Robertson, L. S., and A. E. Erickson. 1978. "Soil Compaction. Symptoms, Causes, Remedies," Parts I, II, III, *Crops and Soils* 30(4,5,6):11–14; 7–9; 8–10.

Rohweder, D. A., W. D. Shrader, and W. C. Templeton, Jr. 1977. "Legumes. What Is Their Place in Today's Agriculture?" *Crops and Soils* 29:11–14.

Sanford, J. O. 1979. "Establishing Wheat After Soybeans in Double-Cropping," *Agronomy Journal* 71:109–112.

Sanford, J. O., D. L. Myhre, and N. C. Merwine. 1973. "Double-Cropping Systems Involving No-Tillage and Conventional Tillage," *Agronomy Journal* 65:978–982.

Sheard, R. W. 1977. "Forages in a Rotation," *Notes on Agriculture*, 8(3):11–13. University of Guelph.

Southwell, P. H., and J. W. Ketcheson. 1978. "Energy and Tillage Practices," *Notes on Agriculture*, 14(1):17–21. University of Guelph.

Voorhees, W. B., V. A. Carlson, and C. G. Senst. 1976. "Soybean Nodulation as Affected by Wheel Traffic," *Agronomy Journal* 68:976–979.

Warncke, D. D., and S. A. Barber. 1974. "Root Development and Nutrient Uptake of Corn Grown in Solution Culture," *Agronomy Journal* 66:514–516.

chapter fifteen

Cultural Energy and Crop Production

In addition to sunlight energy for photosynthesis, successful crop production requires the infusion of supplemental or **cultural energy.** Cultural energy is used directly in crop production to perform seeding, planting, spraying, and harvesting operations; and indirectly to manufacture machinery, pesticides, and fertilizers, and to provide drying and storage facilities and any other inputs associated with crop production.

Cultural energy inputs are not new to agriculture. Historically, slaves, serfs, farm laborers, oxen, mules, and horses performed agricultural-related tasks and represented the major cultural energy input. The coming of the Industrial Revolution meant that powerful, fast, and untiring machines now replaced much human and animal power. With machines, agricultural production flourished because land, once required to produce animal feed, was freed for food production; machines were reliable and speeded up each operation, thereby reducing crop losses by beating the weather through the timeliness of planting and harvesting; and improved housing and storage methods were made available. Mechanization was aided by an abundance of inexpensive fossil fuels, mainly petroleum, used not only to power machines but also to build equipment, transport food, manufacture nitrogen fertilizer and pesticides, operate driers, and provide power for virtually every operation, including trimming the hedge and mowing the lawn. Petroleum has become so much a part of agricultural production that observers commented that agriculture appeared to have become a method of converting fossil fuels to food (Handler 1970; Odum 1971).

The impact of such a massive infusion of cultural energy was demonstrated dramatically in rapid changes in North American agriculture: after 1950, the number of farms declined by 44%; the average farm size increased 240% to over 162 hectares (400ac); and farm population declined from 23.3% in 1940 to about 4.8% in 1980. From feeding himself and four others at the turn of the century, one farmer in 1980 now can feed over 55 people. The once onerous tasks associated with farming could essentially be completed from the control cab of a machine or by pressing an electric switch. Reduced labor requirements of farming and a high degree of farm specialization allowed many farm operator-owners to seek off-farm jobs, more out of necessity to pay for the cultural energy inputs than out of boredom. Consumers around the world benefited, and North American consumers during the sixties and seventies spent an unbelievably low 16 to 20% of their disposable income on food.

But contrary to popular opinion, North American food production costs are high. The money consumers spend on food is a low percentage of their disposable income, but only because per capita income is high. In India, Pimentel et al. (1973) suggested that about 77% of personal income goes toward food, but incomes are low there. Recognition that food is a human necessity and that no one in a land of abundance should go hungry because of food costs, along with depressed prices associated with overproduction, created a "cheap food" attitude in North America. Not surprisingly, therefore, over 50% of North American farmers held off-farm jobs (Steinhart and Steinhart 1974), which generated a greater income than farming.

The volume of food produced per hectare through the infusion of cultural energy, the need to alleviate the precarious world food situation, and the desire for better life styles predictably caused world reaction to encourage the adoption of the North American mechanized style of agriculture (FAO 1970). Wheat and rice cultivars were developed to respond to high nitrogen and fertility levels, and irrigation and pest control inputs were widely distributed in the sixties. These upright-leaved, semidwarf, photoperiodically insensitive, lodging-resistant, and high-yielding cultivars provided the vehicle and incentive toward a cultural energy-intensive style of agriculture. The result was unprecedented food increases in many previously food-deficient countries, and the phenomenon became known as the "green revolution."

The hopes aroused in the sixties by the green revolution were dashed in the early seventies when threats of petroleum shortages appeared and fuel prices soared. The once-abundant petroleum supplies were in short supply, and cultural energy inputs required by the green revolution style of agriculture could not be met. It was the rising cost of production without an accompanying rise in food prices, not a genetic failure within the crop or a lack of farmer know-how, that forced the abandonment of intensive cultural energy inputs (Wilson 1978). North American farmers, perhaps the most dependent on cultural energy for agricultural production, became concerned about the high degree of cultural energy used in crop production. It became evident that agriculture could not continue on a path of adding more cultural energy, that fossil fuels for mechanized agriculture were no longer inexpensive, and that agricultural production must be placed on a long-term, self-sustaining basis. Biswas and Biswas (1976), p. 206 stated: "Ultimately food is a net product for our ecosystem," and they emphasized that instead of "more energy," the aim of agriculture should be "more efficient use of energy."

This is not to say that agricultural productivity cannot be sustained or increased, or that the world population cannot be fed. But increased food production through the use of scarce and expensive petroleum can only lead to a dilemma. Borgstrom (1973) expressed concern about the continued dependence on petroleum and about what a disruption in the flow of petroleum energy could mean. Petroleum inputs or cultural energy use in agriculture must be curtailed. Alternate energy forms will be found, but until that time, efficient use of cultural inputs must be employed.

This chapter is an attempt to determine how much cultural energy is used and with what efficiency is it used. What are the most energy-intensive inputs, and can efficiency be increased? Can cultural energy inputs be reduced without jeopardizing output?

CULTURAL ENERGY EFFICIENCY IN CROP PRODUCTION TO THE FARM GATE[1]

Photosynthetic efficiency measures how completely the energy in the visible light spectrum is transformed into economic yield. Cultural energy, on the other hand, is represented by all the inputs associated with row width, seeding rates, and fertilizer practices that are used to help intercept sunlight, and to harvest, dry, and store that captured energy. Only a small portion of the cultural energy contained in fertilizer materials is transformed into plant tissues; all of the other cultural energy inputs serve to *help* crops convert sunlight energy into food energy. It is possible, therefore, to differentiate cultural energy from the energy associated with photosynthesis and to calculate a **cultural energy efficiency.** The concept of efficiency is closely related to the ratio of input to output of energy in crop production to the farm gate. If output of energy exceeds input, the process is described as "efficient," and if input exceeds output, it is "inefficient." To identify and measure all the cultural energy inputs is a monumental task, and the objective here is to outline several approaches to allow comparisons and an evaluation of the validity of the technique(s).

DETERMINING CULTURAL ENERGY EFFICIENCY

One method of calculating the cultural energy efficiency in crop production is to compare the total energy inputs with the crop yield or output. The resulting "energy budget" has two advantages: (1) it does not become outdated like an economic budget with changing fuel or machinery prices because it is based on physical inputs alone; and, (2) by identifying and quantifying all the inputs, the largest items can be identified.

There are pitfalls and weaknesses with such a system.

1. Emphasis on the energy content of the economic yield ignores the vitamin, mineral, or protein content as well as the aesthetic aspects of

[1]Production to the farm gate considers energy inputs in producing food and feed commodities on the farm.

food. Also, foods with a low energy value cannot be accurately evaluated. However, comparisons are valid among systems producing the same crop commodity (e.g., corn with corn), although less valid when comparing cultural energy inputs among unrelated commodities (e.g., corn with tomatoes).

2. Comparison of cultural energy ratios (output/input) should be used cautiously. Changes in crop management should not be made without an investigation into the economic consequences. A cultural energy-efficient farm may not have a high output. Without a high output the food needs may not be met, the system may not pay for fixed costs (land, taxes, income), and the enterprise may not be profitable. The immediate needs of every human being for food must be considered. Changes resulting from cultural energy-efficiency considerations must be gradual, and they must sustain output as well as maintain long-term productivity.

Calculation of cultural energy inputs must include direct and indirect components. Direct energy is the energy used on the farm such as fuel in the tank of the machines used for crop production. It does not include heating the farmhouse or cooking food required in everyday living. Indirect energy includes energy expended on manufacturing materials, constructing machines, manufacturing pesticides and fertilizers, and transporting these inputs. Cultural energy also includes inputs to extract iron ore, and to obtain rubber latex and other materials.

Because of the difficulty in precisely defining and estimating the amount of indirect energy inputs into crop production, the system used by Pimentel et al. (1973) of limiting direct and indirect energy inputs to one step back from the crop production level is adopted here. Thus, the inputs for farm machinery would include the making of steel, including rolling, casting, and general assembly, but not the energy inputs of mining iron ore, coal, or other resources.

To investigate cultural energy crop production relationships in the United States, Pimentel et al. (1973) selected the corn crop and used national average figures to determine the size of the inputs for selected years between 1945 and 1970 (Table 15–1). Consumption of nitrogen fertilizer showed the largest increase over the 25-year period, and from an average of 8 kg/ha (7 lb/ac) in 1945, it increased to 125 kg/ha (112 lb/ac) in 1970, an increase of 1563%. Machinery, expressed in kilojoules (kJ), increased 233% in the 25-year period or nearly 10% per year. It was during this period that corn drying switched from air-dried cribbed corn to forced hot air with an accompanying energy increase of 1197%. But cultural inputs also had a favorable effect on corn yields; and in the 25-year period, increased from 2132 to 5080 kg/ha (31 to 76 bu/ac) or 238%. Cultural energy inputs per hectare (acre) of corn are shown in Table 15–2. Total energy inputs in kilojoules (Kilocalories) increased 313%, or faster than corn yields increased. The efficiency (output/input) declined from

TABLE 15–1. Average inputs per hectare (and acre in parentheses) in corn production during selected years in the U.S.

Inputs per hectare (acre)	1945		1950		1954		1959		1964		1970	
Labor (hours per crop)	57	(23)	44	(18)	42	(17)	34	(14)	27	(11)	22	(9)
Machinery (kJ + 10³) (kC + 10³)	1,867	(180)	2,594	(250)	3,112	(300)	3,631	(350)	4,357	(420)	4,357	(420)
Fuel (litres) (gallons)	168	(15)	190	(17)	212	(19)	225	(20)	234	(21)	247	(22)
Nitrogen fertilizer (kilograms) (pounds)	8	(7)	17	(15)	30	(27)	46	(41)	65	(58)	125	(112)
Phosphorus (kilograms) (Pounds)	8	(7)	11	(9.8)	13	(12)	18	(16)	20	(18)	35	(31)
Potassium (kilograms) (pounds)	6	(5)	11	(10)	20	(18)	34	(30)	32	(29)	67	(60)
Seeds for planting (kg*) (lb)	10.7	(3.8)	12.5	(4.5)	15.7	(5.7)	18.8	(6.8)	20.7	(7.5)	20.7	(7.5)
Irrigation (kJ × 10³) (kC × 10³)	197	(50)	239	(57)	280	(66)	322	(77)	353	(84)	353	(84)
Herbicides (kilograms) (pounds)	0	(0)	0.06	(0.05)	0.11	(0.10)	0.28	(0.25)	0.43	(0.38)	1.12	(1.00)
Drying (kJ × 10³) (kC × 10³)	104	(10)	311	(30)	622	(60)	1,037	(100)	1,245	(120)	1,245	(120)
Electricity (kJ × 10³) (kC × 10³)	332	(32)	560	(54)	1,037	(100)	1,452	(140)	2,106	(203)	3,216	(310)
Transportation (kJ × 10³) (kC × 10³)	207	(20)	311	(30)	467	(45)	622	(60)	726	(70)	726	(70)
Corn yields (kilograms) (bushels) (3-year average)	2,132	(31.7)	2,383	(35.5)	2,572	(38.2)	3,387	(54.0)	4,265	(63.5)	5,080	(75.6)

Source: Data from Pimental et al. 1973.
*Reflects the increased plant population per hectare (and acre) over the years.

3.70 to 2.82 or 30.3%. Specifically, an efficiency of 2.82 achieved for U.S. corn production in 1970 means that for each unit of cultural energy used in corn production, 2.82 units of energy were returned. Whether or not this is an acceptable depends on one's perspective. The following should also be noted:

1. Nowhere in the world and never in history have national averages reached the level of yield per hectare (acre) found under such a system of agriculture, and it may be reassuring to have a positive return, although marginal.

2. In an era when the price of a hectolitre (bushel) of corn can purchase a barrel of oil, an energy efficiency of near unity may be an acceptable one. However, in an era when a barrel of oil is ten times or more the value of a hectolitre (bushel) of corn, the relationship may be less than satisfactory.

3. Corn production with a marginal cultural energy efficiency of 2.82 leaves little scope for further on-farm processing through livestock or for transportation, processing, packaging, and distribution of corn products beyond the farm gate without expending more energy than is contained in the product itself. If the goal of food production is to be positive, the inescapable conclusion is that serious attention will have to be paid to the more significant elements in cultural energy aspects of agricultural production and processing, especially if cheap food policies are to be pursued.

4. The increased reliance on cultural energy inputs between 1945 and 1970 was not viewed with concern until it became obvious that energy is not an unlimited resource (Newcombe 1976). Although the precariousness of the energy situation and certainly the cost of energy are abundantly clear, mechanized agriculture cannot immediately or abruptly withdraw cultural energy inputs. In countries making a transition from a labor-intensive system to mechanization, and where labor is abundant and cheap and unemployment high, it is not a labor-saving device that is required but a labor-using device. In Tables 15-1 and 15-2, labor was the only input that declined in terms of hours per hectare (acre) for every year selected between 1945 and 1970, in the United States.

5. The decline in efficiency from 1964 to 1970 (Table 15-2) may be interpreted as signaling that the law of diminishing returns is limiting the yield increase with each additional joule (calorie) of cultural energy input. The more intensive the mode of agricultural production, the more energy-expensive it becomes per unit of production. If farmers are to remain in business, energy costs for food production must be reduced and a more efficient form of agriculture must be developed or food prices will rise markedly.

6. Many may not view the marginal efficiency of agricultural production with any concern. Consideration should be given however as to whether a higher level of efficiency is more profitable.

TABLE 15–2. Cultural energy inputs (thousands of kilocalories) per hectare (and acre) of corn for selected years in the United States based on Table 15–1

Input	1945 (hectare)	(acre)	1950 (hectare)	(acre)	1954 (hectare)	(acre)	1959 (hectare)	(acre)	1964 (hectare)	(acre)	1970 (hectare)	(acre)
Labor	130	(12.5)	102	(9.8)	96	(9.3)	78	(7.6)	62	(6.0)	51	(4.9)
Machinery	1,867	(180)	2,594	(250)	3,112	(300)	3,631	(350)	4,357	(420)	4,357	(420)
Fuel	5,637	(543.4)	6,388	(615.8)	7,140	(688.3)	7,516	(724.5)	7,892	(760.7)	8,268	(797.0)
Nitrogen fertilizer	610	(58.8)	1,307	(126.0)	2,353	(226.8)	3,573	(344.4)	5,054	(487.2)	9,760	(940.8)
Phosphorus fertilizer	110	(10.6)	158	(15.2)	189	(18.2)	252	(24.3)	284	(27.4)	489	(47.1)
Potassium fertilizer	54	(5.2)	109	(10.5)	523	(50.4)	627	(60.4)	705	(68.0)	705	(68.0)
Seeds for planting (hybrid input considered)	352	(34.0)	419	(40.4)	495	(18.9)	594	(36.5)	653	(30.4)	653	(63.0)
Irrigation	197	(19)	239	(23)	280	(27)	322	(31)	353	(34)	353	(34)
Insecticides	0	(0)	11	(1.1)	34	(3.3)	80	(7.7)	114	(11.0)	114	(11.0)
Herbicides	0	(0)	6	(0.6)	11	(1.1)	29	(2.8)	44	(4.2)	114	(11.0)
Drying	104	(10)	331	(30)	622	(60)	1,037	(100)	1,245	(120)	1,245	(120)
Electricity	332	(32)	560	(54)	1,037	(100)	1,452	(140)	2,106	(203)	3,216	(310)
Transportation	207	(20)	311	(30)	467	(45)	622	(60)	726	(70)	726	(70)
Total inputs	9,600	(925.5)	12,535	(1,206.4)	16,359	(1,548.3)	19,813	(1,889.2)	23,595	(2,241.9)	30,051	(2,896.8)
Corn yield (output)	35,554	(3,427.2)	39,737	(3,830.4)	42,874	(5,443.2)	56,468	(5,443.2)	71,108	(6,854.4)	84,702	(8,764.8)
Output/input (efficiency)	3.70		3.18		2.62		2.85		3.01		2.82	

Source: Data from Pimental et al. 1973

7. Cultural energy efficiencies indicate that crop production does not waste energy. It takes 1300 m³ (39,300 cubic feet) of natural gas to produce a tonne (ton) of nitrogen fertilizer. Five tonnes (5.5 tons) of this fertilizer could help produce enough energy and protein for 280 people a year at minimal dietary levels. However, a person living alone in a large house heated with natural gas may use this same amount of energy. Thus, societal values will have to determine whether or not cultural energy efficiencies are acceptable or not.

8. The lack of precise and absolute input values associated with the complex undertaking of estimating cultural energy efficiency may lead to a conclusion that the estimate is incorrect. The cultural energy efficiencies calculated by Heichel (1974) are therefore of considerable interest. Many crops such as soybeans, perennial forages, fruits, and vegetables are not produced strictly for their energy value but for protein, vitamins, minerals, or other features. To avoid the pitfall of underestimating the energy contained in crops not noted or produced for their energy content, Heichel (1974) related the annual consumption of energy from mineral fuels to the gross national product and determined the consumption of energy accompanying the production of a dollar of goods and services for 1900 to 1970. In this manner, dollars were translated into energy values without explicit knowledge of the portion of the cost of machinery, fertilizer, herbicides, or depreciation that was directly attributable to the consumption of fossil fuels, other energy sources, or labor. Grain corn in Illinois was found to have a cultural energy efficiency of 4.4 (Table 15-3), at a yield level of 5236 kg/ha (78 bu/ac). At similar output levels, the values are 4.4 (Heichel 1974) and 4.3 (Pimentel et al. 1973). The two measurements for corn at least are comparable and provide a degree of confidence in them.

Striking differences can be observed in cultural energy use efficiency among different crops and systems (Table 15-3). In labor-intensive systems in New Guinea and the Philippines, efficiencies of 16.4 for vegetables and 17.3 for rice, respectively, were found. Highly mechanized rice production in Louisiana had an efficiency of 1.4, although the yield was three times greater than in New Guinea or the Philippines. In a developing country, it appears that the yield of energy is small, but the caloric gain is large. Alfalfa and corn silage were shown to have values greater than 5 because in these crops the entire biomass is harvested. It should also be noted that despite an average increase of 2.6 times more cultural energy from 1915 to 1969, energy efficiency increased from a mean of 4.1 to 4.4.

Both reports (Heichel 1974; Pimentel et al. 1973) are based on average national values. Every farmer feels he is better than average, and he may wish to calculate cultural energy efficiency values for his own farm or crop. As an outline for calculating cultural energy efficiencies, the calculation for a representative corn farm in Ontario is shown below in greater detail. By

TABLE 15–3.   Cultural energy inputs (Thousands of kilo-Joules or kilo-calories) per hectare (acre) annually of various crops

Crop, Location, and Year	Yield (kg/ha)	(lb/ac)	Total Annual Cultural Energy Input (Kilo-Joules)	(kilo-calories)	Efficiency Output/Input
Grain corn (Iowa, 1915)	1756	1564	16,557	1596	4.8
Grain corn (Pennsylvania 1915)	1756	1564	23,673	2282	3.3
Grain corn (Illinois, 1969)	5236	4664	53,281	5136	4.4
Corn silage (Iowa, 1915)	17,065	15,200	34,120	3289	5.9
Corn silage (Iowa, 1969)	27,837	24,794	62,431	6018	5.3
Vegetable garden New Guinea, 1962			3330	321	16.4
Rice (Philippines, 1970)	1359	1210	3434	331	17.3
Rice (Louisiana, 1970)	4095	3648	123,824	11936	1.4
Sorghum (Kansas, 1970)	3211	7064	27,211	2623	5.3
Soybeans (Missouri, 1970)	1820	1621	30,769	2966	2.6
Oats (Minnesota, 1970)	1761	1569	23,902	2304	2.5
Alfalfa (Missouri, 1970)	6274	5588	22,958	2213	5.1
Peanuts (North Carolina, 1970)	2282	2033	120,048	11572	1.4
Sugar beets (California 1970)	5172	4607	91,799	8849	0.8 (including processing)
(refined sugar)					1.2 (excluding processing)
Sugar cane (Hawaii, 1970)	9079	8087	27,377	2639	2.5 (including processing)
(refined sugar)					4.6 (excluding processing)

Source: Data from Heichel 1974.

calculating values pertinent to a specific operation, energy-intensive inputs can be spotted. The following energy inputs were considered by Stevenson and Stoskopf (1974).

The Farm

An 80-hectare (200 acre) corn crop was considered as the main farm enterprise, and a yield level of 5645 kg/ha (90 bushels/acre) was considered.

Machinery.   The equipment involved in the production of this corn crop included:

- 2 tractors (68 kW and 38 kW) (90 and 50 horsepower)
- 1 plow (5 furrow—40 cm (16 in) bottoms)

- 1 cultivator
- 1 corn planter
- 1 sprayer
- ½ ownership in a four-row combine
- 1 stalk chopper
- 1 dryer (forced air)
- 3 wagons and gravity boxes
- accessory equipment
- storage facilities
- 1 farm truck

A small percentage of the farm truck could be credited to operations relative to transporting seed, fertilizer, or dry corn to the elevators. The total weight of this line of equipment and facilities is in the order of 18.2 tonnes (20 tons) or on 80 hectares (200 acres), 227 kg/ha (200 lb/ac). The energy network input based on the manufacture and construction of a 1530 kg (3366 lb) automobile was estimated by Berry and Fels (1973) as 86,788,000 kJ per tonne (18,700,000 Kcal per ton). Using the same conversion factor, the energy required to produce the equipment for one hectare (acre) is:

$$0.227\text{t/ha} \times 86,788,000 \text{ kJ} = 19,745,100 \text{ kJ/ha} (1,900,000 \text{ Kcal/acre})$$

Assuming a ten-year life span for this equipment, the energy per hectare (acre) expressed on an annual basis becomes:

$$1,974,510 \text{ kJ/ha/yr} \qquad (190,000 \text{ Kcal/acre/yr})$$

Including repairs at 6% of the original machine, 6% or 118,669 kJ (12,800 Kcal) for ten years is added in, or 11,867 kJ/yr (1280 Kcal/yr). Thus the total energy equivalent of the farm machinery is:

$$
\begin{array}{ll}
1,974,510 & (190,000) \\
\underline{11,867} & \underline{(1,280)} \\
1,986,377 \text{ kJ/ha/yr} & (191,280 \text{ Kcal/ac/yr})
\end{array}
$$

Labor. About ten hours of labor are required per hectare (4 hr/ac) for corn production (Fisher 1970), and this includes preparatory time such as machinery maintenance and all associated production aspects. A farmer performing moderately hard work during these operations might consume about 21,000 kJ (5000 Kcal) of food energy per day. About 7,140 kJ (1700 Kcal) are required for body maintenance, and only two-thirds of the remaining are directed to corn production. The energy directed to corn production would be:

21,000 (intake) $-$ 7,140 (maintenance) $=$ 13,860 (3300 Kcal)
13,860 $\times$ 2/3 (work) $=$ 9,240 kJ/day (2200 Kcal/day)
$$\frac{9,240}{8 \text{ hr}} = 1,155 \text{ kJ/hr (275 Kcal/ac)}$$

Thus, if a total of ten hours is required per hectare (4 hrs/acre), the labor input is 11,550 kJ/ha (1,100 Kcal/ac).

Fuel.   Between 67 (15) and 84 litres (18 gallons) (mean of 75.5) (16.5 gallons) of fuel are required to produce a hectare (2.47 acres) of corn (Schwart 1974; Constien 1974). Thus:

1 litre of gasoline $=$ 40,249 kJ (1 gallon of gasoline $=$ 43,603 Kcal)
75.5 $\times$ 40,283 $=$ 3,041,367 kJ (293,000 Kcal/ac)

Fertilizer.   Fertilizer was applied at the rate of 112 kg (100 lb) N, 56 kg (50 lb) P_2O_5, and 67 (60 lb) K_2O per hectare (acre), and energy values reported by Biswas and Biswas (1976) were used as follows:

- 1 kg nitrogen $=$ 83,160 kJ to produce and process (1 lb N $=$ 8400 Kcal)
- 1 kg phosphorus $=$ 14,045 kJ to mine and process (1 lb phosphorus $=$ 1520 Kcal)
- 1 kg potassium $=$ 9,702 kJ to mine and process (1 lb potassium $=$ 1050 Kcal)
- 112 kg N/ha $=$ 83,160 $\times$ 112 $=$ 9,313,920 kJ (840,000 Kcal/ac)
- 56 kg P_2O_5/ha $=$ 14,045 $\times$ 56 $=$ 786,520 kJ (76,000 Kcal/ac)
- 67 kg K_2O/ha $=$ 9,702 $\times$ 67 $=$ 650,034 kJ (63,000 Kcal/ac)

Seed.   The energy contained in the seed although small, is included as an input. (1 kilogram of corn has 16,632 kJ) (1 lb $=$ 1800 Kcal). At a seeding rate of 20.7 kg/ha (18.5 lb/ac), the energy equivalent is:

16,632 $\times$ 20.7 $=$ 344,282 kJ (1800 $\times$ 18.5 $=$ 33,300 Kcal)

An additional 90% was included by Pimentel et al. (1973) to cover the effort in producing hybrid seed. Thus:

Total seed energy $=$ 653,562 kg J/ha (63,270 Kcal/ac)

Insecticides and Herbicides.   Pesticides are usually manufactured from a petroleum base and have a high energy content estimated at 20,790 kJ per kilogram (2250 Kcal/lb) including production and processing (Pimentel et al.

1973). Applications totaling 11 kg per hectare (9.8 lb/ac) of total material amount to 228,228 kJ/ha (22,050 Kcal/ac).

Irrigation.   An estimated 782,890 kJ (29,711 Kcal) are required to apply one cm per hectare (one inch per acre) of irrigation water (Pimentel et al. 1973), but higher and lower costs may be associated depending on the method of application (Newcombe 1976). Very little corn is irrigated in Canada; Pimentel et al. (1973) estimated this to be about 3.8% in the United States, and this item is omitted as a cultural energy input in these calculations. Irrigation is a highly energy-intensive input. A litre (gallon) of water weighs 1 kg (10 lb), and based on a requirement of 23.3 million litres (5.1 million gallons) to produce a 5645 kg/ha (90 bu/ac) corn crop (Addison 1961), the energy cost of moving that quantity of water is about 147 million kJ (35 million Kcal). Steinhart and Steinhart (1974) considered that water use at this level would result in irrigation being the single largest user of energy in crop production.

Drying, Ensiling, or Crib Storage.   Corn produced in the northern extremity of its area of adaptation may have a high moisture level at harvest time, and hence the energy input for drying is high. In Ontario, the average propane requirement for drying is 156 litres per hectare (14 gallons/acre). Each litre has 28,872 kJ (31,278 Kcal per gallon) and the total energy required for drying corn per hectare (acre) is:

$$25,872 \times 156 = 4,036,032 \text{ kJ } (437,892 \text{ Kcal/acre})$$

Cultural energy requirements for ensiling high moisture corn or for crib storage is much lower than for artificial drying.

Electricity.   To determine electrical energy consumption, Pimentel et al. (1973) divided total agricultural electrical usage in the United States by the number of cultivated corn hectares, which grossly overestimates the electrical energy inputs into corn production because many farms use electrical energy for a variety of uses. The main electrical inputs into crop production would be associated with grain-conveying equipment and possibly some grain driers. Therefore, an arbitrarily chosen figure of 5% of total agricultural usage was taken for this example. This amounted to 3527 kJ per hectare (340 Kcal/acre).

Transportation.   The value of 726,180 kJ (70,000 Kcal/ac) used by Pimentel et al. (1973) was adopted.

Energy Produced.   Only the economic or grain yield is considered in this calculation. One kilogram of corn contains about 16,632 kJ (1800 Kcal per/pd). At a yield of 5,645 kg/ha (90 bu/ac), the energy produced is:

$$5,645 \times 16,632 = 93,887,640 \text{ kJ } (1800 \times 90 \times 56 = 9,072,000 \text{ Kcal})$$

Total Cultural Energy Inputs. A cultural energy efficiency of 4.38 was obtained (Table 15-4) for the Ontario corn crop. If the cultural energy efficiency of 2.82 (Pimental et al. 1973) (Table 15-2), obtained at a yield level of 5,080 kg/ha, is calculated at a 5,645 kg/ha (90 bu/ac) yield level, the efficiency value obtained is 3.94. The two values of 4.38 and 3.94 can be considered comparable.

DETERMINING THE HIGHEST CULTURAL ENERGY INPUTS

Of interest to crop producers is the identification of the most energy-intensive inputs. For the Ontario field these were, in order of magnitude: nitrogen fertilizer (43.5%), drying (18.8%), fuel (14.2%), and machinery (9.3%)—which together accounted for 85.7% of all inputs. In the calculations by Pimental and his associates, nitrogen fertilizer was the highest component; but because only 30% of the corn produced in the United States was dried and because the average moisture level was reduced from 26.5 to 13%, drying was a more modest input than in Ontario. In the United States, pesticides accounted for 0.8% and in Canada, 1.06%, of the total energy input (Table 15-4).

TABLE 15-4. Cultural energy inputs per hectare (acre), percentage of total inputs represented by each item, and cultural energy efficiency for a representative corn crop in Ontario in 1974

Input	Cultural Energy Input		Percent of Total
	(kJ/ha/yr)	(Kcal/ac/yr)	
Labor	11,550	1,100	.05
Machinery	1,986,377	191,280	9.27
Fuel	3,041,367	293,000	14.19
Fertilizer N	9,313,920	840,000	43.45
P	786,520	76,000	3.67
K	650,034	63,000	3.03
Seed	653,560	63,270	3.05
Pesticides	228,228	22,000	1.06
Irrigation	0	0	0
Drying	4,036,032	437,892	18.83
Electricity	3,527	340	0.01
Transportation	726,180	70,000	3.39
Total input	21,437,295 kJ	2,057,882 Kcal	100
Total output (5645 kg/ha yield) (90 bu/ac)	93,887,640	9,072,000	
Efficiency (output/input)		4.38	

Source: Stevenson and Stoskopf 1974.
N = Nitrogen
P = Phosphorus
K = Potassium

In a study of vegetable production on a hectare of land in Hong Kong, utilizing the most modern farming technology, Newcombe (1976) found that fertilizers accounted for 44.5%, irrigation 37.3%, and machinery that consisted of one rotary hoe accounted for 9.9%—for a total of 91.7% for these three items. Cultural energy efficiency was 0.13. On a larger-based paddy rice and sweet potato operation, cultural efficiency was 1.25, a value similar to that in Australia (Gifford and Millington 1975). Newcombe compared these efficiency values with a double-cropped rice and broad bean program (on a hectare of land) in China during the preindustrialization period of 1935–1937. He criticized the trend toward increasing energy inputs and compared the situation with vegetable production in California under a system of intensive technology where efficiencies ranged between 0.24 and 0.42 (including postfarm gate transportation) (Cervinka et al. 1974).

TABLE 15–5. Comparison of cultural energy relationships on grain farms in western Canada in 1975

Farm Description	Largest Cultural Energy Inputs (%)		Cultural Energy Efficiency
Wheat and fallow (650 ha). Swift Current, Saskatchewan	Machinery and fuel Transportation Fertilizer	— 32 — 18 — 5*	4.56
Grain and fallow (113 ha). Three Hills, Alberta	Fertilizer Machinery and fuel Heating (farm shop)	— 22 — 15 — 16	1.91
Grain and fallow (2355 ha). Vermillion, Alberta	Fertilizer Machinery and fuel Transport	— 38 — 28 — 15	2.48
Irrigated wheat, potatoes, sugar beets, and fallow (377 ha). Taber, Manitoba	Fertilizer Irrigation Machinery and fuel	— 28 — 18* — 12	1.67
		Mean	2.66

Source: Data from Timbers and Jensen 1978.

*The use of commercial fertilizers for small grains on the Great Plains of North America is a recent development; the response of the small grains to fertilizer is variable in this area because the soil moisture levels have a greater effect on grain production than does fertilizer. When small grains are grown on fallow land, they rarely need nitrogen fertilizer. When the crop is irrigated, cereal grains do respond to added nitrogen. Under the dry-land conditions at Swift Current, low levels of fertility meant a low cultural energy input and hence a high efficiency; under irrigated conditions at Taber, cultural energy inputs in terms of fertilizer and irrigation were high, and although higher yields were obtained, cultural energy efficiency was drastically reduced.

In Israel, a country dependent on external fuel sources, 39% of the cultural energy input into crop production was represented by irrigation (Stanhill 1974). Irrigation is often regarded as a key to giant increases in crop production or to make unproductive areas flourish. The report by Stanhill (1974) shows that in Israel, where the average amount of water applied is 78 cm/year (30 in), with water carried a distance of 30 kilometres (18.6 miles), irrigation is a questionable value, especially if irrigation water is to be supplied through desalinization. The cultural energy efficiency in 1974 was found to be 1.06, and if desalinized water for irrigation is used, a large energy deficit in crop production on the farm could result. Considerable interest and hope exist for successful crop production schemes that will increase the cultural energy efficiency.

The dry-land prairie grain farmer of the North American midwest is considered one of the most cultural energy-efficient food-producing areas of the world, despite relatively low yields and extensive use of summer fallow. An energy evaluation of selected Canadian prairie farms was conducted to identify the total distribution of cultural energy inputs on various types and farm sizes (Timbers and Jensen 1978) (Table 15-5).

For apples, peaches, asparagus, and potatoes in Ontario (Lougheed et al. 1975), a mean cultural energy efficiency of 1.52 was obtained (Table 15-6). Fertilizer and fuel were found to be the two highest inputs.

TABLE 15-6. Cultural energy inputs, expressed as percent of total for fruit and vegetable production in Ontario

	Percent of Total Cultural Energy Input				
Input	Apple	Peach	Asparagus	Potato	Mean
Labor	1.3	2.0	1.4	0.3	1.25
Machinery	7.2	12.9	5.9	3.9	7.5
Fuel	28.9	51.4	23.6	15.0	29.7
Pesticides	12.7	9.9	1.9	6.8	7.8
Fertilizer (total)	48.1	23.0	64.6	66.8	50.6
N	41.5	19.4	49.5	—	—
P	1.5	1.2	9.0	—	—
K	5.1	2.4	6.1	—	—
Herbicide	1.8	0.8	2.6	—	—
				7.2 (irrigation)	
Yield (kg/ha)	22,428	6,728	1,993	27,800	
Efficiency (output/input)	1.97	0.28	0.12	3.72	1.52

Source: Data from Lougheed et al. 1975.

N = Nitrogen; P = Phosphorus; K = Potassium.

IMPROVING ENERGY EFFICIENCY

Generally, the more intense the agricultural production system, the greater the cultural energy expended and the lower the cultural energy efficiency. Until inexpensive and reliable energy forms are found to replace petroleum, a greater efficiency of cultural energy usage is a worthwhile goal. Greatest attention should be directed to the highest cultural energy inputs. Newcombe (1976) estimated that 40 to 50% of the energy invested in the Hong Kong food system in 1971 could be conserved without a significant decline in crop production, although he suggested there is little economic incentive to achieve such energy conservation. The greatest savings were found to be related to an increase in labor associated with irrigation. Heichel (1974) suggested gains in energy efficiency are most likely to be realized by producing food in efficient cropping systems (sugar cane) rather than inefficient ones (sugar beets), and by developing new systems (reduced tillage, rotations, manure use) requiring less energy to produce food. Genetic improvements in photosynthetic efficiency, the use of crops that require little processing before consumption, more extensive use of plants as food sources of protein and energy, the use of wastes as energy sources, and the development of frugal cultural practices are all needed to stimulate energy efficiency and slow the onset of a fuel-related crisis in food production.

Crop producers are interested in steps they can take to improve crop production, increase profits, and increase cultural energy efficiency. The following suggestions should be considered for achieving these goals.

Nitrogen Fertilization

Nitrogen is a vitally important plant nutrient, and starving the plant of nitrogen and other fertilizer inputs cannot lead to cultural energy efficiency. Nitrogen must be applied at a rate and in a form that will meet crop needs (Table 15–7). The values shown in the table may vary with previous crop and past fertilization programs.

TABLE 15–7.   Suggested nitrogen fertilizer application rates for selected yields in Ontario as an indication of nitrogen requirements

Yield Objective				Nitrogen (N) Required	
Shelled Corn		Silage Corn			
(t/ha)	(bu/ac)	(t/ha)	(tons/ac)	(kg/ha)	(lb/ac)
5	80	25	11.5	80	71.4
6	96	30	13.5	100	89.3
7	112	35	15.8	120	107.1
8	130	40	18.0	140	125.0

Cultural energy associated with inorganic nitrogen sources is related to the large amount of fossil fuel energy used to manufacture it. An obvious method, therefore, of reducing cultural energy inputs is to find alternate nitrogen sources.

On livestock farms, manure is a source of nitrogen and other fertilizer nutrients, and these nutrients can substitute for appreciable amounts of fertilizer (Table 15–8). For the most effective use of plant nutrients, manure should be applied in the spring and preferably be covered with soil the same day to prevent loss of ammonium nitrogen to the air. When manure is covered immediately after application, it is estimated that 25% more nitrogen will be provided to the crop than when it is not covered. Most of the nitrogen loss to the air occurs in the first week after application.

Proper manure storage can reduce loss of plant nutrients. For maximum conservation of plant nutrients, manure should be stored in a way that saves the urine and with as little exposure to the air as practicable.

Costs of transporting and spreading manure are high relative to its value for plant nutrition, and hence it often must be spread close to the source. Hauling and spreading manure within a radius of 0.8 to 1.6 kilometres (1.3 to 1.0 miles) and applied at a 22 t/ha (10 tons/ac) rate, Pimentel et al. (1973) estimated a cultural energy input of 4,133,780 kJ per hectare (398,475 Kcal/ac).

TABLE 15–8. Reductions in fertilizer application where manure is applied

Solid Manure*		Liquid Manure**		Subtract from Fertilizer Requirement (kg/ha) (lb/ac) Nitrogen***				
(t/ha)	(tons/ac)	(m³/ha)	(gal per ac)	F & W	Spr.	Spr C	P_2O_5	K_2O
		Cattle or mixed livestock manure						
20	9.0	50	4500	25 (22)	50 (45)	60 (54)	20 (18)	90 (80)
30	13.5	75	6750	40 (36)	75 (67)	90 (80)	30 (27)	135 (121)
40	18.0	100	9000	50 (45)	100 (89)	125 (112)	40 (36)	180 (161)
		Swine manure						
20	9.0	30	2700	25 (22)	50 (45)	60 (54)	25 (22)	35 (31)
30	13.5	45	4050	40 (36)	75 (67)	90 (80)	35 (31)	50 (45)
40	18.0	60	5400	50 (45)	100 (89)	125 (112)	50 (45)	70 (63)
		Poultry manure						
4	1.8	10	900	20 (18)	45 (40)	60 (54)	25 (22)	22 (20)
10	4.5	25	2250	50 (45)	110 (98)	140 (125)	60 (54)	44 (39)

Note: Some of the nitrogen and most of the phosphorus not available to the crop in the year of application are in organic form and become available gradually in later years.

*10 tonnes/ha = 4.5 t/acre.

**One cubic metre per hectare = 90 gallons per acre.

***F & W denotes fall- and winter-applied manure.

Spr denotes spring-applied and not covered immediately, including surface applications after seeding.

Spr C denotes spring-applied manure covered after application.

This manure could provide the equivalent of 65 kg N/ha (58 lb/ac) which when applied would have a cultural energy value of 5,405,400 kJ (521,053 Kcal); and if 4.5 litres (1 gallon) of fossil fuels were required for chemical fertilizer application, total cultural energy involved would be 5,405,400 + 181,120 = 5,586,520 kJ (521,053 + 17,459 = 538,512 Kcal/ac). If manure was substituted for chemical fertilizer, there would be a substantial 1,452,740 kJ per hectare saving in cultural energy (140,037 Kcal/acre).

Nitrogen fertilizer inputs can be reduced by planting legumes. When sod containing perennial legumes such as alfalfa, trefoil, and clover is plowed under, these legumes supply an appreciable amount of nitrogen to the following crop (Table 15-9). The cultural energy cost of seeding a legume was estimated by Pimentel et al. (1973) for fuel and seeds as 933,660 kJ per hectare (90,000 Kcal/ac). If the legume crop produced 112 kg/ha (100 lb) of nitrogen, this would be equivalent to 9,313,920 kJ (897,813 Kcal) or a saving of 8,380,260 kJ (807,813 Kcal). Not included is the beneficial effect of legumes or manure on soil structure and erosion.

The switch from crops with high fertilizer demand, such as corn, to crops with a low fertilizer demand, such as soybean legumes, could reduce fertilizer demand. However, these changes would be possible only if the demands for corn products were modified, such as the demand for corn for livestock. Texturized protein meat substitutes would be less expensive than meat, but the cultural energy involved in producing texturized soybean products would need investigation.

Interest in the use of sewage sludge for fertilizers and soil conditioners has been stimulated by the U.S. Resource Conservation & Recovery legislation of 1976 (Sikora et al 1980). Sewage sludges contain nutrients essential for plant growth, but industrial sludges are contaminated with heavy metals, which can become toxic to plants or can lead to undesirable metal accumulation in plant parts. Disease organisms may also restrict land application of sludge.

Application of sewage effluents over long periods in Europe of about 90 t/ha have been reported (Carlson and Menzies 1971), although some evidence exists that copper and zinc concentrations may be causing yield reduction on sewage farms that have been operating for 70 to 100 years. The extent that

TABLE 15-9.   Adjustment of nitrogen requirement where sod containing legumes is plowed under

Type of Sod	For All Crops, Deduct from Nitrogen Requirement	
	(kg N/ha)	(lb N/ac)
Less than one-third	0	0
One-third to one-half	55	49
One-half or more legume	110	98

farmers use sewage sludge will depend on the cost or cultural energy of available nitrogen, phosphorus, and potassium and the possibility of harmful contaminants. Processing of sewage to destroy odor, bacteria, or disease agents may prove to be too energy-intensive in terms of the nutrients contained. It seems certain that agricultural land will have to be used more and more for urban waste disposal, and if this effluent can be applied with a cultural energy advantage, it will be used more readily. Newcombe (1976) noted that night soil, animal manure, and compost, which were formerly an essential feature of nutrient-recycling in Hong Kong, have come to be regarded as pollutants, but he insisted that the nutrients contained in such products must be recycled through the soil. Soils receiving sewage and the crops produced on these soils must be carefully monitored for trace element concentrations.

Machinery and Fuel

Minimum tillage, defined as the minimum soil manipulation necessary for maximum crop production under existing soil and climatic conditions (Lindwall 1980), should be practiced to the fullest extent possible. The extension of minimum tillage to its ultimate is zero-tillage (the least soil disturbance possible), but in areas with cold and heavy soil conditions, zero-tillage may be unrealistic at this time. In many parts of the world with light-textured and warm soils, zero-tillage has been commercial practice for nearly two decades. More than five million hectares (over 12 million acres) of crops such as corn, sorghum, soybeans, and winter wheat are grown with zero-tillage systems in the United States. Hectares of winter cereal and oilseed crops in the United Kingdom and in Australia and spring cereals in Canada seeded with zero-tillage methods are large and expanding (Lindwall 1980).

Cultural energy inputs in crop production may be reduced by the wise use of manure (Fig. 15–1) or minimum tillage practices (Fig. 15–2).

Cultural energy savings associated with minimum or zero-tillage need close examination as increased levels of herbicides, pesticides, and fertilizers are required. Also crop residues are often burned to facilitate seeding, to allow the soil to warm up, and to reduce insect and disease problems, although some conservation of crop residue is necessary to prevent soil erosion and to maintain soil fertility. A complete analysis of the precise costs and benefits under each farm condition is needed to determine the merit of each tillage system.

A wise choice of crops, machines, and tillage operations may reduce energy inputs. Cultural energy input requirements for tillage operations vary appreciably with soil type, crop sequence, and soil moisture content, as well as with design of the tillage implements, land topography, depth, and speed of tillage. The proportion of total cultural energy that tillage represents varies markedly among crops (Table 15–10) (Southwell and Rothwell 1977).

Figure 15–1. Liquid manure is an integral part of an agro-ecosystem. The value of the nutrients in the manure must be considered in terms of energy expended in hauling, soil compaction under wet conditions, odor, and runoff. (Courtesy Ontario Ministry of Agriculture and Food.)

TABLE 15–10. Proportion of total energy inputs used in tillage operations for major Ontario crops

		Percent of Energy Used in Tillage	
Crop	*Tillage Operations*	*Percent of Direct Fuel Use*	*Percent of Total Energy Input*
Grain corn	Plowing and discing	11.8	5.3
Barley	Plowing, discing, and harrowing	42.4	17.1
Oats	Plowing, discing, and harrowing	42.4	22.4
Soybeans	Plowing and 2 discings	55.7	27.3
Winter wheat	1 discing	17.4	4.9
White beans	Plowing, 2 cultivations and 1 interrow cultivation	70.4	34.6

Source: Data from Southwell and Rothwell 1977.

Figure 15-2.   Seeding into soil with the least soil disturbance is being practiced here to reduce energy inputs and reduce soil erosion. Crop producers must determine root bed and seedbed requirements in their area for successful crop production. (Courtesy Ontario Ministry of Agriculture and Food.)

The choice of implement to achieve the desired level of tillage can influence cultural energy inputs (Table 15-11) (Southwell and Rothwell 1977). Table 15-12 shows cultural energy inputs for five tillage systems on three soil types in Indiana (Griffith, Mannering, and Richey 1977). In terms of cultural energy, conventional tillage techniques (fall plow; disc twice; plant) require an average of 1,372,491 kJ (32,678 Kcal) (with 1 litre (gallon) of fuel at 40,249 kJ (43,603 Kcal) compared to zero-tillage with a mean value of 309,917 kJ (73,790 Kcal), a saving of 1,062,574 kJ (252,994 Kcal) for fuel. The possible need for more fertilizer and more herbicides and pesticides and a possible yield reduction may reduce this cultural energy saving for zero-tillage. However, this does not mean that zero-tillage should not be practiced because it is an excellent method for soil and water conservation.

TABLE 15-11.   On-farm measurements of fuel consumption in tillage operations in Ontario, 1975 and 1976

Tillage Operation	Diesel Fuel			
	Average		Range	
	(litres/ha)	(gal/ac)	(litres/ha)	(gal/ac)
Moldboard plowing	15.7	1.4	9.4–20.9	0.8–1.9
Tandem discing	6.4	0.6	5.4– 8.1	0.5–0.7
Cultivating	4.6	0.4	3.8– 6.4	0.3–0.6

Source: Data from Southwell and Rothwell 1977.

TABLE 15–12. Cultural energy requirements for tillage and planting operations on three Indiana soil types

Tillage System	Diesel Fuel Requirements (litres/hectare) and (gal/ac)					
	Sandy Loam		Loam		Silt Loam	
	(litres/hectare)	(gal/ac)	(litres/hectare)	(gal/ac)	(litres/hectare)	(gal/ac)
Fall plow; disc twice; plant	32.0	2.8	34.8	3.1	35.5	3.2
Fall plow; cultivate with planter attached	26.9	2.4	32.5	2.9	31.0	2.8
Fall chisel plow; cultivate with planter attached	21.9	1.9	25.5	2.3	22.5	2.0
Disc twice; plant	13.6	1.2	14.2	1.3	13.8	1.2
Zero-tillage	8.2	0.7	9.5	0.8	5.4	0.5

Source: Data from Griffith, Mannering, and Richey 1977.

Other cultural energy savings exist in tillage operations, even where no reduction occurs in the extent of land preparation. Vyn, Daynard, and Ketcheson (1979) suggested five steps for achieving cultural energy savings.

1. *Avoid excessive speed.* Increased speed causes greater soil acceleration (i.e., the speed at which soil is "thrown" by the tillage implement) and increased soil strength (resistance to tool movement through soil) ahead of the tillage tool, which requires an increase in implement draft. High speed during moldboard plowing in the fall can cause excessive soil pulverization and erosion susceptibility over the winter and in spring. In the spring, however, the creation of many fine aggregates during plowing is desirable to prevent excessive desiccation of surface soil, and to minimize the need for further cultivation.

   Discs perform well at a certain speed of travel, and a faster or slower speed will result in unacceptable performance. Too fast a speed causes soil ridging and too slow a speed means inadequate breakup of large aggregates. With cultivators and chisel plows, a minimum speed is essential to cause adequate soil shattering.

   In general, an increase in implement width is preferable to an increased speed in order to maximize fuel economy and tractor efficiency, all within practical limits. Too large an implement for the tractor increases drive-wheel slippage. Too slow a tractor speed can increase the torque in relation to speed and draft.

2. *Minimize unnecessary drive-wheel slippage.* If slippage is 25%, the outer circumference of the tire travels 4 km (2.5 miles) while the tractor itself moves only 3 km (1.9 miles). A 25% loss in drawbar power results, and fuel inefficiency occurs. Slippage in the magnitude of 10 to 15% is needed to derive traction, but beyond this, slippage is undesirable. Slippage can be reduced by weighting of tractor wheels or by transferring some of the draft of the implement into a downward force on the tractor, as automatically occurs in a three-point-hitch hookup. Wheel weighting does not permit an increase in fuel efficiency in direct proportion to the reduction in slippage since the extra weight means more energy is needed for tractor propulsion.

   Dual tires perform better than single tires with equal vertical axle loads (McLeod et al. 1966; Clark and Liljedahl 1969). Dual tires reduce sinkage on soft soil and may reduce slippage in response to the decreased rolling resistance on soft soils. Power-takeoff units are a means of reducing implement draft and wheel slippage.

3. *Combine tillage operations where possible.* Where the rolling resistance to tractor wheel movement is high, as on previously tilled and soft soil, over 50% of the fuel used may be required simply to

Figure 15–3. Cultural energy inputs may be reduced by combining tillage operations. (Courtesy Ontario Ministry of Agriculture and Food.)

propel the tractor. Tractor fuel consumption is not appreciably higher whether one or several implements are pulled simultaneously. Thus, from the standpoint of labor as well as of fuel consumption, combining field operations is desirable (see Figure 15–3).

4. *Reduce tillage depth.* No yield advantage exists to plowing deeper than 15 to 20 cm (6 to 8 in) (Table 14–16), although wide-bottom plows have a practical limitation as to the shallowness of depth that they can plow effectively. An increase of over 50% in fuel consumption can be expected as the depth of primary tillage is increased by 50%.

5. *Schedule tillage operations.* For tillage operations on wet soils, the surface tension between soil and implement is high, the draft increases dramatically, slippage is greater, and compaction increases.

Draft requirements are also high for plowing when clay (fine-textured) soils are too dry. Crop producers may have to schedule tillage without any control of weather conditions, but operations that can be postponed (e.g., plowing cereal stubble in August versus later dates) should be postponed.

Winter frost action is equivalent or superior on fall-plowed clay and clay-loam soils to the many extra litres per hectare of petroleum fuel required to obtain a finely aggregated seedbed structure with spring-plowed land. Chemical weed control in the fall is superior to extra fall cultivation or discing, which produces a finely aggregated seedbed structure and allows for erosion during winter and spring periods.

Crop Drying

The removal of excessive moisture is expensive and has a high cultural energy requirement. Corn is the major crop that requires drying to a safe storage level of 15% moisture. Crop producers should consider various possibilities.

1. Although it is generally true that full season corn hybrids outyield early maturing hybrids, the increase in yield potential may be offset by the added drying costs of less mature corn. Corn producers should pay close attention to hybrid selection because the potential exists for combining high yield and lower harvest moisture.
2. Storage as high moisture corn means earlier harvest and no drying costs.
3. Existing corn drying units may be modified to increase drying efficiency. *Dryeration* is the combination of high temperature drying and aeration. Instead of drying the corn to 15% moisture, corn can be removed from the drier at 16 to 18% moisture, and at 60° (140°F) contains sufficient heat to allow the removal of up to two percentage points of moisture using ambient air (aeration) at a low rate after a tempering period.
4. The removal of the lower moisture level in corn is very inefficient, and this is the basis of a reduced cultural energy requirement in dryeration. Do not dry corn below a safe storage level.
5. Savings of 20 to 40% are possible through the use of recirculating hot, dry air with the use of cooling air in continuous flow driers.
6. Low temperature driers with careful management can result in satisfactory corn storage and low cultural energy use.
7. As conventional energy sources become more expensive and with additional research, the use of solar energy as a heat source is possible (Teagan 1975).

In many areas of the world crop drying is required for safe storage. Energy required for drying represents a major cultural energy input in crop production (Fig. 15-4).

Irrigation

In dry-land regions, cultural energy inputs in the form of irrigation may be the largest input into agriculture (Stanhill 1974). In the past, the lack of adequate irrigation systems has limited agriculture in many regions; now, with cultural energy limitations added to the technical problems, energy costs may limit irrigated agriculture even more. The fact that cultural energy may severely curtail irrigation was spelled out by Rawlins (1980), but no solutions to reducing energy inputs were given. Rawlins stated that agriculture cannot

Figure 15-4.   Propane driers are used to dry crops to a level where they can be safely stored. Energy required for drying represents the second largest input in crop production after nitrogen fertilizer inputs. (Courtesy Ontario Ministry of Agriculture and Food.)

afford the luxury of abandoning irrigation; and the U.N. Food and Agriculture Organization (FAO) says that irrigated agriculture must be increased if the world is to be fed. The productivity of irrigated land was reported to be twice that of agriculture as a whole, but under Israeli agriculture, irrigation resulted in a negative energy balance (Stanhill 1974).

Irrigation was practiced extensively in ancient agricultural systems without fossil fuel inputs. Today, mechanization has resulted in an intensive energy requirement. Even with surface irrigation where water is allowed to flow over the land and infiltrate the soil, increased mechanization is occurring. Through the use of pipes with controlled openings or semiautomatic or automatic structures, mechanized surface irrigation results in much better water control, but pumps are needed to raise and carry the water, and this requires cultural energy.

In the late 1940s, light-weight aluminum tubing made sprinkler irrigation practical on a large scale. Manual moving of pipes gave way to wheel-roll systems, causing a dramatic change in agricultural irrigation, but this also increased energy requirements. Center-pivot systems consist of a lateral pipe mounted on towers with one end fixed to a pivot pad with lateral lengths from 50 to 800 m (160 to 2600 feet). The unit rotates around the pivot point and can operate on fairly rough topography. Linear-move systems that move up and down a field rather than rotating in a circle eliminate the disadvantage of center-pivot systems that do not irrigate the corners of a square field.

For some crop producers, an irrigation system is a form of insurance against a dry period. For others, irrigation is a necessity for crop production. Improved cultural energy efficiency can be achieved if the water is applied judiciously; i.e., if the water is applied at critical points that will encourage economic yield; if it is applied to encourage root development; and if it is applied only in sufficient quantities to meet the needs of the plant and to leach out salts accumulated by evaporation.

In many areas of the world irrigation is a necessity for successful crop production and represents a major input of cultural energy (Fig. 15-5).

Pesticides

In field crop production, cultural energy expenditures in herbicides, insecticides, and fungicides are a minor input. The dramatic and unrealistic suggestion to abandon chemical farming altogether should recognize this

Figure 15-5. Irrigation is an energy-intensive input in crop production. In areas totally dependent on supplemental water for crop production, irrigation may be the largest single energy input. (Courtesy Ontario Ministry of Agriculture and Food.)

point. Eliminating pesticides would reduce cultural energy inputs very little but could be expected to have a major impact on output, and hence the cultural energy efficiency could be reduced. If cultural energy in the form of pesticides is reduced or eliminated, the trade-off may be increased cultural practices. Direct-seeded forages are possible only with an accompanying chemical weed control application. Direct-seeded forages can produce a crop in the seedling year. Inability to use direct seeding and the use of a companion crop may result in competitive effects so great that seedling year growth is nominal. Herbicides remove the competitive effects by controlling weeds.

In the past, the best insect and disease control has been offered through genetic control. Genetic rust resistance is an excellent example. If genetic disease and insect control is coupled with pesticide use and cultural inputs satisfactory pest control often can be achieved.

Pesticide inputs can be reduced, however, without adversely affecting yield. A change to a policy of "treat when and where necessary" proposed by Steinhart and Steinhart (1974) could bring a 35 to 50% reduction in pesticide use. Spot spraying, crop rotations, removal of alternate hosts, and other biologic control methods offer considerable promise to reduce pesticide use. Monitoring pest situations by a central agency to trace the presence and spread of organisms and to alert growers when to spray may be more effective in controlling pests than systematic spray programs. A resounding success was achieved when weather-timed sprays were used to control carrot blight (McEwen and Sutton 1977). A comparison of unsprayed, nine weekly sprays, and an average of 4.5 weather-timed sprays showed that the weather-timed spray gave the lowest incidence of disease (Table 15–13). Weather-timed sprays were applied according to predicted weather conditions, and knowledge was accumulated as to the effects of weather on infection of carrot leaves by the blight fungus. When applied on a field scale, two weather-timed sprays reduced the blight to 2% infection at harvest; 9% when sprayed six or more times on a weekly basis; and 35% when not sprayed.

Pest monitoring can reduce cultural energy input and the pesticide load on the environment, but such techniques need to be developed on a wide basis and for more pests and may require a cooperative effort among regional authorities.

TABLE 15–13. Comparison of carrot blight infestation with three spray programs in Ontario over a two-year period, 1975 and 1976

Spray Program	Number of Sprays	Percent Blight
Nonsprayed	0	53.0
Weekly	9	18.5
Weather-timed	4.5	9.5

Source: Data from McEwen and Sutton 1977.

ENERGY INPUTS IN FOOD PROCESSING

Very little food makes its way directly from field and farm to the table, but instead enters into a vast system involving processing, packaging, and transportation. Home storage and preparation follow industrial processing and add an additional cultural energy input. Considerations of the cultural energy inputs following farm production are complex because of the diverse nature of the processing industry involved with washing, boiling, steaming, refining, milling, fermenting, freezing, drying, baking, cutting, canning, and packaging the 10,000 food items found on supermarket grocery shelves in North America.

Production of food on the farm accounts for less than 20% of the total energy used to place the food on the table ready for consumption. The breakdown of energy used to get food to the table (Anonymous 1977) is shown below:

- Production (to the farm gate)   18%
- Processing and packaging   32%
- Transport and distribution   20%
- Household preparation   30%

A change in eating habits toward less highly processed foods could reduce cultural energy for processing—but this is beyond the scope of this text. The point is, however, that savings in cultural energy related to food for consumers is not solely the responsibility of the crop producer. Accusations of high food prices should not be directed solely at the crop producer, but neither are price markups excessive in any one step of industrial processing. Consumers need to examine their own buying habits and preferences. As fossil fuel prices increase, the entire food industry will be subject to mounting pressure for change.

CONCLUSIONS

Change is not new to agriculture and the food industry. Perhaps few industries are as competitive and have experienced as many changes and disruptions over the years as the food industry. Innovations are needed throughout the entire food industry, including changes in eating habits and in attitudes toward food-related industries.

This chapter will prepare crop producers for change in cultural energy inputs with an understanding and insight into areas where changes can be most beneficial. Rather than giving a comprehensive treatise on this new and challenging subject, a technique has been provided here for determining cultural energy efficiency along with suggestions to increase efficiency, and hopefully this will stimulate new ideas and thinking that will lead to even greater innovations.

QUESTIONS

1. Review the reasons why cultural energy efficiency increased for grain corn from 1915 to 1969 despite an increase in cultural energy of 2.6 times (Table 15-3). Project the cultural energy efficiency value into the future and give reasons for your projection.

2. In the late 1960s, the sugar beet industry in Ontario was discontinued in preference to imported sugar. Suggest reasons for this action (see Table 15-3).

3. Comment on oat production from a yield and cultural energy efficiency standpoint compared to other crops. Trace the production (yield and hectares) of oats in your region and project any trend found toward the future.

4. Interpret the meaning of cultural energy efficiency values. Is a value of 2.82 acceptable or otherwise?

5. Calculate cultural energy efficiencies for each of the following:
   (a) A selected crop in your area using the format presented in this chapter but substituting your own yield values and appropriate cultural energy inputs.
   (b) A selected crop assuming that the complete water requirements for the crop are met with an overhead irrigation sprinkler system.

6. Set up a hypothetical crop production system aimed at reducing cultural energy inputs by means of manure and a rotation, and calculate cultural crop production efficiencies for a single year and for a sequence of crops.

7. Suggest additional energy-saving modifications in crop production.

8. Consider the pros and cons of on-farm energy production such as methane from manure wastes, alcohol from grain, or use of wastes for fuel and oil production.

9. Consider energy expenditure on food beyond the farm gate and how it may be reduced.

10. Discuss food production in totally enclosed greenhouses, growth rooms, or food factories in short-season areas with climatic limitations for crop production outdoors. Consider: (a) what inputs are necessary, (b) needed production and management inputs such as photoperiod, CO_2 levels, water and fertility, root media; (c) cultural energy efficiency; and (d) success of existing facilities and problems encountered.
    For additional information see Langhans and Poole 1978; Anonymous 1979; and Stoskopf et al. 1970.

11. Plot the price of oil and a crop commodity during the seventies on the world market.

12. Describe a situation in which fossil fuels are not available for crop production.

REFERENCES FOR CHAPTER FIFTEEN

Addison, H. 1961. *Land, Water and Food.* London: Chapman and Hall Ltd. (284 pp.)

Anonymous. 1977. *Energy and the Food System.* Food Systems Branch of Agriculture Canada. (123 pp.)

Anonymous. 1979. "The New Salad Factories," *Business Week,* January 29, 1979.

Berry, R. S., and M. F. Fels. 1973. *The Production and Consumption of Automobiles. An Energy Analysis of the Manufacture, Discard, and Reuse of the Automobile and Its Component Materials.* Chicago: University of Chicago Press. Cited from Pimentel et al. 1973.

Biswas, A. K., and M. R. Biswas. 1976. "Energy and Food Production," *Agro-Ecosystems* 2:195-210.

Borgstrom, G. 1973. "The Breach in the Flow of Mineral Nutrients," *Ambio* 2: 129-135.

Carlson, C. W., and J. D. Menzies. 1971. "Utilization of Urban Wastes in Crop Production," *BioScience* 21:561-564.

Cervinka, V., et al. 1974. *Energy Requirements for Agriculture in California.* Joint study by the California Department of Food and Agriculture and the University of California at Davis, California. (151 pp.)

Clark, S. J., and J. B. Liljedahl. 1969. "Model Studies of Single, Dual and Tandem Wheels," *Transactions of the American Society of Agricultural Engineering* 12:240-215.

Constien, E. J. 1974. "Energy Requirements for Selected Farming Operations," *Agricultural Engineering,* January issue.

FAO. 1970. *Provisional Indicative World Plan for Agricultural Development,* Parts 1 and 2. Rome: FAO (Food and Agriculture Organization) (672 pp.)

Fisher, G. A. 1970. "What Should It Cost to Produce Corn?" *Proceeding of the Ontario Soil and Crop Improvement Association.* 1970 Convention, Toronto.

Gifford, R. M., and R. J. Millington. 1975. "Energetics of Food Production with Special Emphasis on the Australian Situation," *CSIRO Bull.* 288:1-29.

Griffith, D. R., J. V. Mannering, and C. B. Richey. 1977. Energy Requirements and Areas of Adaptation for Eight Tillage-Planting Systems for Corn," In *Agriculture and Energy,* pp. 261-276, W. Lockeretz, ed. New York: Academic Press, Inc.

Handler, P., ed. 1970. *Biology and the Future of Man.* New York: Oxford University Press. (936 pp.)

Heichel, G. H. 1974. *"Comparative Efficiency of Energy Use in Crop Production,"* New Haven, Connecticut Bull. 739 (26 pp.) Connecticut Agr. Exp. Station.

Langhans, R. W., and H. Poole. 1978. "Commercial Use of Controlled Environments," in *Growth Chamber Environments, Proceed. of a Symp. Twentieth Intern. Hort. Congress, Phytotronic Newsletter* 19:52-56. Sydney, Australia.

Lindwall, C. W. 1980. "Zero and Minimum Tillage: Implementation and Consequences," *Agrologist* 9(1):9-10.

Lougheed, E. C., et al. 1975. "Fruit and Vegetable Production and the Energy Shortage," *HortScience* 10(5):459-462.

McEwen, F. L., and J. C. Sutton. 1977. "Pest Monitoring to Reduce Pesticide Use," *Notes on Agr.* 8(1):20-23. University of Guelph.

McLeod, H. E., et al. 1966. "Draft Power Efficiency and Soil Compaction Characteristics of Single, Dual and Low Pressure Tires," *Transactions of the American Society of Agricultural Engineering* 9:41-44.

Newcombe, K. 1976. "Energy Use in the Hong Kong Food System," *Agro-Ecosystems* 2:253-276.

Odum, H. T. 1971. *Environment, Power and Society.* New York: Wiley-Interscience.

Pimentel, D., et al. 1973. "Food Production and the Energy Crisis," *Science* 182:443-449.

Rawlins, S. A. 1980. "Irrigation in a Future Short of Energy," *Crops and Soils 32* (5):5-7.

Schwart, R. B. 1974. "How Much Fuel to Grow Corn and Soybeans?" in *Energy Crisis, Concerns and Conservation.* Cooperative Extension Service, University of Illinois, Urbana-Champaign, Illinois.

Sikora, L. J., et al. 1980. "Fescue Yield Response to Sewage Sludge Compost Amendments," *Agronomy Journal* 72:79-84.

Southwell, P. H., and T. M. Rothwell. 1977. *Analysis of Energy Ratios on Food Production in Ontario.* Research report by School of Engineering, University of Guelph.

Stanhill, G. 1974. "Energy and Agriculture: A National Case Study," *Agro-Ecosystems* 1:205-217.

Steinhart, J. S., and C. E. Steinhart. 1974. "Energy Use in the U.S. Food System," *Science* 184:307-316.

Stevenson, K. R., and N. C. Stoskopf. 1974. "Energy Efficiency in Crop Production," in *Energy and Agriculture,* pp. 38-47. Proceeding of the Ontario Institute of Agrologists. 15th Ann. Conf. Guelph.

Stoskopf, N. C., et al. 1970. "Glasshouse Replacement Rooms for Growing Plants," *Canadian Journal of Plant Science* 50:125-127.

Teagan, W. P. 1975. *Solar Energy Applications,* pp. 14-19. Proceedings from seminar on solar energy. Woburn, Mass.

Timbers, G. E., and N. E. Jensen. 1978. "Energy Analysis of Selected Farms in Western Canada," in *Future Energy Requirements on Farms and Ranches.* Alberta Institute of Agrologists Seminar, Calgary. (12 pp.)

Vyn, T. J., T. B. Daynard, and J. W. Ketcheson. 1979. *Tillage Practices for Field Crops in Ontario,* pp. 35-44. Published by the University of Guelph in cooperation with the Ontario Ministry of Agriculture and Food and the Ontario Ministry of Energy.

Wilson, F. 1978. "The Failure Was Not Genetic," *Ceres FAO Review on Agriculture and Development* 11(5):49-50.

chapter sixteen

Preservation and Storage

Feed for **ruminant animals** has traditionally been the primary reason for forage crop production. Today, however, perennial forage crops are being considered increasingly for their additional advantages including biological nitrogen fixation; for the protective cover they provide year round; for the beneficial effect on soil aggregation, especially from grass roots; and because the entire biomass can be harvested, which represents the greatest energy utilization. Furthermore, successful cash cropping of forages may allow land not suited to intensive crop production of annual crops to be returned to forages. For these reasons, forage crop production is to be encouraged as a cash crop, but its success will depend on whether high forage production can be coupled with successful harvesting, handling, and marketing without undue losses or spoilage. This is a challenge because forage crops have a high moisture content at harvest, are bulky, and have traditionally been produced on or near the farm where they are consumed.

Grain crops stand in contrast to forage crops. The economic component of grain crops is in the form of concentrated grain energy, and the focal harvesting unit is the **combine.** With minor modifications and adjustments to the combine, this one unit serves as the harvesting machine for many seed-producing crops. Once harvested, the grain can be stored readily at 13 to 14% moisture, and is in a concentrated and clean form that can be readily transferred, shipped, and stored again. Corn is both a forage and a grain crop. However, the elaborate system and machinery required to handle this crop as a forage are in marked contrast to what is required to handle the crop as grain. Systems of handling perennial forage crops vary from loose hay stacked outdoors; to rectangular bales stored indoors or stacked outdoors; to large weather-resistant round bales; stacked free from each other outdoors to storage in a fermented state in silos; to storage in energy-intensive dehydration systems. Moisture levels vary from 20 to 80%. Drying is—achieved by field-curing using the sun's energy—to various levels in between such as in a silo at 65% moisture where drying is not needed. In summary, none of these forage materials can be handled with the ease of grain.

Unlike grain, which can dry naturally on the plant, forages are highly susceptible to weather losses especially if field drying is desired. A large forage biomass may be cut at 80% moisture and dried to 20% in June when rain showers are frequent with each shower causing a lowering of quality. Excessively dry material will lose protein-high leaves. Weather losses can be reduced by the use of elaborate machines and protective storage to speed up the operation, to reduce labor inputs, and to beat the weather. But high and ever-rising energy costs, the very reason for including a forage legume in the first place, have forced up the cost of purchasing, operating, and maintaining machines and storage structures. Ensiling crop products is a satisfactory means of storing feed.

HANDLING FORAGES

For ruminant animals, a pasture provides feed at the lowest cost because the animals do the harvesting, but losses associated with trampling and defecation and fencing costs have caused a switch to confined feed-lot systems. If economic yield is determined by the amount of animal products per unit area, stored feed outclasses other systems of forage feeding (Table 16-1) (Young 1977). More and more interest exists in confinement feeding to reduce losses, to avoid cattle walking in a path to compact soil, to ensure rates of gain through improved water availability, and to reduce energy loss in walking to the field. Stored feeding also comes into play where forage production is seasonal and where long severe winters make stored feed a necessity.

Confinement feeding requires a forage-handling system. Such a system may consist of many combinations of machinery and structures. No one system dominates or has become universally adapted, which indicates that there is no one best method. Machinery is constantly undergoing change and must be made to accommodate many barn styles and feeding systems. Careful planning is essential when new facilities are constructed because a high degree of mechanization characterizes most forage-handling systems.

Field-cured hay continues to be a major method of handling and storing feed. When the crop is cut, a hay-conditioner crushes and crimps stems to

TABLE 16-1. Forage systems compared in Wisconsin

	Forage Dry Matter per Hectare (Acre)				Hectares (Acres) per Cow per Season		Milk	
	Available		Consumed					
	(hectare)	(acre)	(hectare)	(acre)	(hectare)	(acre)	(litres/ha)	(lb/ac)
Strip grazing	6609	5900	4366	3898	0.91	2.25	5271	4690
Green chopping	6530	5830	6329	5651	0.74	1.83	6222	5536
Stored feeding	8024	7164	7324	6539	0.59	1.45	6766	6020

Source: Data from Young 1977.

speed drying to an 18 to 20% moisture level. Losses of 300 to 500 kg of protein and carbohydrate per hectare (268 to 446 lb/ac) are not uncommon with field-drying (Hodgson 1976), and under adverse conditions 100% loss may occur. Field-cured hay may be stored, baled, or chopped and then stored indoors, or stored outdoors in large round bales 450 to 600 kg (990 to 1320 lb) in weight or as compressed stacks. In Indiana, after five months of outside storage, 8 to 11% of the outer surface of compressed stacks and large round bales had visibly deteriorated (Table 16–2) (Lechtenberg et al. 1976). High-density bales or pelleting requires less space for transportation and may be a method of adding flexibility to the forage crop.

TABLE 16–2. Weathering losses and quality changes in hay packages stored outside in Indiana for five months

	Portion Weathered in %	T.D.N. Loss from Outside Storage* (% of dry matter)
Grass hay		
Compressed stack	11.1	7.8
Large round bale	10.8	6.4
Alfalfa hay		
Compressed stack	8.1	7.4
Large round bale	10.7	9.1

Source: Data from Lechtenberg et al. 1976.

*T.D.N. = Total digestible nutrients.

Grass silage has the advantage of requiring less field-drying time and has minimal field losses, but silo losses can occur if it is not properly ensiled. Haylage can be ensiled at a moisture level of 40 to 60% in well-sealed structures. Ensilage made from whole-plant corn is cut and hauled directly to the silo. Silage products have the disadvantage of spoiling quickly when removed from storage and cannot be shipped any distance or restored.

Silage production offers the most weather-independent operation, but careful management and suitable structures are required to minimize storage losses. Tower silos may be completely sealed with a pressure-equalization tube to prevent oxygen from being in contact with the silage. Tower silos may be manufactured from glass-lined steel, poured concrete, concrete stave, heavy plastic, or plastic tubing. Top or bottom unloaders within the silo can carry the silage onto feed conveyors. The capacity of upright silos is shown in Appendix C.

Horizontal silos offer a less expensive storage unit with a fast unloading or self-feeding system, but they require more effort and management to seal out oxygen. Self-feeding systems are wasteful, and a push-button mechanization system is less complete. Front-end tractor loaders may remove the feed from the face of the silo. The capacity of horizontal silos is shown in Appendix D.

To produce a quality product from the silo, careful consideration must be given to crop management. A low-quality field crop will produce low-quality silage. Likewise, poor silo management will produce low-quality silage, so it is important to understand the chemical processes occurring in the silo—to achieve good silage.

Crops can be stored as fermented products in silos or in unprotected large round bales (Fig. 16-1). Large round bales should be stored on a dry site with a space between the bales so they can dry out properly.

PRINCIPLES OF SILAGE PRODUCTION

Almost any crop can be ensiled, but whole-plant corn silage is common because of the high yields and the consistency with which a quality product can be produced. Many producers do not understand the fermentation process but follow a recipe that assures them of a quality product. Corn producers who switched to grass silage and haylage in the belief that this was as easy to make into quality silage as corn have frequently encountered problems.

Figure 16-1. Many areas require stored feed for livestock production. Silos are excellent storage units for forage grasses and legumes, whole-plant corn silage, high-moisture grain, or other crop commodities. The round bales can be stored outside with a minimum of spoilage. (Courtesy Ontario Ministry of Agriculture and Food.)

However, when the problem can be accurately diagnosed, improvements can be successfully made.

Silage is produced through the action of fermentation and is similar to the process of wine-making or apple cider production. Since most people have witnessed the changes that occur in a jug of grape or apple juice, it is of interest to describe the changes that occur in such a container. The steel or concrete of a silo prevent the producer from observing events in the silo.

Bacteria are required for fermentation. Grapes and apples have many wild spores of bacteria, yeasts, and molds on their skins, which are carried into the juice when pressed. Under favorable temperatures, these bacteria will feed and multiply on the sugars in the juice and will readily start the fermentation process. Active fermentation can be observed by the release of CO_2 bubbles, which must be allowed to escape. Fermentation converts sugars into alcohol. As the alcohol level increases, a point is reached where the bacteria can no longer survive, and the process is complete.

Several key factors in wine production are noted below:

1. There are always adequate bacteria present, but control of the type of wine (burgundy, sauterne, muscatel) produced is achieved by adding an active and specific strain of bacteria that can dominate the process.
2. A source of sugar or fermentable carbohydrate is required for bacterial growth and fermentation. Sugar can be added to the grape juice, but any unfermented sugar will make the wine sweet. Fermentation will proceed as long as sugars are present or until the alcohol content is too high for the bacteria to exist.
3. **Anaerobic** conditions must prevail. The release of CO_2 during fermentation replaces oxygen, and if the containers are kept sealed, aerobic bacteria cannot survive. If air is allowed to enter, undesirable organisms develop and the wine may be unsatisfactory.
4. The product of fermentation is alcohol and acids, which stabilize the product by killing the bacteria. For example, pickles preserved in a jar will keep well under the acid conditions of the vinegar. Wine can be stored in sealed containers because of the alcohol content.

The process of making silage is based on the same principles as wine-making. The process is the same whether it occurs in an air-tight, glass-lined silo, or in a horizontal open-topped silo, or in a stack on the ground. It is the same for corn silage as it is for grass silage or haylage.

Stages of Ensiling Process

The ensiling process normally extends for about 21 days and several stages can be observed.

Stage 1. Stage 1 is the **aerobic** stage. The freshly chopped and bruised plant cells will carry oxygen into the silo. Some oxygen will be trapped in the air spaces of the chopped material. The trapped air will be used by respiration to produce heat; and the more oxygen trapped, the higher the temperature will rise. A small temperature rise is desired because it will favor bacterial activity. Since respiration converts food energy to energy and CO_2, respiration must be regarded as a wasteful process. However, some loss of energy by respiration cannot be avoided because CO_2 is required to develop anaerobic conditions. It is important to ensile at a moisture level sufficiently high to cause good packing (60–70% moisture) and to cut the material in lengths that will fit closely and exclude oxygen (10–13 mm 1/3 to 1/2 in).

A · basic principle in making quality silage is to maintain as nearly as possible anaerobic conditions. Fast-filling is a key to air exclusion because the faster the crop goes into the silo and the better it is distributed and packed, the less chance air has to enter. When the silo is full, it is wise to seal the top with a plastic film, if it is an open-top silo.

Essential or desirable bacteria will not feed on fresh plant material until cell respiration stops. Anaerobic conditions prevent yeasts, molds and aerobic bacteria present on the tissues from developing.

Under conditions of limited aerobic respiration, the temperature of the respiring mass will rise to about 29.0 to 40.0°C (85.0 to 104°F) by the 4th day.

Stage 2.   During the first three to four days of silage production, acetic acid-forming bacteria will dominate because such bacteria are favored by temperatures between 18 and 35°C (64.4 and 95°F). Acetic acid ionizes slowly; the pH does not fall rapidly and will be insufficient to "pickle" the silage.

Stage 3.   Anaerobic respiration, essential for successful silage, characterizes Stage 3. On the 4th day after ensiling, when the temperatures have reached 29 to 40°C (85 to 104°F), under moisture conditions of 65 to 75% and with a readily available source of sugar or carbohydrates, desirable lactic acid-forming bacteria dominate.

Stage 4.   Lactic acid formation will rapidly lower the pH to a 3.8 to 4.0 level, and further bacteria activity is restricted usually by about the 21st day. Rate of pH fall is an important feature of preservation to prevent undesirable bacteria (clostridia) from developing. The pH levels and the rate of change for orchard grass silage are shown in Table 16–3 (Sprague and Brooks Taylor 1977). The amount of lactic acid required for preservation will equal about 1.5% of the dry weight in corn silage, about 3% in grass silage, and about 6% in clover and legume silage.

Stage 5.   Under conditions in which the pH does not drop sufficiently due to inadequate lactic acid development because of very high moisture levels or inadequate amounts of fermentable carbohydrates, clostridia bacteria will

TABLE 16–3. Mean pH values from two silos of orchard grass in New Jersey from May 21 to September 1

Time after Sealing in Hours and Days	pH of Forage Mass
0 hours	5.58
5 hours	5.41
10 hours	5.48
20 hours	5.33
40 hours	4.59
60 hours	3.85
80 hours	3.79
5 days	3.83
10 days	3.67
20 days	3.93
60 days	4.28
103 days	4.35

Source: Data from Sprague and Brooks Taylor 1977.

develop, which encourage butyric acid development. Lactic acid is broken down as well as protein, which leads to spoilage, discoloration, and foul-smelling silage (like rancid butter) and results in an unpalatable product that limits animal intake. As lactic acid is broken down, the pH rises to 4.5 and proteolytic bacteria develop, which attack and break down protein. It is the breakdown of protein, called *putrefaction*, that renders the silage unpalatable.

Undesirable clostridia bacteria will seldom be a problem in corn silage, ear-corn silage, or high-moisture grain silage because of an adequate supply of fermentable carbohydrate. The soluble carbohydrate content of green perennial crops is extremely variable and depends on species, stage of growth, and weather and fertilizer application. Glucose and fructose, the major carbohydrate sources for fermentation, are required at a minimum level of 6 to 7% on a dry-weight basis to provide fermentation adequate to lower the pH value to 4.0. Glucose and fructose levels determined in five perennial ryegrass samples in England were generally below the required level (Table 16–4). Only sample No. 5 (Table 16–4) could be expected to produce a quality silage, but because less lactic acid is produced from fructose than from glucose, there is no guarantee of success.

Butyric acid development, favored by temperatures in the 35 to 40°C (95 to 104°F) range, is associated with a high moisture and protein content found in direct-cut alfalfa and ensiled grain residues from the brewing and distilling industries.

TABLE 16–4.   Soluble carbohydrates in perennial ryegrass at time of ensiling (percent dry matter) in England

Sample Number	Moisture Level %	Percent of Dry Matter		
		Glucose	Fructose	Total Fermentable Carbohydrate
No. 1	80.8	2.2	1.9	4.1
No. 2	80.1	1.6	1.3	2.9
No. 3	85.1	1.7	1.0	2.7
No. 4	78.5	2.2	1.9	4.1
No. 5	84.4	2.3	3.8	6.1

Source: Data from Whittenbury, McDonald, and Bryan-Torres 1967.

Material Ensiled Too Wet

The ideal moisture level for ensiling is in the mid-60% range. In whole-plant corn silage, this stage can be identified by kernels in the mid-dent stage at 45 to 50% moisture. In an attempt to obtain as large a yield as possible, producers often select a late maturing, full-season hybrid so that at filling time the corn is immature and wet. Grain development is limited, and moisture content may be over 75%.

Direct-cut grass silage may have a moisture level over 80% (see Table 16–4). Such wet material is heavy, packs solidly, and virtually excludes all oxygen. Aerobic respiration is minimal, and plant cells smother and die. With little or no cell respiration, temperatures remain below 29°C (85°F), the level at which acetic acid-forming bacteria thrive. The drop in pH is not sufficiently rapid or low enough, and clostridia bacteria will develop, forming butyric acid and producing a foul-smelling, undesirable product.

Clostridial activity associated with protein destruction is a major problem with wet silage. It is difficult to state an exact pH value at which clostridia are inhibited since this depends on several factors, particularly the moisture content. Clostridia will tolerate very acid conditions in very wet silage, and the critical pH value will be low for preservation. Butyric acid production, readily detectable by smell in the silage product, is an indication of the extent of clostridial activity.

A certain amount of putrefaction may occur in perennial forage silages (especially grass silage) that have an apparent satisfactory pH level. The probable explanation is that the pH drop was relatively slow and clostridial activity occurred in the early stages of fermentation before being inhibited by pH. Rate of pH fall is an important feature of preservation (Whittenbury, McDonald, and Bryan-Jones 1967).

The problem of overpacking with very wet material may be reduced by a coarser cut. Instead of the recommended 10–13 mm (1/3 to 1/2 in) cut for corn silage, a 16–19 mm (2/3 to 3/4 in) cut will save energy and speed up the operation. Fine chopping of perennial forage crops is necessary, however, to lacerate and bruise all the stems so that sugars are readily available for rapid fermentation.

Farmers experiencing a late season and immature corn at silo-filling time may wait until a fall frost has turned the leaves brown and dry before ensiling. Freezing will lower the moisture content of the leaves, but it is not a desirable practice because it destroys the carotene or vitamin A content in the leaves. Forage yields may also be lowered as a result of the direct loss of brittle leaves and tops. First frosts are likely to kill the leaves (although not necessarily the whole plant) so that photosynthesis is reduced or eliminated, but respiration continues to utilize plant reserves and to reduce dry-matter yields. This can lead to increased lodging from stalk decomposition and to decay as the season progresses, with high ear-drop and harvest losses.

To determine the effects of frost on corn, White, Winter, and Kunelius (1976) harvested corn for silage at approximately ten-day intervals between September 5 and November 15, 1973 and 1974, at Charlottetown, Prince Edward Island. Dry-matter yields of the whole plant and of grain increased up to the first frost in late September. Following the first frost, dry-matter yields of the whole plant declined, whereas dry-matter yields of grain remained constant or increased slightly. In vitro dry-matter digestibility was highest at the time of first frost. White and his coworkers recommended that silage corn be harvested prior to or immediately after being frozen to obtain maximum yield and quality. Postponing silo-filling until wet fall conditions prevail may lead to excessive soil compaction and overly dry material. When overly wet material is ensiled, the excess water seeps or drains out. This watery effluent contains soluble nutrients and is a direct loss to feed value.

In tower silos, seepage losses are dwarfed by the seriousness of the damage that can result to the silo structure. High hydraulic pressures associated with seepage can exceed the tensile limits of the silo walls, causing fractures to form. If the walls are concrete, the acidic nature of the effluent and the expanding action of frosts will cause fractures to enlarge, leaving permanent portals for the penetration of undesirable outside oxygen into the silo interior.

Furthermore, the liquid from the silo that accumulates in the soil adjacent to the silo foundation may cause reduction in soil strength, soil failure, and the collapse of the entire silo structure (Bozozuk 1974). Because of the pressure exerted, seepage is greater in a large-sized upright silo. In a mathematical analysis of the interrelationships among tower silo size, silage moisture content, and seepage loss, Arnold (1974) calculated a series of maximum permissible forage dry-matter percentages to avoid seepage. For a small silo 3 × 10 m, (10 × 32 ft) a 72% moisture level could be tolerated; and for a very large 9 × 30 m (60 × 100 ft) structure, 63% was the maximum moisture level.

With horizontal silos, damage to the structure caused by moisture seepage should be minimal, but significant dry-matter losses are associated with seepage.

Winter freezing may cause a serious limitation for high-moisture silage material. Because of seepage and freezing problems, silage above 70% moisture levels should be considered only for storage in horizontal silos.

Material Ensiled Too Dry

Material ensiled with a high dry-matter level remains light and fluffy; does not pack well; and traps quantities of oxygen, which allows for excessive respiration with accompanying losses of carbohydrates and heat build-up, which can lower quality. Temperatures over 65°C (150°F) are excessive and can result in charring and caramelization of the plant material. Temperatures may reach 138° to 148°C (280 to 300°F). High temperatures destroy the fermentation process because acetic acid-producing bacteria are killed at temperatures above 50°C (122°F); butyric acid-producing bacteria are killed at 70°C (158°F), and lactic acid-producing bacteria are killed at 75°C (167°F).

Too much air and a low moisture level means that molds, commonly present on plant tissues, will flourish under aerobic conditions and will cause the silage to be musty and moldy.

Charring, moldiness, and carbohydrate losses from respiration are common when material ensiled is below 60% moisture. The weight of material in a large 9 × 30 m (60 × 100 ft) silo will aid packing and air exclusion, and material may be ensiled successfully at moisture levels of 60 to 64%. Reduction in feed consumption or even refusal by livestock is an indication that the silage material has suffered heat damage. Palatability may be improved by adding molasses or grain, but respiration losses will be significant.

To help preserve dry material, fine chopping can be attempted, which requires energy and slows up the harvest operation. The forage harvester should be adjusted to the finest chop setting on the unit and the blades sharpened and properly adjusted. Rapid filling is encouraged to improve packing and maximum consolidation of material.

Adding water to raise the moisture content is not as easy as may first appear because pithy corn material is not absorbent, and large amounts of water are required to make any appreciable difference. Adding 50 kg (10 gallons) of water to one tonne (ton) of material of a 50% moisture level would change the moisture level by about two percentage points. With a 5½-tonne (6 ton) forage wagonload, 275 kg (60 gallons) of water would be required. To increase the moisture level by ten percentage points, the addition of 250 kg (55 gallons) to a tonne (ton) or 1375 kg (330 gallons) to a load would be required. The water would have to be added at the blower for an upright silo. Thus, the idea becomes rather unworkable for most situations at this moisture level. A general rule of thumb is to add 20 litres (4 gallons) of water per tonne (ton) for each one percentage point increase in moisture desired. There is evidence to show that water added in a silo (not at the blower) tends to concentrate in layers, or flows to the side, and then down the walls, where it is of no benefit.

Adding water to develop a seal on a filled silo is effective if the moisture is distributed and retained over the entire surface. Spoilage is reduced if air infiltration is retarded. A weighted plastic cover is an effective method of reducing air infiltration and is a recommended practice, especially on the large exposed surface area of a horizontal silo.

A novel idea of ensiling residue corn stover after the ears have been picked, and with moisture levels as low as 42 to 50%, was studied by Fulkerson (1978). He combined kale, a crop capable of producing 10–11,000 kg/ha (8900–9800 lb/ac) of dry matter in Ontario but which is characterized by moisture levels as high as 85% (Table 16–5), with corn stover. If kale is blended with corn stover, it produces a silage mass with satisfactory water content and proper conditions for fermentation in a silo. Protein level, feeding value, and palatability are improved over that of corn stover alone.

To increase the moisture level of corn stover from 50 to 70% required approximately 1340 kg (2900 lb) of chopped kale. Forage harvesters equipped with gathering belts will pick up about 60% of the normal 4.5 tonnes (4 tons) of machine-picked corn stover, provided gathering is done immediately after grain harvest. This type of harvester works well to harvest kale. In keeping with good silage practice, chop both kale and stover finely, pack thoroughly, and cover horizontal silos with plastic.

TABLE 16–5.   Cultivar performance of kale, three-year average at Guelph 1975, 1976 and 1977

	Yield kg/ha		Height		Dry Matter	Digestibility*	Protein**
Cultivar	Green	Dry	(cm)	(in)	(%)	(%)	(%)
Gruner Angiliter	70,100	11,040	119	(47)	14.7	81.0	10.8
(lb/ac)	(62,580)	(9857)					
Maris Kestrel	76,700	10,090	71	(28)	13.4	88.0	15.6
(lb/ac)	(68,480)	(9009)					

Source: Data from Fulkerson 1978.

*In vitro digestible dry matter.

**Crude protein.

A feeding trial conducted at Guelph with a stover-kale blend provided gains of 0.66 kg/day (1½ lb/day) on a dry-matter intake of 4.5 kg (10 lb) with beef steers. The ensiled material averaged 72% moisture and had a crude protein of 9.1% and a pH of 4.1. The silage proved to be of top quality, very palatable, and with a high intake.

The lesson of a kale-stover silage production is that a variety of biomass material can be ensiled provided there is sufficient volume. Material such as cannery wastes (tomato seeds, skins, and discarded pulp; apple pomace; citrus pulp; pineapple tops; sugar-beet tops), surplus plant material such as a rye or wheat winter cover crop, oat and pea silage, a late-seeded soybean crop, sorghum, and other materials can produce excellent silage.

MATERIAL LOW IN FERMENTABLE CARBOHYDRATE

Whole-plant corn silage is a popular crop to ensile not only because of high yields (13 to 20 tonnes of dry matter per hectare) (6 to 9 tons/ac) but also because of the ease of producing a quality silage. Whole-plant corn material has an abundance of fermentable carbohydrate in the ears and stalks and a relatively low resistance to pH reduction (low buffering capacity), and butyric acid is not a dominant problem over a wide range of moisture levels. For these reasons it is a reliable crop for silage production.

Perennial forage grasses and legumes are characterized as being high in protein with a low level of fermentable carbohydrate and high in moisture at cutting time (see Table 16–4). These crops have a high buffering capacity so that grass silage requires twice, and legume and clover silage four times, the volume of lactic acid on a dry-weight basis to bring the pH to 4.0. The specific action of bacteria in the silo may cause the buffering action to vary by a factor of two to three times that of the original plant material and explains why butyric acid losses are common. The high buffering capacity and the low sugar content in perennial forages, especially of clovers and legumes, illustrate the difficulties often encountered in ensiling legumes successfully.

Wilting of perennial forage material is generally a necessity. To help provide a source of fermentable carbohydrate, materials can be added that will be quickly converted into fermentable sugars. Ground grain (e.g., barley or corn) contains starch, which can be converted to sugar; and molasses contains a high level of lactic acid, favoring sugar. A general rule is to add 110 to 165 kg (220 to 330 lb) of ground shelled corn, 190 to 220 kg (380 to 440 lb) of ground ear-corn, or 65 to 110 kg (130 to 220 lb) of molasses per tonne of silage.

A number of commercial products are available to provide nutrients for lactic acid development or acid compounds to directly lower the acidity or microorganisms that increase the availability of carbohydrates and other nutrients required by lactic acid-producing bacteria. The wide range of buffering capacity found in silage materials explains why aids to fermentation have given variable and inconsistent results. Concentrations must be sufficient to be effective, and these products cannot compensate for poor management. Some products are aimed more at improved livestock performance than silage production and have given results ranging from zero to highly significant improvements in nutrient preservation, dairy and beef cattle performance, and silage chemical composition (Young 1978).

CHEMICAL CHANGES IN THE SILO

It is obvious that the green material put into a silo undergoes substantial changes during the fermentation process, resulting in noticeable differences in color and odor. The full extent of the chemical changes is complex, although it is usually taken for granted. Chemical changes may lead to losses in feed value and protein losses, some of which are unavoidable, whereas others can be avoided through proper management. Producers should learn to "read" their silage material and benefit from experience. The following chemical changes are noted:

  1. Complex carbohydrate molecules are broken down into simpler

substances, preferably lactic acid. In silos having lactic-acid fermentation, it is estimated that 1 kg (2.2 lb) of organic acid is recovered for every 1.8 kg (4 lb) of carbohydrate lost. The corresponding recovery rate under conditions of undesirable fermentation is 1 kg (2.2 lb) of organic acid for every 2.8 kg (6 lb) of carbohydrate. Because whole-plant corn silage is easily produced as shown by the above relationship, total digestible nutrients (T.D.N.) values for corn silage are higher than for forage silages. The high level of protein in perennial forages compared to corn is also known.

2. A major concern in producing silage from perennial forages is protein destruction by clostridia. Normally, about 50 to 60% of the protein in well-preserved silage is broken down (Whittenbury, McDonald, and Bryan-Jones 1967) into amino acids, the basic components or building blocks of protein. The amino acids are utilized by livestock as digestible protein; but if clostridia activity occurs and butyric acid is produced, losses in protein occur. A reason for making silage is to reduce crude protein losses associated with leaf loss and weathering in field-cured hay. First-cut hay material had an average protein content of 13.2%, whereas the ensiled product averaged 15.7 or 2.5 percentage points higher. Every caution should be taken to maintain this valuable protein advantage in the silo.

If putrefactive bacteria are present because of an inadequate level of acidity and if temperatures above 48° to 49°C (118 to 120°F) are reached, serious tie-up of protein into an unavailable form occurs, and losses in protein can be high. To determine the amounts of protein lost by aerobic heating of silage, a sample of silage spoiled and black in color, obtained at an 8-cm (3 in) depth from the surface of a horizontal silo at Guelph, had only 14% of the protein in a digestible form; at a 23-cm (9 in) depth, the silage was dark brown and 35% of the protein was digestible; and at a 60-cm (24 in) depth (good silage), 66% of the protein was digestible. Color of the silage is an indicator of protein digestibility. The darker the silage, the lower the digestibility.

3. During the initial aerobic stage, the carotene compounds containing vitamin A are lost through oxidation. These losses can be minimized, except in plant material frozen in the field, by proper packing and air exclusion. Vitamins B and D are not affected by the fermentation process, but most green crops are inherently low in these vitamins. Vitamin C, ascorbic acid, is lost through enzyme action within the silage mass. Livestock producers should recognize that silage is low in vitamin content, no matter how well the crop is handled.

4. Minerals may be lost through the leaching action within the silage mass when effluent seeps from silos. Mineral components during the ensiling process may be changed to new forms or combinations but are not destroyed by fermentation. Grass and legume-based silage

had a higher mineral content than hay, possibly because of the greater losses associated with field curing.

5. Color changes in the silage mass are obvious. Fresh green material becomes a yellow-brown during fermentation by the change of the green chlorophyll pigment to a brown pigment due to magnesium removal. A yellow-brown color of silage indicates good fermentation and preservation. Color changes are associated with the loss of carotenoid compounds through oxidation. Excessive heat results in caramelization of the sugars, which produces an undesirable black color.

6. Energy losses occur in silage production. The energy required to produce a desirable but minor period of respiration to increase temperatures from 29° to 40°C (85 to 104°F) accounts for an unavoidable 1 to 2% loss of dry matter. At a 17,000 kg/ha (15,000 lb/ac) dry-matter yield of corn in the field, respiration losses would account for 170 to 340 kg/ha (150 to 300 lb/ac). Losses due to yeasts and molds can amount to 10% or 1700 kg/ha (1500 lb/ac), depending on the amount of oxygen present. Evidence of energy losses can be seen by examining corn silage. Corn ensiled at 65% moisture when the kernels are dented will have firm kernels in the silage, and the kernels will remain yellow (assuming a yellow-kerneled corn was grown). In silage less well-preserved, kernels will be brownish-yellow, and the loss of energy from such kernels will render them soft. In silage that is brown in color, kernels will be very soft because of the high percentage of feed value lost. Livestock may relish silage with soft kernels, but its feed value is greatly reduced, and expensive supplements will be required if livestock production is to be high. Because feed costs account for 60 to 80% of the total livestock production costs, such losses should be avoided if feeding programs are to be profitable.

Stoppages in harvest due to adverse weather, machinery failure, or weekend breaks will expose the surface silage to aerobic respiration, excessive heating, and dry-matter losses. Spoilage due to delays in filling can be detected in the silo as a brown layer with soft mushy kernels. Clostridia bacteria that produce butyric acid can result in dry-matter losses up to 5% of the total dry matter.

Clearly, heating of ensiled material must be avoided if energy and protein losses are to be avoided. Caramelized silage will have a sweet aroma and will appeal to livestock but will have a reduced feed value. Temperature measurements at Guelph in a horizontal silo demonstrated the value of a plastic cover to exclude oxygen. Temperatures of 28°C (82°F) were recorded in a horizontal silo covered with plastic and sealed at the edges with soil, tires, or silage; 35°C (95°F) temperatures were measured when the silage was covered with plastic but not weighted down; at the open face area, with no covering, temperatures reached 60°C (140°F).

SILO SIZE AND STORAGE RECOMMENDATIONS

When silage is unloaded from the silo, it is loosened from its compact state and exposed to oxygen, which allows dormant yeast organisms to become active, causing losses. A sufficient volume of feed must be removed daily from the silo once it is opened to minimize spoilage losses. Table 16-6 indicates the amount of silage that should be removed from different sizes of silos, assuming 1 m³ weighs 750 kg (1 cubic foot weighs 52 lb). Kilograms (pounds) of silage vary with depth, but the total kilograms that must be removed daily to prevent spoilage in the silo remain constant.

TABLE 16-6.   Minimum feeding schedule from silos of various size

Silo Diameter in		Kilograms (pounds) of Silage per cm. (in) of Depth		Kilograms (pounds) to be Removed Daily*			
Metres	Feet		(pounds)	Winter	(pounds)	Summer	(pounds)
3.6	12	106	580	424	933	636	1400
4.2	14	144	812	577	1269	866	1905
4.8	16	189	1066	754	1659	1131	2488
5.4	18	239	1348	954	2099	1431	3148
6.0	20	295	1664	1178	2592	1768	3890
6.6	22	356	2008	1422	3128	2133	4693
7.2	24	424	2392	1697	3733	2545	5599
7.8	26	496	2798	1985	4367	2977	6549
8.4	28	576	3249	2304	5069	3456	7603
9.0	30	662	3734	2646	5821	3969	8732

*Based on removal of 5 cm. (2 in) in winter and 7.5 cm (3 in) in summer.

Information on the capacities of upright and horizontal silos are shown in Appendices C and D.

SUMMARY OF CAUSES FOR STORAGE AND HARVEST LOSSES

Comparisons of storage losses with field and harvest losses for five perennial forage programs were reported by Porter (1978) (Figure 16-2) for work at Michigan State University. The horizontal axis represents dry matter at harvest, with low dry matter (direct-cut) on the left and high dry matter (baled hay) on the right. The vertical axis is the percent dry matter lost. The crosshatch area shows field and harvest losses: the dotted area shows storage losses. When perennial forages are handled as baled hay, the greater dry-matter loss occurred in the field. When ensiled, the greater loss occurred in storage. When perennial forages were handled as baled hay, with over 80% dry matter and under good conditions, about 30% of the dry matter produced is lost (Figure 16-2).

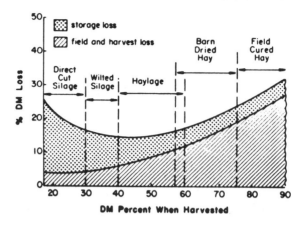

Figure 16–2. Comparisons of estimated field and harvest loss and storage loss.

ORGANIC ACID PRESERVATION

The knowledge that a low pH was an effective method of preservation led to the concept of organic acids (propionic, acetic, or propionic-acetic mixtures) applied to wet material to stop mold and bacteria growth, to stop respiration, and to kill the embryo of seeds to prevent germination of moist grain. The material is thus preserved with no dry-matter loss because biological activity is inhibited due to restricted microorganism activity.

Organic acid storage is useful as an emergency storage method for growers with a small volume of material when yields exceed the normal storage capacity and/or when fuel for hot-air drying is scarce or too expensive, or for farmers temporarily lacking storage facilities because they are uncertain of their long-range plans (Anonymous 1978).

BASIC CONCEPTS OF FORAGE CORN YIELD

Dry-matter yield of the whole plant reaches its maximum about two weeks before the grain reaches its maximum weight, as shown in Table 16–7 (Daynard and Hunter 1975; Stevenson 1976). The study showed that corn reaches its maximum dry-matter yield at approximately 66% whole-plant moisture and could be harvested for silage as early as 70% moisture with a negligible effect on dry-matter yield. Avoidance of seepage represents the best practical guide in choosing the earliest possible date for ensiling corn. Once the corn crop has dried naturally to a sufficiently low moisture to avoid seepage (65–68% for vertical silos), the crop will have reached its maximum dry-matter yield and feeding value. Grain moisture content varies with whole-plant moisture percentage in a predictable manner, and as a consequence, grain moisture is a useful indicator of the proper stage for ensiling. Sixty-five to 68% whole-plant moisture corresponds to grain moisture content of approximately 45%.

TABLE 16–7.   Effect of harvest date on silage and grain yield (average of two years) at Guelph in 1970 and 1971

Date	Percent of Maximum Yield	
	Grain	Whole-Plant
Sept. 4	60	93
10	71	96
14	80	98
18	87	100
26	93	100
Oct. 4	97	100
10	100	99
18	100	97

Source: Data from Daynard and Hunter 1975, and Stevenson 1976.

During the period of maximum whole-plant dry weight and maximum grain yield, carbohydrates are transferred from vegetative parts to the developing grain, but daily photosynthesis does not exceed daily respiration, and total plant dry matter does not increase. These findings suggest that a corn hybrid can be selected for silage production that is 200 corn heat units later maturing than that required for grain; or stated otherwise, areas 200 corn heat units too cool for grain corn production can be used for silage production with little effect on whole-plant dry-matter yield. This is an important concept in the fringe areas of agricultural production because it extends considerably the effective production area of the high-producing corn crop. The system of corn heat units developed by Brown (1975) offers an accurate selection technique to select corn hybrid cultivars in climatically restricted areas. Corn development values were established for daily minimum and maximum values for a defined growing period, and on the basis of daily temperature regimes, corn heat units can be summed and a rating applied to each hybrid by comparing it with standard check hybrids. The effective northern limit for grain corn production is about 2400 corn heat units—the maturity rating for the earliest maturing commerical hybrids available.

The suggestion that silage corn may be of a later-maturing hybrid cultivar than that for grain is contrary to the system in North America during the fifties and sixties that recommended top-grain hybrid cultivars for silage corn production. This was done on the assumption that grain development was closely associated with feed quality of the silage. Although a higher grain content may mean a small increase in dry-matter digestibility, a later-maturing large-stature hybrid cultivar may produce a higher total dry-matter yield. The assimilates in the stalk serve as a source of readily fermentable carbohydrate.

Corn produced in regions with a low number of corn heat units can be expected to have a yield lower than in regions with a higher number of corn heat units, but corn may be competitive with other crops grown under a continental climate (hot summers, cold winters), which is characteristic of most agricultural areas of Canada. As whole-plant corn is harvested for silage at an increasing degree of immaturity, the major changes in composition involve a reduction in the proportion of plant dry matter, which is grain, as well as an increase in percent moisture.

The impact on feeding quality of corn silage with a reduced grain content is difficult to assess because comparisons usually involve silages that differ in water content as well as in percent grain. Trials at Cambridge, England (Bunting 1975) suggest that the importance attached to high grain content as an essential requirement for yield and quality of silage is exaggerated. Sterile plants compared favorably with their fertile counterpart in digestibility and content of crude protein. Elimination of the grain sink prevents starch formation, and food energy is stored as sugars. Content of sugar in the stem has been found to be positively associated with resistance to lodging induced by stalk-rot fungi (Mortimore and Ward 1964). Bunting (1975) suggested that elimination of the grain sink did not reduce yields. This is in contrast to the findings of Moss (1962), who observed that barren plants produced in the United States had a lower photosynthetic rate than fertile plants. Bunting concluded that dry-matter yield in corn is much less dependent on grain formation in northern Europe than it appears to be in the United States. On the "sink-source" hypothesis, dry-matter yield in the United States appears to be limited by sink capacity; whereas at the higher plant densities appropriate for forage corn production in northern Europe, yield is limited by photosynthetic production.

Recommendations for selection of corn hybrids for forage (silage) production and cultural practices must take into consideration the heat units available. In regions with sufficient corn heat units, in order to achieve 30% or more whole-plant dry matter, a number of production practices are recommended to increase yield without any notable effect on silage quality. The first such practice is to increase the leaf area and leaf area duration. This can be accomplished by the use of hybrids 100–200 corn heat units later in maturity for silage than for grain. Full-season hybrids produce larger leaf areas per plant. In areas with 2900 corn heat units or less, where evapotranspiration rates are not likely to cause drought stress, plant populations 20% higher than those for grain production may be considered (Daynard 1977). Table 16–8 shows the yield in fringe areas of corn production when grown under high plant populations (Hunter 1978). The second recommended practice is the use of vegetatively vigorous hybrids or tall leafy types. These include hybrids with rapid leaf area expansion in the spring.

TABLE 16–8.   Forage yields from corn at three Canadian sites

Province	Plant Density (pph)	(ppa)	Dry-Matter Yield (t/ha)	(t/ac)
Quebec	12,000	4800	6.4	2.9
	25,000	10,100	11.9	5.4
	50,000	20,200	17.7	8.0
	75,000	30,300	18.1	8.1
Ontario	35,000	14,200	10.9	4.9
	50,000	20,200	11.7	5.3
Alberta	35,000	14,200	12.0	5.4
	70,000	28,300	15.3	6.9
	115,000	46,000	17.1	7.7

Source: Data from Hunter 1978.

There is a limit to all these possibilities in grain development because sufficient grain development and maturity must be attained to permit the crop to reach a minimum of 30% dry matter before the first autumn frost.

In areas with a minimum corn heat unit availability, such as 2000 corn heat units, 30% dry matter is an impossible or unreliable objective, and some level of feeding quality below optimum must be expected in some years with ensiled forage corn. Certain production practices must be adopted to accelerate grain maturity, including:

1. Selection of earliest maturing hybrids.
2. Avoidance of ultra-high plant densities.
3. Use of fertility and tillage techniques that accelerate the rate of early plant development.
4. Use of spring tillage practices that promote soil warming.

A dry-matter objective of 25% may be a realistic figure in fringe areas of corn production, a value used in Britain (Hough 1975). Daynard (1977) estimated that 25% whole-plant dry matter may be attained 100–200 corn heat units before 30% dry matter is normally achieved, approximately 300 corn heat units before 35% dry matter is normally achieved, and about 500 corn heat units before normal grain maturity (i.e., a grain moisture level of 30–35%). As an approximation, therefore, 25% dry matter might be expected to occur 300–

400 corn heat units earlier than the corn heat unit rating for a given early maturity (e.g., 2600 corn heat unit) hybrid. Even with the handicap of a short season, corn silage can be competitive with alternative forages in terms of digestible energy intake, harvested yield of digestible energy and production cost per unit of digestible energy. Also, in short-season areas, horizontal silos may be preferred to avoid seepage and freezing problems.

ESTIMATING MOISTURE CONTENT

Estimating moisture content of whole corn plants from the morphological development of the kernels is a reasonably accurate method. At 45% grain moisture (associated with 65% whole-plant moisture), kernels are dented (Figure 16–3).

Farmers can also learn to identify a 65% moisture level in perennial forage. A sample can be twisted in the hands and its compactness and rate of untwisting can be associated with a certain moisture level. Farmers are well advised, however, to obtain a moisture meter for a quick and accurate evaluation and until experience is obtained (Figure 16–4). If access to a meter is not possible, moisture levels can be determined using materials available on a farm. Use the following system:

1. Obtain a representative sample of the material to be determined for moisture content.
2. If a scales is not available, construct a balance from two containers. On one side place 100 spikes or large nails; in the other container counterbalance the spikes with the plant sample.
3. Dry the plant sample in an oven at up to 93°C (200°F) until the sample stops changing weight.

Figure 16–3. An ear of corn in the early dent stage. The moisture in these kernels is over 45%, too wet for proper silage fermentation. (Courtesy Ontario Ministry of Agriculture and Food.)

Figure 16–4. A moisture metre is useful to determine with accuracy the proper stage for ensiling. (Courtesy Ontario Ministry of Agriculture and Food.)

4. Reweigh the dried sample by removing spikes to counterbalance the dried sample. The number of spikes removed to balance the scales will be the moisture percentage in the sample.

Moisture levels in whole-plant corn silage can vary according to the size of the upright silo. The following values represent the upper limits for moisture in relation to the size of the upright silo:

- Small (3.7 × 10.7 m) (12 × 35 ft)—68 to 72%
- Medium (6.0 × 18.3 m) (20 × 60 ft)—64 to 68%
- Large (9.1 × 30.5 m) (30 × 100 ft)—60 to 64%

Whole plant corn silage should be over 60% moisture for good preservation in a horizontal silo of any size.

SILO GAS

When crops are produced under conditions of abnormally high levels of nitrogen fertilizers, or under environmental stress conditions such as drought that restrict growth but not nitrogen uptake, or when the plant material is ensiled immature and nitrogen is stored in the stalks and leaves, nitrogen dioxide or silo gas may form and be trapped in upright silos. Silo gas is extremely dangerous, and at high concentrations can cause death to anyone encountering the gas. Irritation to the eyes, throat, and lungs can occur at lower concentrations and can result in respiratory problems days or weeks after exposure.

Silo gas is produced from nitric oxide and oxygen in the air. The chemical pathway leading to silo gas production is as follows:

$$\text{Nitrate } (NO_3) \rightarrow \text{Nitrite } (NO_2) \rightarrow \text{Ammonia } (NH_4)$$

Nitrite (NO_2) in wet silage conditions and at low pH forms nitrous acid (HNO_2):

$$\text{Nitrous acid } (HNO_2) \rightarrow 2NO \text{ (nitric oxide) } + \text{ nitric acid } (HNO_3) + H_2O$$
$$\text{(water)}$$
$$2NO + O_2 \rightarrow 2NO_2 \text{ (nitrogen dioxide or silo gas)}$$

Note that oxygen is required for silo gas production and is most likely to form and be trapped in partially filled uncovered silos in the first few hours after ensiling.

In high concentrations, silo gas is reddish to yellowish-brown, with a bleachy odor; in lower concentrations, it can be colorless and odorless. Because silo gas is heavier than air, it tends to hang over the surface of the silage and in upright silos may spill down the chute into the feed room. The gas can be produced within a few hours after filling, but the greatest danger is 12 to 60 hours after filling, with the danger period extending up to three weeks. In its life-threatening form, nitrogen dioxide is relatively rare, but entry into a silo should always be done with caution and common sense.

HIGH-MOISTURE GRAIN

Immature grain harvested at 20–35% grain moisture requires a substantial energy input if heat-dried to a 15% moisture level. High-moisture grain can be stored as a fermented silage product at a moisture level of 25 to 32% as a feed commodity. As feed it is comparable (Table 16–6) to standard dry corn when both are expressed on a comparable dry-matter basis. Corn can be stored as whole-grain, ground-shelled, or ground-ear corn in which the entire ear including the cob is chopped and ensiled.

High-moisture grain storage offers producers a method of removing an early cereal crop days before it can be combined as dry grain, in order to allow for a second relay crop. Winter cereals can be removed very early and afford such an opportunity. A second late crop may be ensiled at a high-moisture level.

CONCLUSIONS

To take advantage of mechanized forage-handling systems that minimize field losses, there has been a shift toward greater use of ensiled feeds. Whole-plant corn silage, noted for its ease and reliability of producing a quality feed, has led the way in extending silage production to perennial forages and other crops. Silage production is not a sideline; it is often considered the main ration component, with other feeds being supplementary.

The success with corn silage grew out of an almost foolproof recipe without a clear understanding of how silage was produced or how it could be improved. Production of a quality product from crops less easily ensiled demands a complete understanding of the chemistry that occurs in the silo. Silage production cannot be taken for granted.

As forage production is expanded to take advantage of the soil-conserving properties associated with perennial grasses and legumes, effective methods of harvesting, handling, and storage of these crops will be vital to their success.

QUESTIONS

1. Why is it essential to select a corn hybrid with the proper maturity for silage production?

2. Complete the blanks in Table 16-9 below.

TABLE 16-9.

Item	Immature	Overmature and Dry	Ideal
1. Moisture		Less than 65%	
2. Packs	Solidly		
3. Air			Moderate
4. Oxidation		Too much	
5. Temperature			29–40°C (85–104°F)
6. Acids		—	
7. pH change		Slight	
8. Faults			—
9. Cause of loss		Oxidation	—
10. Odor	Putrid		
11. Range of losses			

3. The object of storing green crops as silage is to preserve the material with a minimum loss of nutrients. Calculate in economic terms the value of material lost in item 11 of Table 16-9 using the value of forages found in your area.

4. Calculate the value of hay and silage using the value of protein as a base.

5. It is essential that producers be able to "read" their silage product. Use the following "score card" to evaluate silage samples.

Score Card to Evaluate Silage Samples

Criterion	Possible Score	Your Score
I Color (total 32)		
1. Desirable—slightly brownish.	22–32	_____
2. Acceptable—dark green to yellowish green or yellow to brownish.	15–21	_____
3. Deep brown.	6–14	_____
4. Undesirable—black or white or gray from mold.	0–5	_____
II Odor (total 36)		
1. Desirable—light pleasant odor with no indication of putrefaction.	26–36	_____
2. Acceptable—fruity, yeasty, and musty, which indicates slightly improper fermentation. Slight burnt odor, or sharp vinegar odor.	11–25	_____
3. Undesirable—strong burnt odor indicating excessive heating. Putrid odor indicating improper fermentation. A very musty odor indicating excessive mold.	0–10	_____
III Moisture and texture (total 32)		
1. No free water when squeezed in hand. Well-preserved Silage.	20–32	_____
2. Some moisture can be squeezed from silage, or silage dry and musty.	10–19	_____
3. Silage wet, slimy, or soggy; water can be squeezed from samples. Silage too dry with strong burnt odor.	0–9	_____

(If corn silage, the firmness of kernels should also be considered in III.)

	Score _____

Scoring:
  90 and above—excellent
  80–89—good
  65–79—fair
  Below 65—poor

6. Calculate the total amount of ground-shelled corn required to provide a source of fermentable carbohydrate to a silo about 6×18 m filled with haylage at 50% moisture.

7. Give reasons why corn silage is easier to produce and generally preferred over grass or legume silage.

8. What are the advantages of perennial forage production and grass silage production?

9. State clearly when you would use a preservative in making grass silage and corn silage.

10. Why is silage production so desirable on most farms?

11. Examine a sample of corn silage and from the color and texture describe the condition of the corn under which it was ensiled (frozen, too mature, immature).

REFERENCES FOR CHAPTER SIXTEEN

Anonymous. 1966. *A New Look at Corn Silage.* New Holland, Pa: New Holland Machine Company. (33 pp.)

Anonymous. 1978. "Preserve Grains with Acid," *Crops and Soils* 30(9):24–25.

Arnold, R. L. 1974. "Density—Pressure—Seepage Relationships in Corn (*Zea mays* L.) Silage in Tower Silos. Master's thesis, University of Guelph. (71 pp.)

Bozozuk, M. 1974. "Bearing Capacity of Clays for Tower Silos," *Canadian Agricultural Engineering* 16:13–17.

Brown, D. M. 1975. *Heat Units for Corn in Southern Ontario.* Factsheet No. 75-077. Ontario Ministry of Agriculture and Food. Toronto, Ontario (4 pp.)

Bunting, E. S. 1975. "The Question of Grain Content and Forage Quality in Maize: Comparisons between Isogenic Fertile and Sterile Plants," *Journal of Agricultural Science (Camb.)* 85:455–463.

Cressman, S. G. 1980. *Feed Advisory Program. Average Composition of Ontario Feeds.* Information for Industry Personnel. Ontario Ministry of Agriculture and Food. No. 400/52.

Daynard, T. B. August 1977. "Practices Affecting Quality and Preservation of Whole-Plant Corn Silage." Paper presented at the Symposium, "Corn for Whole-Plant Silage in Canada." Annual Meeting of the Canadian Society of Agronomy, Guelph, Ont. (19 pp.)

Daynard, T. B., and R. B. Hunter. 1975. "Relationships among Whole-Plant Moisture, Grain Moisture, Dry Matter Yield, and Quality of Whole-Plant Corn Silage," *Canadian Journal of Plant Science* 55:77–84.

Fulkerson, R. S. 1978. *Producing Kale-Corn Stover Silage.* Information for Industry Personnel. Ontario Ministry of Agriculture and Food. 4 pp.

Hodgson, H. J. 1976. "Forage Crops," *Scientific American* 234(2):60–75.

Hough, M. N. 1975. "Mapping Areas of Britain Suitable for Maize on the Basis of Temperature," *Agricultural Development Advisory Service Quarterly Review* 18:64–72.

Hunter, R. B. 1978. "Selection and Evaluation Procedures for Whole-Plant Corn Silage," *Canadian Journal of Plant Science* 58:661–678.

Lechtenberg, V. L., et al. 1976. "Big Package Hay Care," *Crops and Soils* 28(5):12-14.

Mortimore, C. G., and G. W. Ward. 1964. "Root and Stalk Rot of Corn in Southwestern Ontario III. Sugar Levels as a Measure of Plant Vigour and Resistance," *Canadian Journal of Plant Science* 44:451-457.

Moss, D. N. 1962. "Photosynthesis and Barrenness," *Crop Science* 2:366-367.

Porter, J. R. December 1978. "Preserving Haylage with Chemstor," *Proceedings of Ontario Forage Conference*, Toronto. (5 pp.)

Sprague, M. A., and B. Brooks Taylor. 1977. "A Technique of Monitoring Acidity of Silage during Preservation and Storage," *Agronomy Journal* 69:727-728.

Stevenson, K. R. 1976. "Corn Silage—Harvesting and Storage" *Proceedings of Ontario Soil and Crop Improvement Association*, pp. 162-163.

———. 1970. "High Moisture Grain Storage Using Organic Acids," *Notes on Agriculture* 6(2):25-26. University of Guelph.

White, R. P., K. A. Winter, and H. T. Kunelius. 1976. "Yield and Quality of Silage Corn as Affected by Frost and Harvest Date," *Canadian Journal of Plant Science* 56:481-486.

Whittenbury, R., P. McDonald, and D. G. Bryan-Jones. 1967. "A Short Review of Some Biochemical and Microbiological Aspects of Ensilage," *Journal of the Science of Agriculture* 18:441-444.

Young, W. S. 1977. "Grazing Systems or Stored Feed," *Notes on Agriculture* 8(3): 23-25. University of Guelph.

———. 1978. *The Use of Aids to Fermentation in Silage Production.* Information for Extension Personnel. University of Guelph: Ontario Ministry of Agriculture and Food.

chapter seventeen

Branch or Tiller Development in Crop Plants

An unresolved question in crop production is whether it is better in a given environment for a plant to produce a few shoots, each bearing a high yield, or whether many small shoots each with a small yield should be developed. The branching or tillering habit is commonly observed and is probably one of the most extensively studied phenomena on an individual plant basis. Critical studies of tillering under cropping conditions in the field are not as numerous as those dealing with single plants in controlled environments. The result is a lack of management studies relating to tillering under field conditions, possibly because of the many interacting factors that influence branching. This has resulted in a complexity and a lack of consistency in specific management practices influencing tillering and grain or forage yield. Although the hopes of a precise, single approach have been dimmed, crop producers must still make decisions about the many crops that exhibit branching and must adopt management practices to promote or restrict branch development—whatever best serves the purpose of increasing yield.

This chapter deals with branching in plants (commonly called *tillering*), lateral shoot development, and suckering or stooling, in terms of the factors influencing such development and in terms of the influence of management practices on branching. It is not possible to resolve the question of branching and yield, but such a discussion does serve to demonstrate the need in crop production of a total management system approach—an integrated adaptation of all the factors influencing yield, both desirable and undesirable. Branching is a biological entity, which means that it does not respond in a predictable manner, and increased nitrogen fertilization or reductions in seeding rate may have little influence on grain yield. There may be no single answer to the problem of the management of branching, yet branching presents a very real challenge to crop producers because it can influence yield. Producers need, therefore, to understand branching in plants.

THE NATURE OF BRANCHING

Branching occurs during the vegetative period of plant development, with branches arising from the buds in the nodes of plants. Branches (tillers) may arise from the buds in the axils of branches in soybeans or forage legumes, or from basal nodes of grasses. Depending on the depth of planting, primary tillers may arise from the lowermost or coleoptile node or from higher nodes. Secondary tillers may arise from the basal nodes of primary tillers, and some tillering may occur during a brief period of time in the seedling stage, after which it declines or ceases at the time of stem elongation and flowering. In perennial crops, tillering may be resumed after ears have emerged. Late tillering may also occur in cereals after heading if light penetration occurs to the basal nodes through lodging or crop removal.

BRANCHING IN PERENNIAL FORAGE CROPS

There can be no question of the essentiality of branching and tillering in forage crops to provide a mechanism for regrowth, persistence, and herbage yield. Tillering is a form of perennation, and it is branching that produces a new crop after each cutting and at the beginning of each season. Branching, therefore, is the very basis of forage yield and more than any other factor should determine the management practices that stimulate crown bud development in perennial forages and tillering in grasses. Each time herbage is removed, considerable tillers are "used-up" as new growth arises from new buds. The turnover may be so rapid that the average tiller may only live for four to six weeks. Management practices such as time and frequency of cutting, fertilizer regimes, a fall rest period, and root reserves must all be considered from the standpoint of how each will influence tillering. Forage management practices are probably based more on dry-matter yield than on a close observation of tiller buds and their development.

It is also not sufficient that the crop or the field be managed so that the population as a whole maintains its tiller population. Plant number per unit area will vary according to the age of the stand. New seedlings have more plants per unit area than older stands. Dense stands reduce tillering per plant; sparse stands allow for greater tillering levels (Donald 1954). Possibly, new and dense stands should be cut at an early calendar date to prevent restrictions on tiller development and to increase herbage yield; whereas sparse stands should not be cut until each plant has reached its full tiller output. Management of each plant may be a consideration, because if an individual plant fails to keep up its replacement rate, it will be crowded out by competition and will die, leaving a gap in the stand, one that will probably be occupied by a weed. Plant physiology is distinguished from crop physiology because the latter normally is concerned only about the entire plant population, instead of focusing on the individual plant as in plant physiology. In forage crop management, however, the tillering capacity of individual plants may have to be collectively considered because tillers are the basis of forage yield.

In the evolutionary scheme of plants, it is possible that tropical conditions favored perennial plant development over annual plant development. Then, as plants migrated out of the favorable tropical regions and were killed by frosts, annuals sprang up from among the perennial population, which retained some of the perennial characteristics such as tillering. Perennation (in annuals) may be expressed as a suppressed or remnant tillering feature. Winter annuals resemble the perennial form and retain a greater tillering habit than other annuals. It is this perennation in annuals with which crop producers must deal. Crossing an annual spring wheat plant to a related perennial species might produce a perennial wheat, but so far efforts have not led to plants that are competitive for grain, and no commercial perennial cereals are available. The management of perennial forages for longevity plus high herbage production is difficult, and adding the perennial feature to cereal crops may be undesirable. The remnant form of perennation in annual cereals that influences leaf area, biological yield, and grain yield is in itself a complex dimension.

Unlike crops for herbage production where tillers are highly prized, in other crops, tillers may be viewed as more of a liability than an asset. A description of the features of tillering in selected crops and the factors affecting branching will help emphasize the importance of tillering in crop production.

FACTORS INFLUENCING BRANCHING

Hormonal Control

The removal of the lead shoot of a geranium plant growing in a house window is an effective method of stimulating lateral bud growth in the leaf axils of the plant. The removal of the lead shoot produces a bushy, short, and multiple-flowered plant.

Apical dominance is destroyed in tobacco plants when topped (removal of the shoot apex) in an effort to maintain vegetative plant development. With apical dominance removed, branches or suckers form in leaf axils, which is considered undesirable. Removal of unwanted suckers can be achieved manually or with the chemical maleic hydrazide.

Tillering in cereal and forage plants is also subject to hormonal influences. If apical dominance is destroyed in a cereal plant by removal of the lead shoot, a cereal plant in the greenhouse under favorable conditions can produce numerous tillers. Under normal circumstances, tillering occurs only in the seedling stage before a lead shoot develops that exerts apical dominance. Newly formed tillers dependent on more advanced culms may be suppressed as well as any further tiller initiation once apical dominance is expressed. An unvernalized winter annual produces a profusion of tillers because shoots that exert apical dominance never develop. In dwarf wheat strains, apical dominance appears to be incomplete, and such plants exhibit a high level of tillering over an extended time period. Late-formed tillers may be green and immature when the first tillers are mature, thereby complicating harvesting and lowering grain quality (Stoskopf and Fairey 1975). The full expression of apical dominance results in more favorable synchronous tiller development.

In perennial forages, apical dominance is exerted in developing inflorescences but is lost as shoots mature or as clipping destroys lead shoot supremacy. Grazing of pasture plants may remove the lead shoot in a manner similar to pinching the apical shoot of the geranium and may stimulate the growth of lateral buds in the branches of legume plants.

Genetic Control

Tillering differences among species and within species can be observed. Winter cereals may have the highest tillering capacity of the cereal plants because they have the benefit of a fall and spring tillering period. More tillers are initiated than develop to maturity; but without an opportunity for fall tillering, such as under very late fall seedings, yield is generally restricted, possibly because of a lack of reserves as such plants go into the winter. Fall tillering appears to influence photosynthetic area, the build-up of reserve food, the extent of root development, winter hardiness, and spring vigor, as well as the degree of spring tillering. In northern regions, fall grazing or clipping invariably reduces subsequent grain yield, possibly because of the detrimental effect on one or more of the above features. In southern regions of the United States where positive assimilation rates can be measured during most of the winter, grazing is a common practice. Wheat yields may be sacrificed under such a situation depending on the relative prices for wheat and livestock products.

Because grain yield does not relate directly to the degree of fall tillering, there is no advantage to seeding fall cereals for grain production earlier than the recommended date. The abundance of leaves associated with the higher tillering of unvernalized fall-seeded cereals may lead to smothering in the winter. Early fall seeding subjects plants to the Hessian fly, which can lead to plant losses the following summer as the developing larvae tunnel in the wheat culms.

Tillering differences between winter and spring types were demonstrated in winter and spring oats in 30-cm (12 in) spacings in Iowa. Winter oats averaged 12.0 tillers per plant, and spring oats, 9.1 (Frey and Wiggans 1957). Fall tillering of winter oats in Ontario will not ensure the survival of this crop under Ontario conditions.

Two-row barley has a higher tillering capacity than six-row barley, but grain yield of the two types may be similar under similar seeding rates and environmental conditions. More tillers develop to maturity in a two-row barley than in a six-row barley. Tillering capacity in a two-row barley must be regarded, therefore, as a positive attribute. Attempts to transfer the high tillering capacity and the large-sized kernels of two-row barley into six-row types have not been successful because the photosynthetic source has not kept pace.

The development of the genetically controlled nontillering (uniculm) mutant barley cultivar Kindred (Donald 1968; Swaminathan, Chopra, and Bhaskaran 1962) lead Donald to propose the development of genetically nontillering types to reduce intra- and interplant competition. Such a uniculm barley, comprised of only one main stem and no tillers, will have no remnant of perenniality, no intraplant competition among developing tillers, and without late tiller development, all plants could exert uniform interplant competition. With these concepts in mind, uniculm barley development was initiated in the mid-1960s at a number of plant-breeding institutions, but there is no good evidence that these cultivars are inherently higher yielding than multiple-tillered types. A weakness of uniculm strains appears to be poor root development, leading to low drought-tolerance and susceptibility to root-induced lodging. In cereals that tiller, the main lateral shoots produce adventitious roots, which help anchor and nourish the plant, an apparent major function of lateral tillers.

In northwestern Europe, MacKey (1966) described a trend to develop decreased tillering wheat lines. On the basis that wheat cultivars released in England were high in yield and showed a reduced tillering capacity, this approach was supported by Bunting and Drennan (1966) and Bingham (1967). Some tillering ability in cereals is assumed to be desirable in order to enable recovery from poor establishment or thin stands due to insects, disease, winterkill, or other injury.

In contrast to the extensive fibrous root system associated with tillers on cereal plants, flax produces a poorly branched taproot system, which may be a weakness for flax production. Branching in flax arises from a lower node 1 or 2 cm (1/3 to 3/4 in) above the soil. The entire plant is supported by one root system. Flax is capable of producing branches from buds below the primary flower. Branching in flax is of minor importance to flax yield under recommended seeding rates.

In grain corn production, the virtual elimination of tillering in commercial corn hybrid cultivars has provided a form of population control considered to be an asset for grain production. Tillers are regarded as an undesirable feature during environmental stress conditions often associated with the reproductive development stage.

For forage corn production, evidence that grain development is not associated with high forage yield and quality (Chapter Sixteen) led to the investigation of tillering corn hybrid cultivars for silage use. Would tillering types increase vegetative growth and forage yield? Up to twelve tillers per corn plant has been reported (Kuleshov 1933).

Comparison of tillering and nontillering hybrid cultivars of corn showed that the tillering hybrid cultivar produced more dry matter than a nontillering hybrid cultivar under low populations and irrigation in southern Alberta (Freyman et al. 1973; Bowden, Freyman, and McLaughlin 1973; Bowden, McLaughlin, and Freyman 1975). Subsequent studies in Alberta (Major 1977), in Ontario (Hunter 1978), and in Quebec (Francis and Gendron 1977) suggest that at low plant populations, tillering hybrids outperform nontillering hybrids. At high plant densities, yield is comparable; but at plant densities recommended for forage crop production, there is no apparent advantage and tillering hybrids are not grown extensively.

As in small seeded cereals, except flax, tillering corn-cultivars have better drought resistance as a result of a more extensive root system (Duncan 1975).

Herbaceous forage plants, like other plant species, also exhibit genetic differences in amount and duration of tiller production (Langer 1963; Mitchell 1956). The undeniable importance of tillering in forage grasses has led to a preponderance of the most prolific types to be used for forage production.

Light Control

Light of favorable intensity and quality stimulates branching (Friend 1965; Matsushima 1967), possibly due to plant hormone degradation or synthesis or to increased availability of food energy. Cultural practices of spaced planting and low plant population facilitate light penetration to lower branch or tiller buds to stimulate their development. Under the reverse situation of high plant density, light penetration is limited and branch development restricted. Seed growers wishing to multiply a limited quantity of seed should utilize this concept. Seed produced on tillers is genetically identical, and in such cases tillers are considered an asset for seed multiplication.

Early seeding is associated with high tiller production. Seasonal increases in far-red illumination suppress branching (Tucker 1976) and are a factor related to low tillering in late-seeded cereals. Day length is also a factor. In plants having a specific day-length requirement, photoperiod governs whether or not an apical or lateral bud will develop into a flowering shoot.

Harrowing of pastures is recommended to spread animal droppings; in addition it may stimulate growth, possibly because dead plant tissue or manure is dislodged from the crown of the plants to allow light penetration to stimulate tiller development. When a field is cut at harvest, it is the increased light penetration that stimulates tiller growth.

The use of the rotary hoe on soybeans 15 cm (6 in) in height, to destroy weed seedlings, is regarded as a severe treatment; but the loss of leaves, the possible loss of apical buds and apical dominance, along with a reorientation of leaves, and even the loss of individual plants will allow light to penetrate to lateral branch buds; and recovery is rapid, dramatic, and possibly complementary to yield.

Forage producers should recognize the importance of light and tiller development. Under conditions of a closed leaf canopy, tiller development is reduced or halted by the low light intensity at the base of plants. Could this be an indication of a suitable time for cutting to reduce seasonal production cycles? In swards of timothy and meadow fescue cut before or at ear emergence, tillering has been shown to decline from spring to summer (Langer 1959), but plots defoliated every four weeks exhibited this trend only in a dry season. As noted in Table 8–16, highest productivity scores for perennial ryegrass were achieved under ten cuts per season and exceeded productivity scores at three or five cuts even though total dry-matter yields were reduced with increased cutting regimes. Brougham (1959) found that in a ryegrass-clover pasture, tiller population was generally greater when the sward was allowed to grow 18–23 cm in height and then grazed to 2.5 to 7.5 cm (1 to 3 in) than under a more intensive or lenient management system.

Moisture, Temperature, and Soil Fertility

Whether tillering is governed through hormonal, genetic, or other mechanisms, the full expression of branching is determined predominantly by the supply of minerals and assimilates (Aspinall 1963). Tiller survival and the production of an inflorescence are paramount to yield. The degree of inter- and intraplant competition for minerals, assimilates, and water will determine how many tillers will reach maturity (Bunting and Drennan 1966). Bingham (1966) found that a shortage of water available to the plant generally reduced branching, probably because it reduced photosynthesis, reduced nutrient uptake, and altered the level of plant hormones.

Temperature increases accelerate the speed of chemical and physical plant processes. Late-seeded crops including wheat, barley, rice, and soybeans have shown a reduction in branching (Friend 1965; Pfund 1974; Chi, Tsai, and Huang 1974; Shibles, Anderson, and Gibson 1975). High temperatures were associated with a high level of tiller mortality. In barley, tiller production and survival were higher at a 10:6°C (50:43°F) than at a 24:15°C (75:59°F) day:night temperature regime (Cannell 1969a and b).

Any deviation from the optimum conditions for plant growth is likely to be reflected in a decrease in branching. In general, the earliest formed tillers have a better chance of surviving and to form a large head than those developing later, although the shoots that die are not necessarily the last to be produced (Langer 1967). Early-formed tillers have a better chance of surviving and producing ears because they develop before the onset of high temperatures, which increase apical dominance.

Most alfalfa crops in the cool and humid continental zone of North America are harvested three times during the summer growing period, usually when flowers are visible on only 10% of the tillers. Under this three-cut system, dry-matter yields are greater than at subsequent harvests (Smith et al. 1966; Moore 1967; Daigger, Axihelm, and Ashburn 1970). Alfalfa regrowth was found to be related to temperature, with greater regrowth occurring at 20:15°C (68:59°F) day:night temperatures than at a 30:25°C (86:77°F) temperature regime (Peason and Hunt 1972). Contrary to popular opinion, which credited water stress and poor management with the low second- and third-cut yields, response to temperature appears to be a factor determining alfalfa regrowth, responsible at least in part for the decline in harvest yield from first to third growth under field conditions.

Crop producers cannot modify temperature conditions in the macroclimate, but they can manage plants so that alfalfa regrowth coincides with cooler temperature periods. Plant breeders can select alfalfa, which can "tolerate" higher temperatures, to help level out seasonal variation in forage production. Perhaps a first cut taken well in advance of visible flower buds may allow for regrowth to occur at an earlier, cooler period in the season and help level out seasonal variation in forage production. Further, research on longevity and seasonal yield is needed.

CONCLUSIONS

Branching is the key to herbage yields and an essential feature for regrowth in forages. In other crops, branching cannot be unequivocably stated as being desirable or undesirable.

In cereals, the lack of a commercial uniculm cultivar suggests that tillers exert a positive value; yet only a small proportion of the tillers produced actually survive to produce grain.

In corn, under most circumstances, efforts should not be aimed at increasing tillers because a large production of tillers does not result in the formation of many ears, and tillers have a lower yield than the main stem.

It is generally assumed that the tillers that die without producing ears are deleterious as regards grain production, and they waste assimilates, water, and nutrients that could otherwise contribute to grain yield (Donald 1968). Experiments with radioactive carbon dioxide show that young tillers of cereals import assimilates from the main shoot (Lupton 1966; Quinlan and Sagar 1962); but at a later stage in plant development, nongrain-bearing tillers of wheat may export assimilates to the rest of the plant (Lupton and Pinthus 1969). Other studies with wheat, however, have indicated that assimilates move from the flowering to the nonflowering tillers at all times (Dimova and Popova 1973).

If it is accepted that some of the resources are passed to the fertile tillers as the sterile tillers senesce, unproductive tillers must be viewed as undesirable because water transpired or light absorbed is essentially irreversible. Under droughty conditions the loss of water through unproductive tillers may be particularly significant.

In sorghum, because it is a perennial, a high degree of tillering is exhibited. When main stems have set seed, the sorghum plant continues to tiller. Because of this feature, sorghum can be used for grain and pasture.

Conclusions drawn from one species or cultural practice may not be applicable to other crops. The annual crop canola has heavily branched stems. Similar branching in soybeans may lead to an excessive leaf area index, but because the seed-bearing pods in canola are the main photosynthetic area, branching generally does not lead to an excessive leaf area. Under favorable conditions, therefore, branching habit in canola must be viewed as a highly desirable feature.

In rice a good relationship was demonstrated between tillering and yield when grown at normal densities (Yoshida 1972). In wheat and corn, however, low tiller and ear number were not consistently correlated with either low or high yields (Bingham 1969; Duncan 1975). The question of tillering from the standpoint of using this attribute to maximum advantage in crop production thus remains largely unresolved.

QUESTIONS

1. Forage fields cut two weeks after the recommended date often have slow aftermath recovery.
   (a) Compare and explain reasons for quick or slow regrowth under conditions of early or late cutting.
   (b) Explain differences in regrowth between grasses and legumes.
2. What management practices should accompany forage fields cut frequently in order to encourage tiller development to help level out seasonal variation in forage yields?
3. Under what circumstances, if any, is high tillering in cereals beneficial?
   (a) When would you select a cultivar noted for its high tillering capacity over one with a lower capacity?
   (b) Would you prefer a forage grass species with a high or low tillering capacity?
4. In the development of hybrid corn, pollination is controlled by removal of the pollen-producing tassels on the female plant. What effect does this action have on tillering or plant branching? Explain.
   (a) What might happen if the ear is removed at an early stage of development?
   (b) Under what circumstances may a plant become multiple-eared?
5. Explain why fall-seeded cereals tiller so profusely during the fall period.
6. List the desirable and less desirable features associated with branching and tillering in crops.
7. What management practices would you employ to promote branching and tillering in a crop?
8. Is there any advantage to achieving the desired plant population in cereals by using a high seeding rate rather than a low seeding rate and relying on tillers to produce a satisfactory leaf area and grain yield?

9. Prepare a library-researched article on fall-seeding date of winter cereals with emphasis on tillering as it affects winter survival, root development, and grain yield. Consider the impact of fall grazing on grain yield for your area.

10. Comment on tillering capacity from the standpoint of competitive ability.

11. Discuss the possible merits of a multitillering corn hybrids for silage corn production.

12. Review the literature on tillering cited for this chapter and note the titles and dates of publication. What features strike you about the literature on tillering?

REFERENCES FOR CHAPTER SEVENTEEN

Aspinall, D. 1963. "The Control of Tillering in the Barley Plant II. The Control of Tiller-Bud Growth during Ear Development," *Australian Journal of Biological Sciences* 16:285-304.

Bingham, J. 1966. "Varietal Response in Wheat to Water Supply in the Field, and Male Sterility Caused by a Period of Drought in a Glasshouse Experiment," *Annals of Applied Biology* 57:365-377.

———. 1967. "Investigations on the Physiology of Yield in Winter Wheat by Comparisons of Varieties and by Artificial Variation in Grain Number per Ear," *Journal of Agricultural Science (Camb.)* 68:441-452.

———. 1969. "The Physiological Determinants of Grain Yield in Cereals," *Agricultural Progress* 44:30-42.

Bowden, D. M., S. Freyman, and N. B. McLaughlin. 1973. "Comparison of Nutritive Value of Silage from a Tillering and Nontillering Hybrid Corn," *Canadian Journal of Plant Science* 53:817-819.

Bowden, D. M., N. B. McLaughlin, and S. Freyman. 1975. "Feeding Value of Silage from a Tillering and Nontillering Hybrid Corn," *Canadian Journal of Plant Science* 55:955-959.

Brougham, R. W. 1959. "The Effects of Frequency and Intensity of Grazing on the Productivity of a Pasture of Short-Rotation Ryegrass and Red and White Clover," *New Zealand Journal Agricultural Research* 2:1232-1248.

Bunting, A. H., and D. S. H. Drennan. 1966. "Some Aspects of the Morphology and Physiology of Cereals in the Vegetative Phase," in *The Growth of Cereals and Grasses*, pp. 20-38, F. L. Milthorpe and J. D. Ivins, eds. London: Butterworth & Co. (Publishers) Ltd.

Cannell, R. Q. 1969a. "The Tillering Pattern in Barley Varieties I. Production, Survival, and Contribution to Yield by Component Tillers," *Journal of Agricultural Science (Camb.)* 72:405-422.

———. 1969b. "The Tillering Pattern in Barley Varieties. The Effect of Temperature, Light Intensity and Day Length on the Frequency of Occurrence of the Coleoptile Node and Second Tiller in Barley," *Journal of Agricultural Science (Camb.)* 72:423-435.

Chi, K. S., C. H. Tsai, and C. S. Huang. 1974. "Growth Behaviour and Grain Production of Three Rice Varieties Exhibiting Different Plant Types in Taipei," *Journal of the Taiwan Agricultural Research* 23:166–175.

Daigger, L. A., L. S. Axihelm, and C. L. Ashburn. 1970. "Consumptive Use of Water and Alfalfa in Western Nebraska," *Agronomy Journal* 62:507–508.

Dimova, R., and D. Popova. 1973. "Inter-Relation between Productive and Unproductive Tillers in Wheat and Rye," *Field Crop Abstracts* 28:86:1975.

Donald, C. M. 1954. "Competition among Pasture Plants II. The Influence of Density on Flowering and Seed Production in Annual Pasture Plants," *Australian Journal of Agricultural Research* 5:585–597.

———. 1968. "The Breeding of Crop Ideotypes," *Euphytica* 17:385–403.

Duncan, W. G. 1975. "Maize," in *Crop Physiology; Some Case Histories,* pp. 23–50, ed. L. T. Evans.

Francis, T. R., and G. Gendron. 1977. "Population Studies of Multitillering versus Nontillering Corn for Silage in Eastern Canada," *Canadian Journal of Plant Science* 57:312.

Frey, K. J., and S. C. Wiggans. 1957. "Tillering Studies in Oats. I: Tillering Characteristics of Oat Varieties," *Agronomy Journal* 49:48–50.

Freyman, S., et al. 1973. "Nutritive Potential of Multitillering Corn Compared with Nontillering Corn for Silage, *Canadian Journal of Plant Science* 53:129–130.

Friend, D. J. C. 1965. "Tillering and Leaf Production in Wheat as Affected by Temperature and Light Intensity," *Canadian Journal of Botany* 43:1063–1076.

Hunter, R. B. 1978. "Selection and Evaluation Procedures for Whole-Plant Corn Silage," *Canadian Journal of Plant Science* 58:661–678.

Kuleshov, N. N. 1933. "World's Diversity of Phenotypes of Maize," *Journal of the American Society of Agronomy* 25:688–200.

Langer, R. H. M. 1959. "A Study of Growth in Swards of Timothy and Meadow Fescue II. The Effects of Cutting Treatments," *Journal of Agricultural Science* 52:273–281.

———. 1963. "Tillering in Herbage Grasses," *Herbage Abstracts* 33:141–148.

———. 1967. "Physiological Approaches to Yield Determination in Wheat and Barley," *Field Crop Abstracts* 20:101–105.

Lupton, F. G. H. 1966. "Translocation of Photosynthetic Assimilates in Wheat," *Annals of Applied Biology* 57:355–364.

Lupton, F. G. H., and M. J. Pinthus. 1969. "Carbohydrate Translocation from Small Tillers to Spike-Producing Shoots in Wheat," *Nature* 221:483–484.

MacKey, J. 1966. *"The Wheat Plant as a Model in Adaptation to High Productivity under Different Environments"* Proceedings of the 5th Yugoslavian Symposium on Research in Wheat, Novi Sad.

Major, D. J. 1977. "Seasonal Dry-Weight Distribution of Single Stalked and Multitillered Corn Hybrids Grown at Three Population Densities in Southern Alberta," *Canadian Journal Plant Science* 57:1041–1047.

Matsushima, S. 1967. *Crop Science in Rice—Theory of Yield Determination and Its Application.* Tokyo: Fuji Pub. Co. Ltd.

Mitchell, K. J. 1956. "Growth of Pasture Species under Controlled Environment I.

Growth at Various Levels of Constant Temperature," *New Zealand Journal of Science and Technology* 38A:203–216.

Moore, C. E. 1967. "Growth Analysis of Vernal Alfalfa throughout the Growing Season under One System of Cutting Management." Master's thesis, University of Guelph.

Pearson, C. J., and L. A. Hunt. 1972. "Effects of Temperature on Primary Growth and Regrowths of Alfalfa," *Canadian Journal of Plant Science* 52:1017–1027.

Pfund, J. H. 1974. "Optimum Culm Number in Barley (*Hordeum vulgare* L. emend. Lam.)," *Dissertation Abstracts International B* 35:640.

Quinlan, J. D., and G. R. Sagar. 1962. An Autoradiographic Study of the Movement of 14C-Labelled Assimilates in the Developing Wheat Plant. *Weed Research* 2:264–273.

Shibles, R. M., I. C. Anderson, and H. G. Gibson. 1975. "Soybean," in *Crop Physiology; Some Case Histories*, pp. 151–189. ed. L. T. Evans.

Smith, D., et al. 1966. *"The Performance of Vernal and du Puits Alfalfa Harvested at First Flower or Three Times by Date,"* *Research Report of the Wisconsin Agricultural Experiment Station* No. 23.

Stoskopf, N. C., and D. T. Fairey. 1975. "Asynchronous Tiller Maturity—A Potential Problem in the Development of Dwarf Winter Wheat (*Triticum aestivum* L.)," *Plant Breeding Abst.* 45:467–472.

Swaminathan, M. S., V. L. Chopra, and S. Bhaskaran. 1962. "Chromosome Aberrations and the Frequency and Spectrum of Mutations Induced by Ethylmethane Sulphonate in Barley and Wheat," *Indian Journal of Genetics and Plant Breeding* 22:192–207.

Tucker, D. J. 1976. "Effects of Far-Red Light on the Hormonal Control of Side-Shoot Growth in Tomato," *Annals of Botany* 40:1033–1042.

Yoshida, S. 1972. "Physiological Aspects of Grain Yield," *Annual Review of Plant Physiology* 33:437–464.

chapter eighteen
Crop Lodging

A barrier to yield increases is the inability of the plants to support their ever-increasing weight of economic yield, and they fall over or lodge. *Lodging* was described as an "abundance disease" by Pinthus (1973), meaning that under highly productive environmental conditions of high soil fertility, good moisture, and high populations to promote a high LAI, lodging may occur, restricting the exploitation of otherwise yield-promoting factors. Cultural practices must aim, therefore, at a balance or compromise between yield-promoting inputs, yet prevent lodging. With the number of interacting factors in crop production, this presents a very real challenge to crop producers.

Lodging (Fig. 18-1) is a serious problem in many crops and may result in yield reductions. Lodging results when forces exerted by wind, rain, or hail permanently displace plant stems from their upright position. Lodging may involve bending of stems at the base, breakage of the stems, or root lodging—the predominant form of lodging in cereals. Lodging-resistant types have a root system that provides relatively better anchorage because of ample horizontal and vertical branching.

The degree of stem lodging, that is, the angle at which stems deviate from the perpendicular, can vary from slight to lying flat on the ground. Wind action on plants held rigidly by dry, firm soil may lead to stem breaking. And under conditions of a limited root system in wet soil, the torque created on the upper part of the plant is transferred to the roots and root lodging results.

Lodging may occur at any stage of plant development, but if lodging occurs in the seedling or vegetative stages, plants may recover their upright position—a negative geotropic response—resulting in goosenecking or curvature of the lower stems. Reduced light on the underside of actively growing stems causes cells to elongate, pushing the stems upright. In corn, a large root system and sturdy young stalks prevent lodging in the early stages of plant development. Lodging in corn is most prevalent late in the season, after physiological maturity has been reached and when the weight of the ear is greatest and plant stalks are weakened by translocation of carbohydrates out of

Figure 18-1. Lodging in oats. Lodging may reduce yield because of poor light interception and reduction in photosynthesis. (Courtesy Ontario Ministry of Agriculture and Food.)

the stalks to the developing sink. Diseased, damaged, or weathered stalks may promote lodging in crops.

Traditionally, lodging-induced losses have been associated only with indirect yield losses, i.e., harvest-related losses, a situation that predominates in grain corn production. Recognition that photosynthesis is the basis of all crop yield and that lodging can upset photosynthesis has led to the realization that direct yield losses occur also if lodging takes place while the crop is metabolically active. Lodging disrupts the highly organized plant leaf canopy, resulting in greater mutual shading and reduced air and CO_2 movement, causing reductions in the net assimilation rate. Lodged stems may disrupt the flow of water, nutrients, and assimilates. The uppermost exposed leaves may provide food energy for a restricted number of grain-bearing heads. Lodging can destroy apical dominance, and if the disrupted leaf canopy allows sunlight to strike basal nodes, lateral branching and tillering may be stimulated after such development has normally ceased. The additional late vegetative growth contributes nothing to grain yield and complicates harvesting.

Lodging reduces grain quality. Lodged plants may allow heads to come in contact with the soil, encouraging mold and decay organisms. Birds can feed readily on the grain of lodged plants. Restricted air movement and high humidity may encourage sprouting, which reduces feed value, milling, and malting quality. Improperly filled kernels are low in quality.

If cereals are used as a companion crop for forage seedling establishment, lodging can destroy underseeded forages.

The effect of lodging on grain yield is dependent on the time of occurrence and the severity of lodging. Lodging during the early vegetative stage when plants can recover their upright position may not be as detrimental to yield as lodging during the early reproductive stage when grain filling occurs and photosynthetic demands are greatest. Juvenile lodging, before stems are differentiated, was observed to have some detrimental yield effects (Gardener and Rathjen 1975). Under conditions of a Mediterranean climate, where spring-type cereals are grown in winter under high fertility conditions, grain yield in barley was depressed by juvenile lodging by a reduction in the number of productive tillers apparently as a result of poor light interception coupled with drastically changed conditions of ventilation, evapotranspiration, and temperature. Lodging that coincides with early stages of reproductive development was found to reduce yields more than at a later stage (Table 18-1).

TABLE 18-1. Reduction of grain yield in cereals due to artificially induced 90° lodging at two growth stages

Crop	Location	Lodging Induced at	
		Heading	15–20 Days After Heading
		Percent Yield Reduction	
Winter wheat	Kansas	27	22
Winter wheat	Illinois	31	20
Spring wheat	Manitoba	34	24
Fall-sown barley	Arizona	40	39
Oats	Iowa	36	23
Oats	Illinois	37	17

Source: Data from Pinthus 1973.

In a study with winter wheat growing through a 5-cm wire mesh and artificially lodged by moving the mesh and forcing and holding the plants in a horizontal position to simulate complete lodging, yield reductions resulted at each of the four stages studied (Weibel and Pendleton 1964). Lodging was induced at heading and each week thereafter (Figure 18-2). When compared to erect plants, reductions in yield due to lodging were as follows: at heading, 31%; at milk stage, 25%; at soft dough stage, 20%; and at hard dough stage, 12%. Lodging lowered weight per hectolitre (bushel) and kernel weight of the harvested grain.

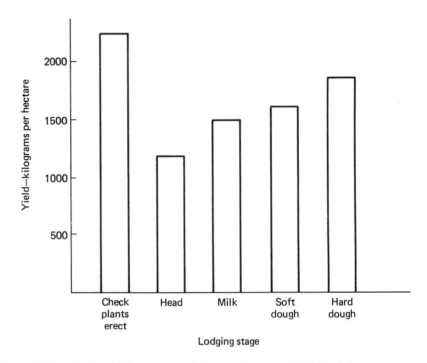

Figure 18-2.   Grain yield response of winter wheat to artificial lodging treatments at four stages of plant development for a three-year average, Urbana, Illinois.

Of greatest interest to producers are management practices that reduce lodging in crops, but because of the differences among crops, selected crops will be dealt with individually in the following pages.

LODGING IN CORN

Hybrid-corn development has resulted in greater stalk lodging resistance than open-pollinated cultivars. The amount of the increase was suggested by Zuber and Kang (1978) to be in the magnitude of two or three times. However, increased fertilizer applications, higher plant populations, and monoculture production have put greater stresses on the plants and have partially offset these gains. Annual stalk rot lodging losses in the United States were estimated at 5 to 25% (Zuber and Kang 1978).

Most lodging losses in corn are a result of mechanical losses. Mechanical corn pickers must gain control of the plant, which is difficult when stalks are broken over below the ear or are leaning at an angle greater than 45°. Lodging in corn normally occurs late in the season, generally after the corn has reached physiological maturity (35% kernel moisture), and it increases in intensity as the corn remains standing in the field (Johnson 1960) (Table 18-2).

TABLE 18–2.   Lodging as determined by date of observation in Ohio

	Hybrid 1		Hybrid 2	
Date	Percent Kernel Moisture	Percent Lodging	Percent Kernel Moisture	Percent Lodging
Oct. 2	27.5	5.5	29.1	18.9
14	25.2	3.0	26.7	26.0
23	22.4	16.3	24.7	35.1
Nov. 3	21.2	17.4	23.6	35.4
13	19.0	28.0	20.7	40.8

Source: Data from Johnson 1960.

To pick up lodged stalks, it is necessary for the gathering points to ride on the ground surface and the gathering chains to operate close to the ground so that they can grasp lodged stalks. Thus, the risk of picking up stones and damage to the machine is increased.

The major causes of lodging in corn include the following:

1. Cultural practices and stress conditions.
   (a) Strong wind and/or heavy rain.
   (b) Fertility imbalance such as potassium deficiency.
   (c) High yield and high plant populations.
   (d) Damage from herbicide 2, 4-D.
2. Stalk rots caused by fungal organisms.
3. Insects—stalk borers, rootworm.
4. Genetically inherited susceptibility or resistance to lodging.

Stalk decay is a major cause of corn lodging and is caused by fungal organisms such as *Gibberella* species commonly present on crop residue and the soil. Corn plants become susceptible to rot organisms when sugar levels are decreased by stresses that interfere with normal plant growth (Mortimore and Ward 1964). Stress conditions include leaf blights, drought, wet weather, frost, hail, unbalanced fertility, and poor soil drainage. Infestation of decay organisms may occur after pollination, long before stalk breakage symptoms occur. Stalk decay develops slowly into a dry, internal stalk shredding, often pinkish-red in color. Stalk rot weakens the stalk and breakage occurs, although lodging may not take place until after physiological maturity when food energy has moved from the stalk into the ear.

To control stalk rot, producers must select hybrid cultivars with a test record of low stalk breakage and with the proper maturity rating. Corn late in maturity, which must remain in the field to dry, may lodge after frost has destroyed the photosynthetic area and assimilates move out of the stalk into the immature ear. Increases above recommended population levels could lead to a breakdown of hybrid cultivars that have genetic resistance to stalk rot if planted at recommended plant populations. Excessive interplant competition under high populations may be expected to limit the total carbohydrate manufactured within any single plant and to predispose plants to stalk rot when the low level of food reserves move into the ear. Although stalk rot symptoms may not occur until after the plants have reached physiological maturity, Mortimore and Wall (1965) concluded that stalk rot had an adverse effect on grain yield in addition to lodging.

In a stalk rot-susceptible hybrid, stalk rot reached high levels at relatively low populations, whereas a stalk rot-resistant hybrid had minor levels at recommended rates (Table 18-3). Stalk rot increased, however, in both hybrids as the season progressed (Mortimore and Wall 1965).

Plants bearing more than one ear were observed to develop stalk rot before single-eared plants (Mortimore and Wall 1965) in the same population. Barren plants or plants producing a nubbin ear do not develop stalk rot.

Since stalk rot is a mature plant disease, timeliness of harvesting is important. The longer the crop is left in the field, the greater the risk of stalk breakage, although this risk is reduced with hybrid cultivars that reach physiological maturity before frost.

TABLE 18-3.   Percentage of corn plants with stalk rot in a susceptible and a resistant hybrid cultivar on three dates at five populations, mean of two years, 1961 and 1962, conducted at Harrow, Ontario

Corn Hybrid	Plants per Hectare	(acre)	Percent of Plants with Stalk Rot			Yield	
			Sept. 26	Oct. 11	Oct. 27	(t/ha)	(tons/ac)
Susceptible	12,350	5,000	0	0	2.0	3.13	1.27
	24,700	10,000	4.5	19.4	42.7	6.66	2.70
	37,000	15,000	15.1	44.8	70.7	7.20	2.91
	49,400	20,000	23.2	49.3	74.6	7.22	2.92
	61,700	25,000	38.5	64.6	81.7	6.74	2.73
Resistant	12,350	5,000	0	0	0	3.54	1.43
	24,700	10,000	0.1	1.0	1.8	6.91	2.80
	37,000	15,000	0.2	0.6	1.6	8.53	3.45
	49,400	20,000	2.7	7.8	12.4	9.08	3.68
	61,700	25,000	1.9	8.4	14.5	9.29	3.76

Source: Data from Mortimore and Wall 1965.

Crop rotation reduced the incidence of stalk rot and root rot (Figure 18-3) (Williams and Schmitthenner 1963). Generally root rot, common on corn seedlings, can be controlled by seed treatment with an approved fungicide.

Plant population experiments have shown that increasing plant number to increase LAI to maximize light interception and photosynthesis can lead to increased yields under conditions of adequate soil fertility and moisture

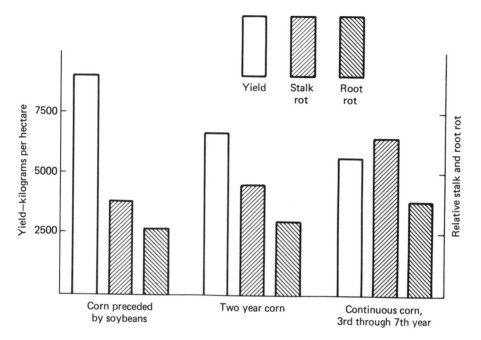

Figure 18–3. Relationship of corn yields to the severity of root and stalk rot. (After Williams and Schmitthenner 1963.)

conditions (Woods and Rossman 1956). Smaller ears and increased lodging usually associated with high plant populations may lead to greater harvest losses. At six Michigan locations, total picker losses averaged 6.0, 6.9, 10.3, and 13.4% of the total yield for average populations of 26,160, 37,800, 48,000, and 58,160 plants per hectare (10,600, 15,300, 19,400 and 23,500 ppa). As high as 35.9% of the plants lodged in one of the three hybrids tested (Table 18-4).

As plant density is increased, plants are under greater stress since there are more plants competing for the moisture, fertility, and light. With dense plant

TABLE 18-4. Average stalk lodging as influenced by hybrid and plant population at six locations in Michigan

Plant Population per Hectare	(acre)	Percent of stalk-lodging			Mean Yield	
		Hybrid 1	Hybrid 2	Hybrid 3	(t/ha)	(tons/ac)
26,160	10,600	9.9	6.6	7.4	4.87	2.19
37,800	15,300	19.4	14.3	14.3	5.20	2.34
48,000	19,400	29.6	24.0	23.6	5.42	2.44
58,160	23,500	35.9	31.7	32.6	4.94	2.22
Average		23.7	19.2	19.5	5.11	11.75

Source: Data from Woods and Rossman 1956.

stands, stress caused by one or more factors such as drought, low fertility, weed competition, insects, or diseases will have a more detrimental effect on grain yield, grain quality, and lodging. It is possible that small increases in grain yield achieved by increased plant density could be lost at harvest time due to increased harvesting losses by lodging.

A balanced soil fertility program can help reduce lodging. Potassium is important for stalk strength because it promotes good cell growth, making the stalks stronger and more lodging-resistant. A shortage of potassium is frequently accompanied by a weakening of plant stalks in cereals, corn, and sorghum (Table 18-5 after Josephson 1962). This weakening may be associated with a decrease in photosynthesis and an increase in respiration, resulting in a low net assimilation rate with insufficient potassium. This seriously reduces the supply of food energy and consequently the growth of the plant. The translocation of sugar from the leaves of cane in potash-deficient plants was reduced to less than half that of the potash-sufficient plants (Tisdale and Nelson 1966). As noted in Table 18-5, a balanced soil fertility regime can reduce lodging. Good cell growth associated with potassium promotes stalk strength. Potassium applied in excess of recommendations made on the basis of soil tests does not reduce lodging. Thus, it is important to maintain a well-balanced fertility program without applying too much or too little of any one element.

TABLE 18-5.   The effect of nitrogen and potassium fertilizer on corn lodging in Tennessee

Fertilizer Added*		Lodged Plants (%)	
(kg/ha)	*(lb/ac)*	*1959*	*1960*
67-40-0	60-36-0	21.8	19.2
134-40-0	120-36-0	25.4	24.6
67-40-47	60-36-42	5.1	1.7
134-40-47	120-36-42	6.0	6.7
67-40-93	60-36-83	1.6	1.9
134-40-93	120-36-83	1.7	3.3
67-40-186	60-36-66	2.0	3.2
134-40-186	120-36-166	1.4	2.4

Source: Data from Josephson 1962.

*In order of nitrogen, phosphorus, and potassium.

When making a final decision on plant population, a producer should consider limiting population size on soils with low fertility, on soils susceptible to drought, on soils imperfectly drained, or under conditions of late seeding.

If the herbicide 2, 4-D is used as an overall spray after the corn is 15 cm (6 in)

tall (leaf extended), rapid cell elongation and root distortion can result, making the plants more susceptible to lodging. To avoid injury the use of drop nozzles is recommended.

Larvae of the European corn borer can cause injury to all parts of the corn plant. Tiny borers hatch from eggs laid on the underside of leaves by the adult moth in the early stages of corn plant development. Initially, the larvae feed on the leaves, making small round punctures or elongated slits. Later, the borers work downward into the whorl of the plant and feed on the developing leaves and eventually tunnel into the stalks, weakening them and causing them to break over. Losses can be reduced but not controlled by plowing to incorporate cornstalks, by using rotations to reduce the number of adults (although they can fly in from adjacent fields), and by planting with low plant populations to avoid thin stalks that break over more readily when the larvae tunnel.

Larvae of corn rootworms feed on roots prior to anthesis and weaken the root system, which promotes lodging. Lodging induced by rootworms occurs early in the season, and as the plants recover, they show a characteristic goosenecking.

Rootworm infestations may vary in intensity and distribution from field to field, year to year, and therefore the degree of rootworm damage is hard to assess. Some corn hybrid cultivars may be able to tolerate considerable feeding damage because they produce new roots quickly in response to damage.

Young corn plants generally have a shallow, spreading root system, and for this reason mechanical cutting of the roots to simulate corn rootworm damage was attempted using a flat-bladed spade (Fitzgerald, Ortman, and Branson 1968). Eight corn hybrids were subjected to root cutting experiments on one, two, or three dates. Yield and lodging were strongly affected (Table 18-6).

TABLE 18-6. Effects of mechanical damage to corn roots on yield and lodging at Brookings, South Dakota

Percent of Roots Removed*	Mean Yield				Percent Lodging	
	1964		1965		1964	1965
	(t/ha)	(t/ac)	(t/ha)	(t/ac)		
0+0+0	9.75	4.39	5.76	2.60	3.1	0.9
50+0+0	8.93	4.02	5.23	2.35	14.4	5.3
25+25+0	8.65	3.89	5.34	2.40	10.6	6.6
50+25+0	8.07	3.63	4.65	2.09	20.9	10.9
75+0+0	7.92	3.56	4.31	1.94	36.3	17.2
25+25+25	7.37	3.32	4.67	2.10	40.0	26.3

Source: Data from Fitzgerald, Ortman, and Branson 1968.

*Straight cuts, perpendicular to each other, were made with a spade on one, two, or three sides of the stalk and 2.5 cm from the stalk on June 9, 15, and 22 in 1964 and on July 20, 26, and August 3 in 1965.

Crop rotation is recommended to control corn rootworms, but if crop rotation is not feasible, a soil-applied insecticide can be applied at planting time.

Under conditions when corn is lodging severely, field losses may be reduced by early harvesting in the form of whole-plant, high-moisture ear corn or shelled-corn silage. Whole-plant silage, at 65% moisture, usually can be harvested before serious lodging occurs.

To assess corn losses in the field, select a number of areas at random after harvest. Eighteen (21) shelled kernels in one square metre (yard) equals 63 kg/ha (1 bushel per acre). One large ear 340 grams (3/4 lb) per 1/100 of a hectare (acre) is equivalent to 63 kg/ha (1 bu/ac) of shelled, 15.5% moisture corn. The length of row in metres for 1/100th of a hectare at seven row widths is shown in Table 18–7. The same values in centimetres can be used to determine a one square metre area for kernel counts. For ear loss measurements, include all rows of the machine swath. For example, 1/100th of a hectare (acre) for a 91-cm (36 in) wide row requires a 110-metre (146 feet) long section of row for a one-row harvester. If a two-row harvester is used, check two parallel rows 55 metres (73 feet) in length, or for a four-row harvester, select four rows 27.5 metres (36.5 ft) long. Add ear loss to kernel loss to obtain total loss per hectare (acre) (Henry and Bereza 1979).

TABLE 18–7.   Dimensions for yield loss test areas

Row Width		Length of Row in Metres (feet) for 1/100 of a Hectare (acre) or Measurement in Centimetres (inches) to Find a Metre Area	
(cm)	(in)	(metric units)	(imperial units)
71	28	141	187
76	29	132	175
81	32	124	165
86	34	117	155
91	36	110	146
97	38	104	138
102	40	99	131

Source: Data from Henry and Bereza 1979.

LODGING IN SOYBEANS

Conditions such as high available soil nitrogen, abundant water, and warm weather that favor rank vegetative growth promote lodging in soybeans. Cultural practices that increase competition for light at high populations and the use of narrow rows result in reduced stem diameters and spindlier plants subject to lodging.

Lodging in soybeans results in a loss of apical dominance, which enhances lateral branch development. Increased vegetative growth in lodged plants reduces air circulation, decreases CO_2 supply, increases humidity, reduces light penetration, and promotes disease. The result is increased flower and pod abortion because less food energy is available for reproductive development. Yield reductions under lodged conditions can be attributed mainly to decreased numbers of pods per plant rather than smaller seed or fewer seeds per pod. Mechanical losses resulting from plant parts missing the cutter bar also occur and may be the only losses attributed to late soybean lodging.

Lodging can be reduced by using a low seed rate and wide rows, but yields may also be reduced. Selection of a lodging-resistant cultivar is essential. The best spatial arrangement for soybeans and the optimum plant populations for highest seed yields are dependent on the particular cultivar as well as on soil productivity.

In studies to determine the yield reductions attributed to lodging, Cooper (1970, 1971) reported yield reductions of 21 to 23% when natural lodging occurred in the early pod-filling stage. Weber and Fehr (1966) found a 13% increase in soybean seed yield when early lodging was prevented. Under conditions of simulated lodging, soybean seed yields were reduced from 11 to 32% in 1972 and from 12 to 22% in 1973 in Indiana (Woods and Swearingin 1977). The most critical time for lodging was found to occur when beans were beginning to develop in pods at the uppermost nodes.

LODGING IN DRY BEANS

Dry beans are usually harvested by a windrow method using puller blades located just beneath the soil surface and putting two adjacent rows into a windrow. Direct harvesting may be possible with types exhibiting an erect plant habit, with pods high off the ground and if plants are lodging-resistant.

The relationship of root size to seed yield and lodging was studied in New York State using black beans (Stoffella et al. 1979). In field studies in both years of the test, four strains produced larger seed yields and lodged less than the two check cultivars (Table 18–8). Uprooting resistance of the four lines

TABLE 18–8. Mean seed yield, lodging scores, and uprooting resistance under field conditions 1977 and 1978 in New York State

Strain	Seed Yield (kg/ha)	(lb/ac)	Lodging Score*	Pressure Required to Uproot Plants (kg)	(lb)
Four strains	3575	3192	1.4	12.5	27.5
Two checks	3293	2940	3.8	10.4	22.9

Source: Data from Stoffella et al. 1979.

*Lodging rating based upon 1 (erect) to 9 (prostrate).

was higher than for the check cultivars. The results suggest that a larger root biomass may be an important component of lodging resistance in beans.

LODGING IN CEREALS

Lodging is a major problem in cereal production and can occur at any stage of plant development. The rapidly dividing succulent meristems of elongating internodes are vulnerable to wind and rain damage. Cereal plants respond to nitrogen fertilizer by developing tall stems and large leaves, conditions inducive to lodging. Cereal cultivars are often described as being strong- or weak-strawed, suggesting that lodging-resistant cultivars do indeed exist and that the quality of the straw is responsible for differences in lodging. Although differences in straw strength do exist, all cereal crops are subject to severe lodging with root lodging being a major factor.

Seeding date and rate can influence the degree of lodging in cereals. Early-seeded spring cereals are lodging resistant because they benefit from early cool temperature conditions which encourage slow growth, cell differentiation, restricted cell elongation, and thus shorter culm internodes. Early seeding generally promotes tillering with an accompanying better root system and root lodging resistance.

Late-seeded cereal crops are subjected to higher temperatures, respond to the longer day length, and will reach head initiation in a shorter time period than with early seedings.

Seeding date along with seeding rate are possibly the best management tools to reduce cereal lodging. Pinthus (1973) reported that numerous studies with cereals in all parts of the world have indicated that lodging may be prevented or reduced by seeding rate reductions. Tillering, however, prevents the producer from exerted a high degree of control of plant population, although tillering does promote crown root formation, which reduces lodging.

Seeding in narrow rows to improve light interception, as discussed in Chapter seven, without any change in seeding rate, did increase the length of basal internodes; it also decreased the diameter and wall thickness of basal culms of wheat (Watson and French 1971). The difference between wide and narrow row spacings in straw diameter was only 6%, however. Narrow rows did increase lodging and may prevent producers from achieving the high plant populations necessary for top yields.

Under conditions of irrigation, it is essential to apply sufficient water during the vegetative stage to initiate a suitable sink and source. Stimulation of excessive vegetative growth almost surely will lead to lodging. One timely irrigation at the beginning of reproductive development can contribute to yield development and minimum lodging. The amount of irrigation water applied will be influenced by the stage of plant development, not just evapotranspiration rates or available rainfall.

Some of the most dramatic yield increases during the sixties were achieved by shortening the lower internodes and thus reducing crop lodging. Semi-dwarf and dwarf cultivars respond well to high nitrogen fertility and irrigation. Similar results were obtained with rice in the Philippines. Reduced lodging by the case of dwarf and semidwarf selections offers a viable tool to producers to increase yield. The significance of straw strength was demonstrated clearly in a series of experiments with two winter wheat cultivars differing in straw strength (Van Dobben 1966). At low yields, the grain yield of the two cultivars was equal, but as the yield was raised by a better nitrogen supply, the weak-strawed cultivar became less productive because of a greater decline in straw strength. Surprisingly, in these experiments no lodging of the crop occurred, so that the results suggest that even bending of the culms, which reduced light interception and photosynthesis, caused a yield depression. Van Dobben (1966) supported this supposition with calculations reported by a coworker, de Wit, in a closed canopy of a grass crop. According to Van Dobben's calculations, in midsummer, a lodged crop produces 50 kg carbohydrates per hectare per day less than an erect crop, which results in a loss in production of about 2 tonnes dry matter per hectare during the postheading period (1786 lb/ac).

In cultivars not completely resistant to lodging, Van Dobben (1966) suggested that grain yields may be increased and harvesting conditions improved by a split nitrogen application. Part of the nitrogen could be applied in the spring and a second application given just before or even during stem elongation. This late application does not influence plant height or tillering, but it prolongs leaf area duration. Systems of this kind are used in several European countries.

The association between short straw and reduced lodging has led to the widespread use of chemical substances such as CCC (2 chloroethyltrimethyl-ammonium chloride) to restrict cell division and elongation and reduce plant height (Appleby, Kronstad, and Rohde 1966). These substances can be applied as a foliar spray or as a seed dressing. Prevention of lodging and subsequent increase in grain yield have been measured, but different responses to environment and genotype have been reported. Dwarfing compounds may be a useful dwarfing tool under emergency situations, but it is much better for the character to be inherited than induced chemically.

The extensive literature on the use of CCC to reduce lodging in cereals was reviewed by Humphries (1968). Yield responses were variable, but the widespread interest in reducing lodging is clearly shown. The ultimate of an applied growth regulator would be the ability to alter plant development in a specified way, although the complexity of associated yield production practices make a positive approach to yield increases somewhat obscure.

Solid-stemmed wheats are known to possess resistance to the wheat stem sawfly, an insect that increases lodging and lowers yield (Kemp 1934). Solid-stemmed wheat yields are usually lower than the yields of hollow-stemmed selections (McNeal et al. 1965). Many thousands of solid-stemmed wheat plants have been evaluated for yield potential, but few have been observed that will produce yields equal to those of commercially grown hollow-stemmed cultivars (McNeal et al. 1965). Mitchell (1970) suggested a possible explanation is that development of a solid pith uses food energy that would have gone into grain development.

As a means of stabilizing performance, Grafius (1966) tested lodging resistance of oat mixtures and concluded that mixtures could improve the overall worth of a cultivar with respect to lodging, test weight, and yield. If the yield increases due to reduced lodging are greater than yield reductions from competition, a blend may be acceptable. However, in the years following the report by Grafius, blends have been slow to develop.

CONCLUSIONS

The problem of lodging must be solved because it creates a very specific barrier to increasing crop yield. To benefit from the anticipated increase in grain yield potential, improved lodging resistance is needed. The combined efforts of breeding for lodging resistance, management practices to maximize sunlight interception, reduction of diseases and insects, and possibly the use of growth regulators will be required to reduce the lodging problem. It may be concluded that a permanent requirement exists for increased lodging resistance, linked to the continuous process of grain yield increases.

QUESTIONS

1. Comment on the statement that crop lodging is an "abundance disease."
2. At what stage in the development of a crop is lodging likely to be most detrimental to yield? Explain.
3. Suggest management practices that crop producers may use to reduce the lodging problems. Can you suggest any in addition to those mentioned in the text?
4. As a crop producer, what features would you recommend to a plant breeder to incorporate that would help you reduce your lodging problem and produce top yield?
5. Some farmers have accepted crop lodging as the inevitable fate of a good crop and regard lodging as an indication that they have produced a good crop. Comment.
6. Review how crop lodging may reduce grain quality.
7. Explain why lodging at heading time was more detrimental than fifteen to twenty days after heading (Table 18-1).
8. Comment on the importance of lodging in forage crops to be used for herbage production.

9. With the use of modern machinery such as a swather to pick up lodged crops, should lodging continue to receive the emphasis and be regarded as seriously as suggested in this chapter?

10. Speculate on the action of growth regulators in addition to the action of CCC as a tool to reduce lodging.

11. Review the value of crop rotations in reducing lodging. Where in the rotation would cereal crops be best suited?

12. Explain the high amount of lodging when nitrogen was applied without any potassium (Table 18-5).

13. What limitations (if any) might exist to the development of further reductions in plant height to reduce lodging? What is the ideal plant height? How short might plants become?

14. Relate branch or tiller development to lodging.

15. Researchers in Iowa (Shaw and Weber 1967) suggested a small degree of lodging enhances soybean seed yields. Speculate on why this was observed.

REFERENCES FOR CHAPTER EIGHTEEN

Appleby, A. P., W. E. Kronstad, and C. R. Rohde. 1966. "Influence of 2-Chloroethyltrimethylammonium Chloride (CCC) on Wheat (*Triticum aestivum* L.) When Applied as a Seed Treatment," *Agronomy Journal* 58:435–437.

Cooper, R. L. 1970. "Early Lodging—A Major Barrier to Higher Yields," *Soybean Digest* 30(3):12–13.

———. 1971. "Influence of Early Lodging on Yield of Soybean [*Glycine max* (L.) Merr.] *Agronomy Journal* 63:449–450.

Fitzgerald, P. J., E. E. Ortman, and T. F. Branson. 1968. "Evaluation of Mechanical Damage to Roots of Commercial Varieties of Corn (*Zea mays* L.)," *Crop Science* 8:419–421.

Gardener, C. J., and A. J. Rathjen. 1975. "Juvenile Lodging in Barley: A Yield-Depressing Phenomenon," *Australian Journal of Agricultural Research* 26: 231–242.

Grafius, J. E. 1966. "Rate of Change of Lodging Resistance, Yield, and Test Weight in Varietal Mixtures of Oats *Avena sativa* L.," *Crop Science* 6:369–370.

Henry, G. H., and K. Bereza. 1979. *Corn Lodging.* Ontario Ministry of Agriculture and Food. Fact sheet No. 79-021. (4 pp.)

Humphries, E. C. 1968. "CCC and Cereals," *Field Crop Abstracts* 21:91–99.

Johnson, W. H. 1960. *Corn Harvesting 1959-60.* Report of the Department of Agricultural Engineering, Ohio Agricultural Experiment Station.

Josephson, L. M. 1962. "Effects of Potash on Premature Stalk Drying and Lodging of Corn," *Agronomy Journal* 54:179–180.

Kemp, H. J. 1934. "Studies of Solid Stem Wheat Varieties in Relation to Wheat Stem Sawfly Control," *Sci. Agr.* 15:30–38.

McNeal, F. H., et al. 1965. "Relationship of Stem Solidness to Yield and Lignin Content of Wheat Selections," *Agronomy Journal* 57:20-21.

Mitchell, R. L. 1970. *Crop Growth and Culture.* Ames, Iowa: Iowa State University Press.

Mortimore, C. G., and G. M. Ward. 1964. "Root and Stalk Rot of Corn in Southwestern Ontario III. Sugar Levels as a Measure of Plant Vigor and Resistance," *Canadian Journal of Plant Science* 44:451-457.

Mortimore, C. G., and R. E. Wall. 1965. "Stalk Rot of Corn in Relation to Plant Population and Grain Yield," *Canadian Journal of Plant Science* 45:487-492.

Pinthus, M. J. 1973. "Lodging in Wheat, Barley and Oats: The Phenomenon, Its Causes and Preventive Measures," *Advances in Agronomy* 25:209-263.

Shaw, R. H., and C. R. Weber. 1967. "Effects of Canopy Arrangements on Light Interception and Yield of Soybeans," *Agronomy Journal* 59:155-159.

Stoffella, P. J., et al. 1959. "Root Characteristics of Black Beans I. Relationship of Root Size to Lodging and Seed Yield," *Crop Science* 19:823-826.

Tisdale, S. L., and W. L. Nelson. 1966. *Soil Fertility and Fertilizer,* 2nd ed., pp. 82-83. New York: The Macmillan Publishing Co., Inc.

Van Dobben, W. H. 1966. "Systems of Management of Cereals for Improved Yield and Quality," in *The Growth of Cereals and Grasses,* pp. 320-334. F. L. Milthorpe and J. D. Ivins, eds. Butterworth & Co. (Publishers) Ltd.

Watson, D. J., and S. A. W. French. 1971. "Interference between Rows and between Plants within Rows of a Wheat Crop, and Its Effects on Growth and Yield of Differently Spaced Rows," *Journal of Applied Ecology* 8:421-445.

Weber, C. R., and W. R. Fehr. 1966. "Seed Yield Losses from Lodging and Combine Harvesting in Soybeans," *Agronomy Journal* 58:287-289.

Weibel, R. O., and J. W. Pendeton. 1964. "Effect of Artificial Lodging on Winter Wheat Grain Yield and Quality," *Agronomy Journal* 56:487-488.

Williams, L. E., and A. F. Schmitthenner. 1963. "Rotation Affects Corn Stalk and Root Rot," *Ohio Farm and Home Research* 48:67.

Woods, D. J., and E. C. Rossman. 1956. "Mechanical Harvest of Corn at Different Plant Populations," *Agronomy Journal* 48:394-397.

Woods, S. J., and M. L. Swearingin. 1977. "Influence of Simulated Early Lodging upon Soybean Seed Yield and Its Components," *Agronomy Journal* 69: 239-242.

Zuber, M. S., and M. S. Kang. 1978. "Corn Lodging Slowed by Sturdier Stalks," *Crops and Soils* 30(5):13-15.

chapter nineteen

Root Growth, Seedling Development and Crop Production

The saying "Out of sight, out of mind" applies to roots of plants. Roots are the unseen and frequently forgotten component of crop production, and perhaps it is no accident that roots are the last topic in this book. There has been less research directed to roots than to other plant parts, perhaps because roots are difficult and awkward to work with; they have no clear criterion of measurement such as dry weight, area, and extent of lateral or vertical growth; and they have no market value except in root crops per se. Approaches to root studies often are purely descriptive with little interpretive discussion, and they are frequently based on pot experiments, or on sand or nutrient culture.

Crop producers seeking effective inputs to increase crop production are becoming increasingly aware that root growth and functions in many instances set the pace of development for the entire plant. There is a growing awareness among crop scientists that root neglect has resulted in a gap in the knowledge of root-related crop production inputs. It is possible that the soil environment in which the plant roots grow, although of far greater complexity than the environment of the above-ground parts, could be subject to greater control than the environment in which the leaves and stems grow. Clearly, roots deserve more attention than they have received in the past.

THE ROOT ENVIRONMENT

The soil environment has a direct effect on root development size and activities and subsequently on the growth and yield of above-ground parts. Soil temperature, air or oxygen content, and the supply of water and fertilizer nutrients are important features of the root environment.

365

Soil Temperature

Favorable soil temperatures to promote rapid seed germination, seedling establishment, and rapid and vigorous root growth are paramount to successful crop production. Many crop producers in the temperate regions of the world have to contend with cool soil temperatures, and they must direct considerable effort into cultural practices such as tillage and soil drainage to promote favorable soil temperatures. Plants vary in regard to soil temperature requirements as demonstrated by three crops that need low, medium and high soil temperatures for germination (Weaver 1926) (Table 19-1).

TABLE 19-1.   Temperatures for germination

Crop	Minimum °C	°F	Optimum °C	°F	Maximum °C	°F
Wheat	4.4	40	28.9	84	42.2	108
Corn	9.4	49	33.9	93	46.1	115
Pumpkin	11.1	52	33.9	93	46.1	115

Source: Data from Weaver 1926.

Soil temperatures are influenced by soil moisture, and because water has a specific heat requirement about five times greater than that of solid soil constituents, to cause equal warming, a wet soil is a cold soil. Heat required to evaporate water is not available to warm the soil. To promote early warming of the soil, drainage, tillage, level of organic matter, and removal of a surface mulch should be considered.

Oxygen Content

Soil has a porous structure filled with water and air. About 20% of the soil volume may be filled with air, but 10% is considered a minimal but adequate level for normal growth of many plant species (Russell 1952; Wessling and Van Wijk 1937; Harris and Van Bavel 1957a and b). Excess water must be removed from the root zone for healthy root development. Oxygen is necessary for seed germination, root growth, root-hair development, root absorption, and for the activity of soil microorganisms. Oxygen is required for normal root respiration, and under anaerobic conditions roots may produce toxins (Clements 1921). Without adequate oxygen, crop plants may turn yellow, wilt, and eventually die. By placing roots in nitrogen gas to exclude oxygen for various time periods (and to simulate conditions of excess moisture), Letey, Stolzy, and Blank (1962) demonstrated that shoot growth was critically reduced and root growth essentially ceased (Table 19-2). Similar results were obtained with green beans and cotton; three days without air reduced growth when compared to plants in the all-air treatment. Exclusion of oxygen has a less adverse effect following a prior extended period with oxygen than with a prior short period with oxygen.

TABLE 19–2.   Response of sunflower seedlings (dry shoot weight) to soil oxygen deficiencies under potted conditions in California

Treatment Sequence	Dry Shoot Weight (grams)	(oz)
14 days with oxygen	3.51	0.12
4 days with oxygen, 1 without, followed by 9 with oxygen	2.98	0.11
7 days with oxygen, 7 without	2.69	0.09
Alternating days with and without oxygen	2.59	0.09
4 days with oxygen, 3 without, 7 with oxygen	1.93	0.07
4 days with oxygen, 5 without, 5 with oxygen	1.61	0.06
7 days without oxygen, 7 with oxygen	0.76	0.03
14 days without oxygen	0.68	0.02

Source: Data from Letey, Stolzy, and Blank 1962.

Based on the results of this study (Letey, Stolzy, and Blank 1962), the following conclusions about soil oxygen can be drawn:

1. Low soil oxygen is most detrimental during early stages of growth following germination.
2. Roots cease growing at low oxygen levels.
3. There is a time lag in recovery of root growth after being stunted by low oxygen supply.
4. If oxygen content is low, the transpiration of plants is reduced, and the plant is not effective in helping itself by "pumping" out water through transpiration. Water removal must be achieved through drainage and evaporation.
5. The number of days that a soil can be low in oxygen without serious crop damage is dependent on the stage of plant and root development when oxygen deficiency occurs.

Water Needs

An appreciation of the importance of an adequate root system to supply the water needs of the plant may be obtained through a realization of the amount of water lost through transpiration. Transpiration is the loss of water vapor from the plant and accounts for about 99% of the water taken up by the plant. The transpiration ratio, i.e., the amount of water required to produce an equivalent amount of plant dry matter for any one crop, may vary among locations according to temperature, humidity, wind, and soil fertility. Transpiration rates generally are low at low temperatures. Under high temperature conditions or conditions inducive to high transpiration rates, plant water losses may exceed the water uptake by the plant and wilting results. As water becomes limited, growth is slowed and yields are reduced. To avoid crop losses from wilting, a large root system is required, especially under dry-land conditions or periods of drought.

Plant water requirements or the transpiration ratio is based on the weight of the above-ground dry matter harvested, and only in the case of root crops is the weight of the underground parts included. The classical work on plant water requirements as reported by Shantz and Piemeisel (1927) gives a relative comparison among plants. The study was conducted at Akron, Colorado, over the seven-year period 1911–1917 under good fertility conditions in pots (Table 19-3). The lowest values were obtained for the millets, sorghums, and corns; the highest values for flaxes and legumes. The water requirement range to produce a unit of dry matter is very great, from a mean value of 280 for 8 cultivars of millet to 1131 units for Franseria, a native plant. In other words, Franseria required five times as much water to produce a unit of dry matter as did the millet cultivars. In general terms, the millets, sorghums, and corns are

TABLE 19-3.   Average water requirement based on dry-matter transpiration ratio and units of dry matter produced for each 1000 units of water consumed, for plants grown at Akron, Colorado, 1911–1917

Plant	Mean Transpiration Ratio	Units of Dry Matter Produced for Every 1000 Units of Water Consumed
Millet (8 strains)	280	3.64
Sorghum (9 strains)	305	3.46
Corn (15 open-pollinated)	349	2.88
(8 single-cross hybrids)	350	2.87
Teosinte (2 strains)	369	2.72
Sugar beet	377	2.65

Plant	Mean Transpiration Ratio	Units of Dry Matter Produced for Every 1000 Units of Water Consumed
Weeds		
Tumbleweed	260	3.85
Pigweed	305	3.28
Russian thistle	314	3.18
Lamb's-quarter	658	1.52
Polygonum	678	1.47
Barley (4 cultivars)	517	1.93
Buckwheat (1 cultivar)	540	1.85
Wheat (Emmer)	517	1.93
Wheat (Durum—6 cultivars)	520	1.93
(Common—16 cultivars)	557	1.81
(2 hybrids)	534	1.88
Cotton (1 cultivar)	568	1.76
Irish potato (2 cultivars)	575	1.77
Oats (4 cultivars)	583	1.72
Native weeds (11 strains)	589	1.87
Turnip, cabbage, rape	615	1.53
Rye (1 cultivar)	634	1.58
Franseria (native plant)	1131	0.88
Native grasses (8 species)	638	1.99
Cucurbits (watermelon, cantaloupe, cucumber, squash, and pumpkin)	676	1.50
Rice (1 cultivar)	682	1.47
Legumes (17 species of pea, bean, soybean, vetch, cowpea, chickpea)	738	1.55
Alfalfa (2 strains)	720	1.42
(8 strains of Grimm)	875	1.14

Source: Data from Shantz and Piemeisel 1927.

most efficient. The small grains—barley, wheat, oats, and rye—required almost twice as much water, and the legumes required almost three times as much as the millets, sorghums, and corns.

The ten alfalfa strains tested had a transpiration ratio of 844 to 1; i.e., it required 844 kg (or pounds) of water to produce 1 kg (or one pound) of dry matter without any consideration of root weight. In Table 11-7, alfalfa at Kemptville, Ontario, under a three-cut system produced 10,321 kg/ha (9215 lb/ac) at 15% moisture. To produce this crop, assuming a transpiration ratio of 844 to 1, would require 8,710,924 kg (17,421,848 lb) of water. One centimetre of rainfall is equivalent to 100 tonnes per hectare, which means that 87.1 cm (34.29 in) of rainfall were required, the average amount of annual rainfall at Kemptville. Since some water would be lost by evaporation, runoff, or percolation through the soil, the evapotranspiration rate at Kemptville may be lower than at Akron, Colorado.

The transpiration ratio is not a measure of the yielding ability of a crop under drought conditions but rather a measure of a physical plant process to supply the water needs for optimum growth. Under conditions of equal dry-matter production, crops with a low transpiration ratio will require less water than those with a higher transpiration ratio. Corn, with a transpiration ratio of 350 may be efficient because it is a C_4 plant. Barley and buckwheat, with transpiration ratios of 517 and 540, respectively, are not capable of yielding as much as alfalfa with a transpiration ratio of 844.

If the soil is not supplied with sufficient water to meet the water needs of the plant, the yield will be restricted. Under conditions of limited rainfall and without irrigation facilities, crops such as sorghum and millets might be considered because they have a low transpiration ratio. A knowledge of the amount of water required to produce crops is essential for crop producers when irrigating crops or calculating irrigation water supplies to produce a crop under dry-land conditions. Crop producers should employ techniques that increase crop efficiency or that stimulate a root system to meet the water needs of the crop.

Fertilization

Fertility can increase water-use efficiency. The water requirement of corn in Table 19-3 is 350 units per unit of dry matter. Kelley (1954) reported on work by Montgomery and Kiesselbach in Nebraska in 1912 where the fertility from manure increased water use efficiency by lowering the transpiration ratio 64% with infertile soils (Table 19-4), accompanied by a 333% increase in dry weight per plant.

In common Bermuda grass, 224 kg N/ha decreased the transpiration ratio from 9268 to 3294 (Table 19-5) at Tifton, Georgia (Burton, Prine, and Jackson 1957). Under the hot, dry conditions of southeastern United States, very high transpiration ratios were found. Coastal and Suwanee Bermuda grass were

TABLE 19–4. Water use efficiency of corn as influenced by soil fertility

Soil Fertility Level	Average Dry Weight per Plant (g) (oz)				Transpiration Ratio	
	No Manure		Manure		No Manure	Manure
	(g)	(oz)	(g)	(oz)		
Infertile	113	3.98	376	13.26	550	350
Medium fertility	184	6.49	414	14.60	479	341
Fertile	270	9.52	473	16.68	392	347

Source: Data from Kelley 1954.

more efficient at all fertility levels than common Bermuda grass, Pensacola Bahiagrass, and Pangola grass. Except for Pangola grass, a similar increase in efficiency was found with nitrogen fertility. A crop producer would choose Bermuda grass cultivars carefully for production under dryland conditions.

TABLE 19–5. The effect of nitrogen fertility on transpiration ratios at Tifton, Georgia, mean of two years 1953 and 1954

Species	Water Used per Unit of D.M.			Percent Improvement of 224 (220) over 56 (50) kg N/ha (lb N/ac)
	56 kg N/ha (50 lb N/ac)	112 kg N/ha (100 lb N/ac)	224 kg N/ha (200 lb N/ac)	
Common Bermuda grass	9268	4437	3294	281
Coastal Bermuda grass	2253	1351	809	278
Suwanee Bermuda grass	1697	1025	641	264
Pensacola Bahia grass	2970	1829	1180	251
Pangola grass	2852	2295	2937	−9.7

Source: Data from Burton, Prine, and Jackson 1957.

*D. M. = dry matter.

PLANT ROOTS AND CROP PRODUCTION PRACTICES

Land Drainage

The amount of water in a soil influences temperature, air, and plant food relationships in the soil, and to keep these elements in proper balance, a system of land drainage may be required to remove excess water, so as to help hasten timely seeding and to promote good root structure and depth (Irwin 1969).

Three types of soil water can be observed: hygroscopic, capillary, and gravitational. A film of hygroscopic water normally adheres so tightly to soil particles that it is of no practical value to roots. Gravitational water flows by the force of gravity after a heavy rainfall or following the melting of snow, and it is gravitational water for which drainage outlets must be provided to allow this water to escape and to help oxygen reenter the soil. Capillary water resists the pull of gravity, and it is the water useful to plants (Figure 19–1). Capillary water is replaced by rainfall, through capillary rise from the water table or by irrigation water. The amount of capillary water held by a soil varies with soil type. Coarse-textured soils have fewer particles to hold capillary water than do fine-textured soils. Coarse-textured soils may not require a system of underdrainage. On soils requiring periodic drainage, it is sometimes believed that underdrainage will cause these soils to become droughty or dried out. Underdraining a soil removes gravitational water only, as soils do not give up capillary moisture, and for this reason there is no danger of overdraining a mineral soil. Organic soils can be overdrained to the extent that they become difficult to rewet if they become too dry.

Drainage of wet mineral soils increases the volume of soil that the roots can use because of the improved physical conditions of the soil. A high water table in the spring promotes shallow root development and stunted top growth. Later in the year during the dry season when the water table drops, plant roots are not able to obtain sufficient moisture and plant food (fertilizer nutrients). A water deficiency causes crop stress, growth is reduced, and yields are depressed. Crops produced on well-drained soil have a more extensive root system, which helps them withstand drought. They can obtain more moisture and plant food and will produce larger yields. Figure 19–2 shows root development in drained and undrained soil. Fertilizer nutrients are taken up by plant roots above the water table. Roots existing in a wet, poorly drained soil will not absorb fertilizer nutrients.

Soil drainage can be accomplished by providing surface outlets, open ditches, a system of subsurface drainage pipes, or a system of unlined channels

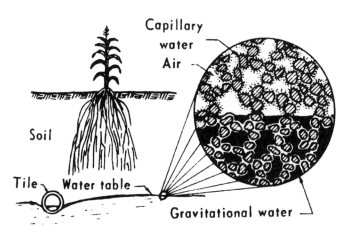

Figure 19–1. Capillary and gravitational forms of soil water.

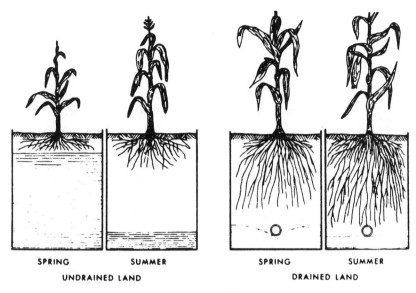

Figure 19-2. Root development of plants on drained and undrained soil. The circle in the drained soil is a perforated pipe, tube, or tile forming a subsurface system of drainage.

formed by pulling a subsoil plow with a mole ball attached. If this ball is pulled through wet clay soil, a temporary drainage system can be formed.

In areas with limited heat units available for corn production, and where wet, cool soils are found in the spring, subsurface drainage will extend the corn growing season. On drained soils, corn can be planted earlier, and this allows a greater potential for physiological maturity to be reached before fall frosts. Also a drained soil provides a greater assurance of soil conditions suitable for fall harvesting before freeze-up.

Root Development and Plant Nutrients

Inadequate land drainage can usually be identified by water standing on the soil surface (Fig. 19-3) and can lead to reduced crop yields.

An increase in crop yields from a balanced fertility program can be demonstrated. Fertility needs should be based on a soil test rather than on observations of top growth because it may be too late to correct fertility deficiencies when deficiency symptoms are already apparent in vegetative plant parts. Soil fertility has a direct and readily observable influence on leaf area, leaf color, lodging, and yield; thus, soil fertility is believed to directly affect plant top growth with a subsequent indirect effect on roots through an increase in available food energy for root growth. The influence of soil fertility on root growth is less clearly understood because of the hidden nature

Figure 19-3.   Inadequate land drainage can delay seeding, root development, and crop yield. (Courtesy Ontario Ministry of Agriculture and Food.)

of roots and because of the complexity of root growth. Taylor and Klepper (1978) listed 24 factors that affect root elongation and development and described the list as incomplete.

The influence of fertilization on root development is not clear. Work of Maizlish, Fritton, and Kendall (1980) confirmed the observations of researchers that nitrogen fertilizers increased root length and number of apices (Table 19-6). A close relationship was found between root weight, above-ground shoot weight, and leaf area (Table 19-7). Maizlish and his researchers speculated that shallow placement of nitrogen fertilizer encourages shallow rooting of maize. And under conditions where subsoil moisture is an important water source, corn drought-susceptibility would be increased by shallow rooting. Further research on the effects of fertility levels and nitrogen placement on root growth and yield is needed.

TABLE 19-6.   Effect of nitrogen on root length and number of root apices of corn grown in a growth chamber, seventeen days after emergence tests were conducted in Pennsylvania

Nitrogen Level (ppm)	Length per Plant Metres	Length per Plant (feet)	Number of Apices per Plant
0	21	70	855
21	54	177	1,439
42	73	239	1,961
105	121	397	2,517
210	148	485	2,963

Source: Data from Maizlish, Fritton, and Kendall 1980.

TABLE 19–7.   Effect of nitrogen level on root dry weight, shoot dry weight, and leaf area seventeen days after emergence tests were conducted in growth chambers in Pennsylvania

Nitrogen Level (ppm)	Root Dry Weight (grams)	(oz)	Shoot Dry Weight (grams)	(oz)	Leaf Area (cm$_2$)	(sq in)
0	0.38	0.013	0.24	0.008	53	8.5
21	0.84	0.03	0.75	0.026	140	22.4
42	1.30	0.06	1.34	0.05	187	29.9
105	1.93	0.07	2.40	0.08	365	58.4
210	2.89	0.10	4.49	0.16	428	58.5

Source: Data from Maizlish, Fritton, and Kendall 1980.

One of the effects of soil fertility is to improve water-use efficiencies of crops. In a study conducted in Wisconsin, under high fertility treatments, a more efficient use of water was found in every comparison (Table 19–8) (Hanks and Tanner 1952). Overall, high fertility resulted in a 35.3% increase in water-use efficiency over low fertility conditions.

TABLE 19–8.   Yield of four crops, per centimetre of water used by the crops, as influenced by soil fertility, in Wisconsin field trials during the years 1949 and 1950

Crop and Year	Fertility Level	Crop Yield per Unit of Water Used by Crop (Includes Evaporation) (kg)	(lb)
Corn	High	36.9	81.2
	Low	26.3	57.9
Oats	High	24.3	53.5
	Low	14.3	31.5
Cucumbers	High	90.5	199.1
	Low	71.0	156.2
Alfalfa	High	64.3	141.5
	Low	47.9	105.4

Source: Data from Hanks and Tanner 1952.

In a greenhouse study, alfalfa and bluegrass were subjected to a cutting and fertility regime to determine if management might improve regrowth (Sprague and Graber 1938). Nitrogen fertility decreased the water required per unit of dry matter produced (Table 19–9). Fertility conditions improved the transpiration ratio in alfalfa and bluegrass 25.4 and 17.3%, respectively.

TABLE 19–9.   Total dry matter produced for alfalfa and bluegrass under high and low nitrogen fertility levels in greenhouse experiments in Wisconsin

Crop	Fertility Level	Total Water Used (g)	(lb)	Total Dry Matter Produced (g)	(lb)	Transpiration Ratio
Alfalfa	High	11,125	24.5	20.61	0.045	539
	Low	10,417	23.0	16.27	0.036	676
Bluegrass	High	6,241	13.8	8.98	0.020	986
	Low	5,314	11.7	5.16	0.011	1,157

Source: Data from Sprague and Graber 1938.

Seeding Practices

It is often assumed that deep seeding encourages deep rooting as well as promoting better anchorage and better nutrient and water uptake. However, in most instances deep seeding does not promote deep rooting, and to understand why, it is necessary to examine seed germination and seedling growth.

Seed germination in the grass family begins with the emergence of the radicle or primary root that grows downward and the coleoptile that grows upward—regardless of how the seed is placed in the soil (Figure 19–4). When the coleoptile, pushed by the elongating epicotyl, breaks through the soil surface, elongation ceases, and the first leaf unrolls from the protective coleoptile and begins photosynthesis (Figure 19–5). Seminal roots arise from the germ, and adventitious roots arise from the first node, as shown for corn in the sequential stages of corn seedling development (Figure 19–5).

Shortly after emergence of the first leaf, photosynthesis begins, and it provides food for growth and development. Reserve food in the endosperm of

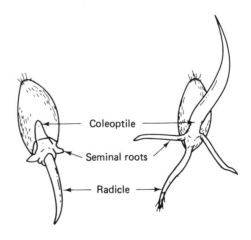

Figure 19–4.   Seed germination in the grass species (wheat).

the seed may remain unused if the seed is planted shallowly. If the seed is planted too deeply, there may be insufficient food reserves in the seed to provide enough energy for emergence of the coleoptile. Grasses are called monocots, and germination and seedling development differ markedly from that of dicots such as in the legume crops.

In dicots such as clover, sugar beet, sunflower, alfalfa, and other legumes, the radicle or primary root pushes downward to develop into a taproot with many branches throughout its entire length. In the bean (Figure 19-6), the radicle or primary root is the first to emerge from the seed, followed by the appearance of an arched hypocotyl hook that pushes upward to the soil surface, pulling instead of pushing and thus protecting the exposed growing point or plumule of the young seedling (diagrams A and B of Figure 19-6). Considerable energy is required to pull the cotyledons of the relatively large bean seed to the surface.

The hypocotyl hook is the first part of the seedling to emerge from the soil, and as the hypocotyl straightens out, it carries the two cotyledons upward into the air, where they turn green and begin photosynthesis. The plumule, located between the cotyledons, gives rise to the first true leaves, which begin photosynthesis; then photosynthesis in the cotyledons ceases and they shrivel and fall off.

Peas, faba beans, and runner beans do not have a hypocotyl that pulls the cotyledons, but instead the cotyledons are pushed throught the soil.

In perennial forage legumes such as alfalfa and red clover, a crown develops just below the soil surface. As the seedlings develop in the first year of growth, the hypocotyl cells become more compact and cause a shortening of the stem, thereby pulling the cotyledonary node down below the soil surface. The buds in the cotyledonary node [Figure 19-6 (D)] eventually give rise to a crown, which may contain up to 100 tiller buds beneath the soil surface.

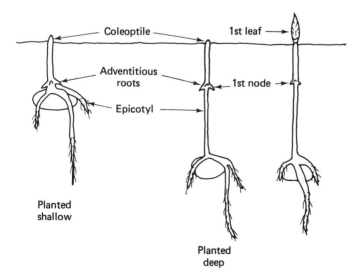

Figure 19-5. Seedling emergence as influenced by depth of seeding in grasses.

Clovers and legume plants are characterized by three leaves. If clovers and legumes are seeded without a companion crop, a properly timed herbicidal treatment for weed control is essential, usually when the alfalfa seedling has developed three trifoliate leaves. Crop producers should learn to recognize the developmental stages in the legume seedling. The first leaf developing between the cotyledons is a unifoliate leaf (see Figure 19-7). Since the cotyledons are not true leaves, the unifoliate leaf stage is not to be confused with the trifoliate stage. Note that as the alfalfa seedling develops, the height of the cotyledonary leaves diminishes as the cotyledons are pulled to the soil surface to develop a crown.

Deep seeding may not promote deep rooting. The radicle or primary root system emerges and serves to supply water and nutrients to the developing seedling. In alfalfa, roots have been traced to a depth of over 9 m (30 ft) with

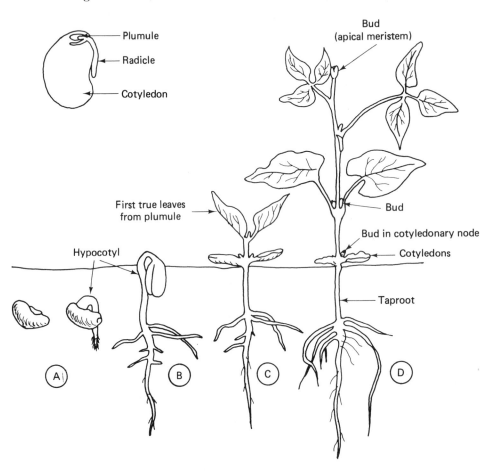

Figure 19-6. Seedling development of field bean.

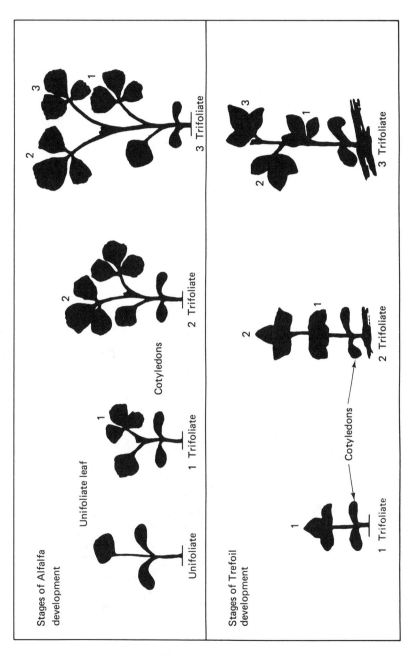

Figure 19–7. Seedling leaf development in legumes.

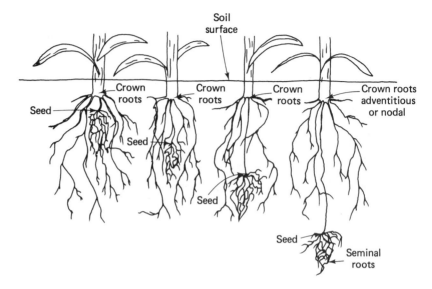

Figure 19–8.   Depth of secondary or crown root development in corn seeded 5, 10, 15 and 25 cm deep (2, 4, 6 and 10 in deep).

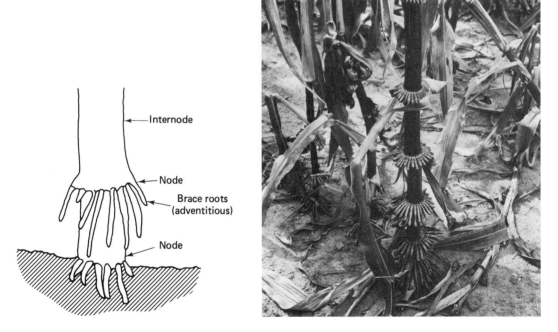

Figure 19–9.   Brace roots in corn. (Courtesy Ontario Ministry of Agriculture and Food.)

surface seeding. In the grasses, roots generated by the radicle never become large and are generally unimportant, and it is the adventitious or nodal roots that become the main functional roots. In wheat, oats, and barley, considerable disagreement exists as to the significance of each of the two root systems for plant growth (Brouwer 1966). In corn and forage grass species, it is believed that the seminal roots degenerate and cannot be found six weeks after germination begins, and it is the adventitious or crown root system that supports the plant. Deep seeding does not promote deep rooting in such crops (Figure 19–8). The epicotyl elongates until the coleoptile reaches light, and the depth of the basal node is independent of seeding depth, so rooting depth is reasonably constant.

In corn, adventitious roots also arise from higher nodes once the stem has begun to elongate and are referred to as *brace* or *anchor roots* (Figure 19–9).

Seedbed Preparation and Root Development

A fine, firm seedbed is essential for good contact with capillary moisture. Seeds placed in soil that is too loose and dry will not come in contact with sufficient moisture to germinate. With cereals, a common practice is to mechanically roll or cultipack the soil after seeding to make the soil firm and to ensure good seed-soil contact. Experiments to confirm these observations were conducted in Czechoslovakia by Stranak (1968) by sowing spring cereals into loose loam soil and later compacting the soil by rolling. Three main treatments were compared:

1. Sowing into loosened soil without rolling.
2. Compacting soil with a light roller.
3. Compacting soil with a heavy roller.

Rolling was always completed within five days of sowing. As bulk density increased with rolling, yields increased (Table 19–10).

TABLE 19–10. Effect of soil density on cereal yields—experiments conducted in Czechoslovakia with three types of treatment, 1959 to 1963

Seedbed Treatment	Bulk Density (g/cm^3)	(oz/in^3)	Yield Expressed in Percent of Lowest Bulk Density Soil
Loose	1.01	0.58	100
Dense	1.22	0.70	118.3
Very dense	1.51	0.87	126.4

Source: Data from Stranak 1968.

Based on the favorable response of cereals to compact soil, studies were conducted with direct seeding into soil without plowing and with chemical weed control. Highest yields were obtained without plowing and confirmed several years of observation at the Elora Research Station in which winter wheat was seeded directly into unworked soil with a degree of success.

Although a firm seedbed assures good contact of soil and moisture, excessive soil compaction may be unfavorable to root growth because of a lack or excess of moisture, inadequate oxygen, and/or mechanical impedance to root penetration. Research on the subject indicates a difficulty in showing conclusively which property affects the plant. Low yields in a compact soil may be associated with unfavorable soil temperatures or with root impedance and related small roots.

Mechanical impedance was found to be the physical property associated with reduced growth and yield of corn in Iowa (Phillips and Kirkham 1962). The test was conducted on a clay soil having 27% sand, 31% silt, and 42% clay at three compaction treatments. Normal compaction was that associated with usual corn culture; moderate compaction was produced by driving the wheels of a tractor back and forth twenty times on the same strip the width of the rear tractor wheel until the entire plot was compacted prior to planting; and severe compaction was achieved by driving a tractor back and forth on the plow sole twenty times and then applying an additional twenty passes with the tractor on top of the plowed and cultivated soil. Average bulk densities achieved ranged from 1.15 to 1.46 g/cm³ (0.66 to 0.84 oz in³) (Table 19–11).

Seedling emergence values for the normal, moderate, and severe compaction treatments were 85, 46, and 44%, respectively. Reductions in plant height; in uptake of N, P, and K; and in grain yield are shown in Table 19–12. In every instance, compaction reduced plant height; total levels of N, P, and K in leaves; and grain yield. However, soil fertility effectively reduced the impact of compaction. The reason for these observations was related to mechanical root impedance. No differences in the percent of soil oxygen, soil temperature at 10 cm (4 in), or soil moisture were measured, and these did not affect plant

TABLE 19–11. Average bulk density (g/cm³) (oz/in³) at four depths, mean of two years 1957 and 1958 at Ames, Iowa

		Average Bulk Density per Compaction Treatment					
		Normal		Moderate		Severe	
Depth							
(cm)	(in)	(g/cm³)	(oz/in³)	(g/cm³)	(oz/in³)	(g/cm³)	(oz/in³)
0– 7.6	0– 3.0	1.15	0.66	1.32	0.76	1.32	0.76
7.6–15.2	3.0– 6.0	1.33	0.77	1.46	0.84	1.45	0.84
15.2–23.0	6.0– 9.0	1.36	0.79	1.39	0.80	1.41	0.81
23.0–30.5	9.0–12.0	1.33	0.77	1.34	0.77	1.35	0.78

Source: Data from Phillips and Kirkham 1962.

TABLE 19–12. Relative corn plant heights; uptake of N, P, and K per twenty leaves sampled at silking time; and relative grain yields for differing compaction-fertility treatments, averaged over a two-year period where measurements of the normal compaction, added-fertility treatment are taken as unity (1.00). Values represent a two year mean 1957 and 1958 at Ames, Iowa

Compaction Fertility Treatments	Relative Corn Plant Height			Relative Uptake Per 20 Leaves*			Relative Grain Yield
	June 28	July 8	July 18	N	P	K	
Added fertility							
Normal	1.00	1.00	1.00	1.00	1.00	1.00	1.00
Moderate	0.66	0.70	0.74	0.81	0.77	0.81	0.82
Severe	0.64	0.64	0.67	0.76	0.73	0.77	0.72
Existing fertility							
Normal	0.73	0.77	0.78	0.66	0.79	0.81	0.70
Moderate	0.43	0.45	0.50	0.59	0.62	0.62	0.51
Severe	0.41	0.42	0.47	0.53	0.56	0.56	0.45

Source: Data from Phillips and Kirkham 1962.

*N = nitrogen; P = phosphorus; K = potassium.

growth and yield. Oxygen contents averaged 13% and ranged from 10 to 12% to 16 to 18% for all treatments.

The results of this study (Phillips and Kirkham 1962) can be summarized as follows:

1. Soil compaction reduced corn stands.
2. Soil compaction resulted in reduced corn yields after stands had been equalized on compacted and uncompacted soils.
3. Compaction reduced yields about equally at both fertility levels.
4. Compaction increased mechanical impedance to root growth.
5. In June and July, soil temperatures were the same in all plots. In areas with cool spring temperatures, May temperatures may be a factor in production in compacted soils unworked in early spring.
6. No differences in soil moisture were measured.
7. Yield reductions were believed to be related to mechanical impedance.
8. Interrow cultivation in the compacted plots increased yields.
9. Grain yields were directly related to root volume when measured in compacted or uncompacted, and fertilized or unfertilized plots.
10. Production of corn in the year following the last compaction, and after one plowing and freezing and thawing over the winter, apparently restored the soil to a nearly normal state. No differences in grain yields among former treatments could be found.

Interrow Cultivation

A knowledge of root development is important to crop producers in terms of interrow cultivation and fertilizer practices. Foth (1962) studied root development of corn and found that root growth consisted of a series of overlapping stages associated with top growth development. Early root growth occurred largely in a downward-diagonal direction, followed by extensive lateral growth (Table 19-13). Lateral growth was completed a week or two before tassel emergence and resulted in a marked uniformity in root density in the upper 30 to 38 cm (12 to 15 in) of soil. Extensive growth of roots below 38 cm (15 in) occurred near tasseling time. When kernels were in the milk stage, all root growth stopped. On day 23, corn plants measured 30 cm (12 in) from the base to the tip of the longest leaf. At 37 days, this measure was 91.4 cm (36 in), and on the 54th day, tassel emergence was beginning.

It has generally been assumed that interrow cultivation will result in detrimental root pruning. Troughton (1962) stated: "The greater the proportion of the root system that is removed by pruning, the greater is the decrease in the subsequent growth of the shoot" (p. 58). With the advent of chemical weed control, cultivation for weed control was no longer necessary, but many

TABLE 19-13. Vertical and lateral distribution of roots from base of corn plant at eight dates after planting in Michigan in 1960

Location from Base of Plant		23	37	41	47	54	67	80	100
					Grams of Roots				
Vertical (cm)	**(in)**								
0– 7.6	0–3	0.13	0.42	0.83	1.33	2.42	3.28	5.31	5.63
7.6–15.2	3–6	0.12	0.68	0.94	1.46	1.91	1.80	2.56	2.73
15.2–22.9	6–9	0.02	0.57	1.03	1.56	1.75	1.07	1.78	1.75
22.9–30.5	9–12		0.38	0.86	0.94	1.55	1.02	1.11	1.19
30.5–38.7	12–15		0.11	0.31	0.50	0.67	0.52	0.49	0.43
38.7–45.8	15–18		0.02	0.13	0.17	0.17	0.38	0.30	0.31
45.8–53.3	18–21			0.06	0.08	0.06	0.33	0.27	0.28
53.3–61.0	21–24				0.05	0.03	0.19	0.21	0.23
61.0–68.6	24–27				0.04	0.02	0.11	0.18	0.10
68.6–76.2	27–30						0.09	0.09	0.07
76.2–91.4	30–36						0.05	0.07	0.06
Lateral (cm)									
0– 7.6	0–3	0.15	0.93	1.18	2.20	3.63	4.95	6.43	7.23
7.6–15.2	3–6	0.10	0.72	1.21	1.99	2.08	1.84	3.30	2.92
15.2–38.7	6–9	0.02	0.34	0.87	1.29	1.60	1.33	1.14	1.72
38.7–53.3	9–12	0.06	0.19	0.90	0.65	1.27	0.72	1.50	0.91
Total lateral		0.27	2.18	4.16	6.13	8.58	8.84	12.37	12.78

Source: Data from Foth 1962.

farmers debated the merits of interrow cultivation and often continued the custom perhaps largely out of tradition. Studies involving deliberate root pruning have reduced top growth, but few studies are available of the effect of interrow cultivation under actual field conditions.

Spencer (1941) reported a significant increase in lateral root development in corn from root pruning damage, resulting from a proliferation of regenerated roots. At a late stage of root pruning damage, however, regeneration may compete with developing grain for available carbohydrate. Spencer found a consistent increase in root dry weight when ear formation was prevented, indicating a decrease in intraplant competition for available food energy.

Root studies on corn and soybeans in a review by Taylor and Klepper (1978), suggest that root pruning damage potentially could be extensive because of the shallow development of lateral roots in these crops (Table 19–14). Regardless of row width, soybeans showed a preponderance of lateral roots (72–73%) in the top 10 cm (4 in) of soil.

TABLE 19–14. Percentage of corn and soybean roots 66 and 44 days after seeding, respectively, at various soil depths. Tests were conducted in Iowa in 1975

Depth (cm)	(in)	Percentage of Corn Roots at Various Depths	Percentage of Soybean Roots at Various Depths Under Two Row Widths	
			25-cm (10 in) Wide rows	100-cm (39 in) Wide rows
5	2	7	22	41
10	4	14	50	32
15	6	24	15	10
20	8	17	6	6
30	12	15	3	8
40	16	8	2	1
50	20	6	1	2
60	24	4	1	0
Below 60	24	5	0	0

Source: Data from Taylor and Klepper 1978.

Mitchell and Russell (1971) showed that the soybean root system by 21 days after planting consisted primarily of secondary lateral roots that had developed from an enlarged upper 10- to 15-cm (4 to 6 in) segment of the taproot. Lateral roots grew horizontally for 40 to 50 days and after reaching 35 to 40 cm (14 to 16 in) lateral growth (in 76-cm (30 in) wide rows), turned downward, providing extensive deep penetration to the soil profile.

Essentially three stages of root development (Mitchell and Russell, 1971) were observed:

1. During early vegetative growth, the taproot developed downward along with shallow horizontal lateral roots. One month after

planting, lateral soybean root development had spread 35 cm (14 in)
laterally within 2.5 cm (1 in) of the soil surface.

2. During flowering and pod formation, roots developed to 76 cm (30 in) in depth.
3. During seed maturity, deep penetration of several lateral roots occurred to a depth of 122 to 183 cm (48 to 72 in) under favorable conditions.

Because of the high concentration of lateral roots in the upper soil zone, Russell, Fehr, and Mitchell (1971) anticipated that interrow cultivation would damage a significant amount of soybean roots (Figure 19-10). Six cultivation treatments were evaluated on four soybean cultivars in three Iowa State environments. Treatments were combinations of cultivation 6.4 cm and 11.4 cm (2½ and 4½ in) deep, spaced 15.2m, 22.9m, or 30.5 cm (6, 9, or 12 in) on each side of the plants in the 68-cm (27 in) wide rows. Interrow cultivation was conducted at four stages of soybean development.

1. The first cultivation occurred when three trifoliate leaves were fully unrolled.
2. The second cultivation took place when seven to eight trifoliate leaves were unrolled and 40 to 60% of the plants were flowering. More

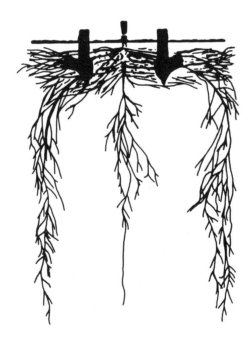

Figure 19-10. Diagrammatic representation of a fully developed soybean root system cultivated 11.4 cm (4.5 in) deep and 15.2 cm (6 in) from the row showing potential for significant root-pruning damage.

than 50% of the total root development (Mitchell and Russell 1971) was observed to occur after this stage.

3. The third cultivation treatment was imposed when pods in the lower half of the plants were well formed and up to 3.8 cm (1.5 in) in length.

4. The fourth cultivation treatment consisted of a double cultivation applied at the three trifoliate stage and when the lower pods were well formed. Weeds were controlled chemically.

Only small yield reductions were observed and on the overall average no real yield difference was found (Table 19-15).

The reasons for the surprising lack of a yield reduction from root pruning damage may be explained as follows:

1. The shallow 6.4 cm (2½ in) cultivation resulted in minimum root pruning damage

2. Cultivation during vegetative growth is not seriously detrimental. Cultivation is most detrimental at late stages of plant development,

3. Interrow cultivation may have improved soil conditions for water infiltration, CO_2 diffusion, reduced evaporation, and warmer soil conditions. Prihar and Van Doren (1967) found increases in water infiltration and slight reductions in evaporation on cultivated corn plots. Van Doren and Triplett (1965) suggested that interrow cultivation may be beneficial on soils that tend to crust, especially after heavy rains.

4. Early cultivation, before lateral roots had extended, may have resulted in little actual damage from root pruning.

5. An increase in root area from regeneration of severed roots may have contributed to the yield increases observed in early cultivations.

Fertilizer Placement

Not all the fertilizer nutrients applied to a field are used by the plants. Some nitrogen goes back into the atmosphere through nitrification and subsequent denitrification. Other nitrogen is lost through volatilization and by leaching

TABLE 19-15. Mean grain yield (kg/ha) (bu/ac) of four soybean cultivars with and without interrow cultivation (mean of all treatments and all stages of development) conducted in Iowa in 1968 and 1969

Treatment	Grain Yield	
	(Kg/ha)	(bu/ac)
Cultivated	2924	43.5
Not Cultivated	2942	43.8

Source: Data from Russell, Fehr and Mitchell 1971.

action, especially in coarse-textured soils. Inorganic nitrogen is "tied up" by soil microorganisms that need energy to decompose organic matter. Under conditions of inexpensive and plentiful fertilizer nutrients, often an excess was applied to assure an adequacy of supply; but under conditions of expensive and scarce fertilizer nutrients, a greater efficiency must be sought. Proper placement of fertilizer nutrients may be an effective means of promoting a desirable root morphology and increasing the efficiency of fertilizer use, thereby reducing the quantities of fertilizer needed.

In the Philippines and Southeast Asia fertilizer efficiency was increased by 10-cm (4 in) deep placement of nitrogen in lowland rice production (Anonymous 1976). Ten-centimetre deep placement resulted in 68% being used by the rice crop as compared to 28% when broadcast and incorporated by harrowing (Table 19–16).

Drilling seeds above bands of fertilizer so that seedling roots quickly come in contact with nutrients is an example of how proper fertilizer placement can be used to encourage rapid root development, reduce fertilizer consumption, and increase efficiency. However, not all the fertilizer requirements can be met with banding because excessive levels can cause "burning" of the seedlings. Techniques of injecting nitrogen into the soil after seedlings have developed is another way of providing a separate application during the time when the plants are actively absorbing nutrients.

TABLE 19–16. Yield of rice (t/ha) (tons/ac) comparisons with two levels of nitrogen placed 10 cm deep or top-dressed in the Philippines in 1974

| Fertilizer Rate | | Yield When N applied | | | | Efficiency of N use (kg rice/kg N) | |
| | | 10 cm deep | | Top-dressed | | 10 cm deep | |
(kg N/ha)	(lb N/ac)	(t/ha)	(lb/ac)	(t/ha)	lb/ac	(4 in)	Top-dressed
60	53.6	8.0	3.6	5.8	2.6	53	23
100	89.3	8.4	3.8	6.6	3.0	38	21

Source: Data from Anonymous 1976.

Band application of phosphorus is particularly important during periods of low soil temperature when less efficient utilization is found and when a deficiency is likely to occur. Band application is important on soils low in available phosphorus, particularly if the soil has a high phosphorus-fixing capacity. Localized applications of fertilizer at planting are commonly referred to as *starter* or *planting fertilizers.*

Band seeding of up to 30 kg P/ha (26.8 lb P/ac) placed 5 cm (2 in) directly below the surface-placed seed increased seedling growth as much as fivefold in five field experiments with alfalfa (Figure 19–11), birdsfoot tiefoil, and bromegrass (Sheard 1980). Inclusion of nitrogen in the phosphorus band was

Figure 19-11. Proper placement of fertilizer, 5 cm below the surface-placed seed can result in a fivefold increase in alfalfa seedlings. (Courtesy Ontario Ministry of Agriculture and Food.)

of minor importance in increasing seedling vigor. Band seeding of phosphorus for forage seedling establishment is considered of singular importance for early seedling growth of cool season forage species, even on soils testing high in phosphorus.

Broadcast applications of fertilizer imply large amounts of nutrients. Phosphorus is very immobile in the soil, and potassium is only a little more mobile. For this reason both are usually incorporated with the soil. Nitrogen is very mobile and moves vertically or horizontally with water movement.

The effect of nutrients on the growth of the roots of plants is somewhat obscure. Fertilizers may have both a direct and an indirect effect, usually increasing top growth to a greater degree than root growth; they may also make plants more competitive with late developing weeds and may increase disease resistance.

Irrigation

When water is applied at the proper time, a limited number of irrigations gives as good results as a greater number. Much of the water used for irrigation has required considerable energy to lift it, convey it, or pressurize it, and proper use of water is an economic necessity. Depending on the natural rainfall and water-holding capacity of the soil, attention should be directed to the stage of crop development in determining the time to irrigate. Water should be applied when about 50 to 60% of the readily available soil moisture has been consumed. The amount and frequency of water to be applied depend to a large extent on how the root development will be affected. Keeping the surface soil too moist during early plant development may promote a shallow rooting habit, and the crop may later suffer from drought unless watered very frequently. Deep-rooted plants generally survive drought better than shallow-rooted plants because more water is accessible to deep-root systems. When crops are subject to occasional drought, deep-rooted species often yield more than those with shallow roots. Pearson (1966) found that deepening the rooting zone enhanced plant water supply. But because roots deep in the soil are farther from the plant stem, it has been suggested that deep roots may not be as effective as shallower roots in applying water to tops (Kramer 1969). In the taproot cotton plant, however, Taylor and Klepper (1971) found that roots at the 150–180 cm (59 to 71 in) depths absorbed as much water per centimetre of length as did roots higher in the soil. In the fibrous rooted corn plant, Taylor and Klepper (1973) found that roots deep in the soil were effective because they were young roots in moist soil.

A system of limiting irrigation water to sugar beets was studied in a semiarid climate (Winter 1980). Irrigation schedules and amounts of water applied were varied to produce a wide range of water-stress severity, duration, and timing. Sugar beets made efficient use of limited rainfall and soil profile water, even when subjected to periods of water stress as long as five months. Irrigation water was most efficiently used when water application was adequate to maintain a nearly full canopy with no periods of major water stress or excessive water. One irrigation applied prior to onset of major water stress in July promoted high water-use efficiency. Sugar-beet response to limited irrigation compared favorably to that of grain sorghum. No unusual production problems were observed with limited irrigation. The work of Winter in Texas indicates that sugar beets can be efficiently grown under a wide range of irrigation levels. Sugar beets readily adapt to limited irrigation because they utilize deep-stored soil water and quickly recover growth following water stress (Table 19–17). Similar results were found in 1977. Treatments used did not affect sucrose percentage of the sugar beet root for either year of the study. Average percent sucrose was 15.2 and 16.2 in 1976 and 1977, respectively. Perhaps somewhat surprisingly, 48.9 cm (19.3 in) of irrigation water produced a yield of 60.5 t/ha or 85% of the 66.5 t/ha yield produced when 73.4 cm (28.5 in) of irrigation water were applied. The efficiency of water use varied little but was highest with four and five seasonal irrigations rather than with six and eight. If sugar beets can be efficiently grown under such a wide range of irrigation levels, studies of root morphology and timeliness of applications may lead to more effective irrigation methods in other crops.

TABLE 19–17. Sugar-beet water use, root yield, and water-use efficiency with ten irrigation schedules in Texas in 1976

Number of Seasonal Irrigations	Total Irrigation Water		Root Yield		Total Water-use Efficiency Tonnes of Sugar Beets per cm
	(cm)	(in)	(t/ha)	(tons/ac)	
0	0	0	33.2	14.9	0.522
2	30.5	12	49.7	22.4	0.554
3	41.4	16	56.7	25.5	0.562
4	48.9	19	60.5	27.2	0.564
5	52.8	21	66.3	29.8	0.582
6	60.7	24	64.1	28.8	0.536
8	69.6	27	67.4	30.3	0.557
8	73.4	29	66.5	29.9	0.535

Source: Data from Winter 1980.

A delay in applying irrigation water may promote deep roots. Sound crop production is associated with a good root to top ratio because a good root system is associated with good top development. Lowering soil moisture to increase rooting depth may lead to an improved root-top ratio. Consideration of root development may lead to an understanding of observed differences in crop yields that may be obscure otherwise.

Too much irrigation water may extend the vegetative growth period, making plants more susceptible to disease or late droughts. Also the associated delay in reproductive development may lead to lower economic yields.

ROOT MORPHOLOGY AND YIELD

An extensive root system is associated with drought resistance in wheat (Hurd 1974), but effective breeding programs are difficult because of the hidden nature of the root system. Little evidence of cultivar differences in the root growth of semidwarf and tall winter wheat cultivars was found (Lupton and Oliver 1974).

A detailed study of root morphology in black runner beans in relation to yield and lodging was conducted by Stoffella et al. (1979). They found that the total root biomass consists of the following root sections shown in Figure 19–12.

1. Adventitious roots that arise from the stem tissue of the hypocotyl.
2. Basal roots arising from the basal region of the hypocotyl.
3. The primary or taproot.
4. Lateral roots arising from the taproot.

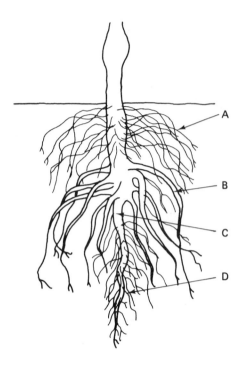

Figure 19–12.   Root morphological components of black beans. A) Adventitious
                root; B) Basal root; C) Taproot; D) Lateral root.

Four lodging-resistant lines were found to have higher basal root weights
than two lower yielding, more lodging-susceptible check cultivars (Table 19–
18). The high yielding, lodging-resistant strains had the highest total root
weight with the basal root weight comprising 58% of the total weight. This
may be advantageous if beans are direct-combined rather than pulled and
windrowed for harvesting.

TABLE 19–18.   Root biomass components (in grams) for two black bean cultivars and
                four selected lines grown in the field at Ithaca, New York in 1978

Black Bean Type	Adventitious Root Weight (g)	Basal Root Weight (g)	Taproot and Lateral Root Weight (g)	Total Weight (g)
Four strains	0.054	0.602	0.387	1.043
Two checks	0.065	0.296	0.250	0.611

Source: Data from Stoffella et al. 1979.

SEEDLING DEVELOPMENT AND CROP PRODUCTION PRACTICES

Seedling Emergence and Seeding Depth

Since photosynthesis is the key to crop production, producers are interested in having seedlings emerge and begin photosynthesis as soon as possible after seeding. The rate of seedling emergence is dependent on seeding depth, soil temperatures, and amount of reserve food energy stored in the seed (seedling vigor). Large seeds can emerge from a greater depth than small seeds, but generally there is no advantage to deep seeding even with large-seeded crops.

In cereals, the length of the coleoptile and epicotyl is genetically controlled. In dwarf and semidwarf wheat selections a short coleoptile normally is found, and poor seedling emergence is a major shortcoming of these wheats if seeded too deeply (Allan et al. 1961). Slow coleoptile growth along with a short coleoptile length was found to exist in barley (Liptay and Davidson 1971) as well as in many short wheat lines, and this means that cereals should be planted at a depth of 3.8 to 5.0 cm (1.5 to 2 in). If dwarf and semidwarf wheat lines are planted deeper than 5.0 cm (2 in), a genetically limited coleoptile length may prevent emergence; in such wheat lines seeding depth is critical, as shown by the work of Feather, Qualset, and Vogt (1968) in Table 19–19. When the coleoptile reaches a genetically determined length, it stops elongating regardless of whether it has reached the surface. The leaf emerges from the protective coleoptile but has little ability to penetrate soil and, without light for photosynthesis, eventually dies. Variations in coleoptile length are found among species and cultivars within species. Crops with short coleoptiles such as timothy must be seeded shallowly.

TABLE 19–19: Coleoptile length and seedling emergence planted at three depths for five spring wheat selections at California in 1968

	Seeding Depth					
Parameter	2.5 (cm)	(1.0 inches)	7.6 (cm)	(3.0 inches)	12.7 (cm)	(5.0 inches)
Coleoptile length (cm) (in)	1.53	(0.60)	2.14	(0.45)	2.18	(0.86)
Percent emergence	93.1		21.5		0	

Source: Data from Feather, Qualset, and Vogt 1968.

Unnecessarily deep seeding that delays emergence may also reduce crop yield. Spring wheat seeded at four depths at Swift Current, Saskatchewan over a four-year period showed equal emergence at 5.1 and 7.6 cm seeding depths; but possibly because of the delay in emergence, the 7.6 cm seeding depth had a lower grain yield (Table 19–20).

TABLE 19–20. Influence of seeding depth on spring wheat performance at Swift Current, over a four-year period 1970 to 1973

Seeding Depth (cm)	(in)	Number of Seedlings per 3-m Row	(per yd of row)	Yield (kg/ha)	(bu/ac)	Yield as Percent of 5.1 cm Depth (2.0 in depth)
2.5	1.0	93.3	85.6	1468	21.8	93.6
5.1	2.0	101.7	93.3	1568	23.3	100.0
7.6	3.0	101.7	93.3	1403	20.9	89.5
10.2	4.0	86.0	78.9	1213	18.1	77.4

Source: Unpublished report from the University of Saskatchewan.

In the small-seeded perennial forage crops, seeds should be covered shallowly because of the small cotyledons and low reserve food energy available for germination. Forage seeds are often dropped on the soil surface with a cultipacker-roller, which pushes soil over the seeds and assures good soil-seed contact.

As in cereals, Grabe and Metzer (1969) reported that the elongation of the hypocotyl was controlled genetically; and they classified 25 soybean cultivars as short, intermediate, and long, based on their ability to emerge from 10 cm of sand at 25°C (77°F). Burris and Fehr (1971) classified a short hypocotyl as ranging from 8.2 to 11.3 cm (3.2 to 4.4 in); intermediate as 11.0 to 17.5 (4.3 to 6.9 in); and long as 13.3 to 18.1 cm (5.2 to 7.1 in). In a field study, Fehr, Burris, Gilman (1973) investigated the ability of six soybean cultivars to emerge at different planting depths, seeding rates, and soil temperatures (Table 19–21).

TABLE 19–21. Percentage emergence for soybeans seeded 5 and 10 cm deep on three dates in two years in Iowa

Planting Date	Percentage Emergence	
	5 cm Deep (2 in)	10 cm Deep (4 in)
1970		
May 1	76.5	74.5
May 26	95.0	52.0
June 18	95.5	26.5
1971		
May 1	61.5	30.0
June 9	64.0	43.0
June 25	47.5	35.0
Mean	73.5	43.5

Source: Data from Fehr, Burris, and Gilman 1973.

In large-seeded legumes such as soybeans, proper seeding depth is important because the deeper the seed is placed in the soil, the more energy is required to pull the cotyledons to the soil surface; and the deeper the seed, the longer it takes to emerge, increasing the risk that soil crusting from heavy rains may occur, restricting hypocotyl emergen_e. Uniform planting depth is also important so that mass action can break any soil crusts and so that seedling competition is uniform.

Hatfield and Egli (1974) measured temperature effects on soybean hypocotyl elongation and found that the time to reach a 5-cm length decreased as the soil temperature increased from 10° to 32°C (50 to 90°F). The optimum temperature range for hypocotyl extension was concluded to be between 25° and 32°C (77 and 90°F).

Under very dry surface soil conditions with late seeding in a relay cropping program in Illinois, Stucky (1976) reasoned that deep seeding would favor emergence if seed was in contact with deeper soil moisture. Under the dry soil conditions of this test, no advantage to deep seeding was found (Table 19-22). With flax in Saskatchewan, seeding depth reduced yield and plant stand (Table 19-23).

TABLE 19-22. Percent emergence and grain yield of three soybean cultivars at seeding depths in Illinois, 1973 and 1974

Planting Depth (cm)	(in)	Number of Plants per Metre of Row* at emergence				Grain Yield (kg/ha)	
		1973	(yd)	1974	(yd)	1973	1974
5	2 in	20.0	18.3	15.2	13.9	1751	1819
7.5	3 in	23.3	21.4	10.7	9.8	1507	1788
10	4 in	17.0	15.6	—		1359	—

Source: Data from Stucky 1976.

*Fine—silty soil.

TABLE 19-23. The effect of seeding depth on yield and plant stand—Mean of six cultivars at Saskatoor in 1976 and 1977

Planting Depth (cm)	(in)	Seed Yield (kg/ha)	(bu/ac)	Plant Stand per Metre of Row	(yard)
2.5	1	1679	26.8	75	69
5.0	2	1701	27.1	55	50
7.6	3	1509	24.1	34	31
10.0	4	1293	20.6	25	23

Source: Unpublished data work conducted by G. G. Rowland and K. M. Panchuk.

To evaluate the effects of seeding depth and soil temperature on corn seedling emergence, Alessi and Power (1971) varied soil temperatures in the growth room. From 4 to 24 days were required to achieve 80% emergence depending on soil temperatures and seed depth. Temperature had a much greater effect on rate of emergence than did seed depth. Increasing soil temperature from 13.3° to 26.7°C (56 to 80°F) reduced the time for 80% emergence.

In field experiments in North Dakota, Alessi and Power (1971) found that corn under adequate soil moisture conditions required 8 to 13 days to reach 80% emergence. About one additional day was required for emergence for each 2.5-cm (1 in) increase in planting depth, an important consideration in short-season areas.

Seed Size and Yield

Under conditions of deep seeding, a large seed is important to provide sufficient food energy for seedling emergence. Under shallow seeding, seed size may be of less consequence, and seed uniformity may be of more importance.

Seed size in corn varying from 23 to 39 g/100 kernels (8.1 to 13.8 oz/1000 kernels) of a single-cross hybrid had no effect on the rate or extent of emergence (Hunter and Kannenberg 1972). The deeper the seed was planted and the lower the temperature, the longer the period to emergence (Table 19-24). Percent emergence also decreased slightly with increased depth of planting. Only the date of planting affected field yield performance, with the May 19 seeding giving the greatest grain yield. Plants from small seed were shorter than those from large seed.

In four barley cultivars, Kaufmann and McFadden (1963) found that plants from large seeds had more tillers and higher yields than small, medium, or mixed large, medium and small seeds over a three-year period. Kaufmann and Guitard (1967) found that in the greenhouse, large-seeded barley produced a superior rate of seedling growth and size of the first two leaves. In spring wheat, Austenson and Walton (1970) found that the size of the individual seeds accounted for 2.5 to 4.5% of yield variation.

Seed size and seed uniformity are crop production variables that can be controlled easily by sieving and grading the seed. As pointed out by Kaufmann and McFadden (1960), yield differences between large and small barley seeds were accentuated under increased competition caused by closely spaced plants.

CONCLUSIONS

As yield levels are increased, eventually water, nutrients, solar radiation, CO_2, or some factor becomes limiting. A concentration of effort has been devoted to plant spatial arrangements to intercept 95% of the incoming solar radiation.

TABLE 19–24. Mean number of days to 50% emergence, final percent emergence, plant height on June 30, and grain yield for large and small seed lots of a single-cross hybrid for three planting dates (May 9, 19, and 29) and three depths of planting at Elora, Ontario in 1967

Planting Depth (cm)	(in)	Days to 50% Emergence		Final Percent Emergence		Plant Height June 30				Grain Yield			
		Large Seed	Small Seed	Large Seed	Small Seed	Large Seed (cm)	(in)	Small Seed (cm)	(in)	Large Seed (kg/ha)	(bu/ac)	Small Seed (kg/ha)	(bu/ac)
2.5	1	13.3	13.3	94	95	70.2	(27.6)	67.7	(26.7)	6905	(110)	6717	(107)
5.0	2	14.9	14.8	91	95	69.7	(27.4)	64.6	(25.6)	7049	(112)	6999	(112)
7.5	3	16.1	15.9	90	93	67.0	(26.4)	63.9	(25.2)	7012	(112)	6767	(108)
Mean		14.8	14.7	92	95	69.0	(27.2)	65.4	(25.7)	6989	(111)	6828	(109)

Source: Data from Hunter and Kannenberg 1972.

Nutrients may be applied even to excess, and as a deficiency is discovered, attempts are made to eliminate that factor. Root development and morphology in relation to crop production practices may limit yields, and attention must be directed to improving our understanding of this complex entity.

QUESTIONS

1. Outline the extent of interrow cultivation in your area. How many times is interrow cultivation practiced? What is the depth of cultivation? How close to the plants does cultivation occur? Speculate on interrow cultivation on root damage and yield.

2. Do crop producers cultivate row crops when herbicidal weed control is adequate?

3. List reasons why crop producers practice interrow cultivation.

4. What is the total annual rainfall in your region? Without irrigation, what is the upper limit of crop production of crops commonly grown in your area, assuming complete water utilization and using transpiration values from Table 19-3? Is moisture a limiting factor in your region?

5. What factors determine the water-use efficiency of crops?

6. Discuss the fact that open-pollinated and hybrid corn cultivars have similar transpiration ratios (Table 19-3).

7. Comment on the bulk density values in Tables 19-10 and 19-11 in terms of limits to crop production.

8. What is the usual seeding depth for selected crops in your area? Record the time period from seeding to seedling emergence.

9. How would you determine if seeds were placed too deeply? Is there any evidence of improper seeding depth in your region?

10. How important are prop or brace roots to corn crop production?

11. Calculate the amount of water used to properly irrigate selected crops in your area. Outline how water efficiency can be improved.

12. Comment on root size and competitive ability of plants.

13. Some weeds have low transpiration ratio and yet are very competitive in a crop. Explain.

14. Comment on minimum or zero-tillage techniques and the ideal requirements of a seedbed.

15. How important is the root bed versus the seedbed in crop production?

16. Can moisture stress be beneficial?

17. Review soil compaction and root development.

18. Consider the problem of measuring roots. What can you conclude about the number and quality of root studies?

19. Comment on the statement, "When crop producers learn to manipulate

the root environment to their advantage, a major step in crop production will be achieved." Speculate on additional possible ways the root environment may be modified to the possible advantage of crop production.

20. Suggest research studies that might be conducted on how to manage roots for improved crop production.

REFERENCES FOR CHAPTER NINETEEN

Alessi, J., and J. F. Power. 1971. "Corn Emergence in Relation to Soil Temperature and Seeding Depth," *Agronomy Journal.* 63:717-719.

Allan, R. E., et al. 1961. "Inheritance of Coleoptile Length and Its Association with Culm Length in Four Winter Wheat Crosses," *Crop Science* 1:328-332.

Anonymous. 1976. "Root-Zone Placement Stretches Scarce Agricultural Chemicals," *The IRRI Reporter.* Published by the International Rice Research Institute Philippines. Issue 2.

Austenson, H. M., and P. D. Walton. 1970. "Relationships between Initial Seed Weight and Mature Plant Characters in Spring Wheat," *Canadian Journal of Plant Science* 50:53-58.

Brouwer, R. 1966. "Root Growth of Grasses and Cereals," in The *Growth of Grasses and Cereals,* pp. 153-166, F. L. Milthorpe and J. D. Ivins, eds. London: Butterworth & Co. (Publishers) Ltd.

Burris, J. S., and W. R. Fehr. 1971. "Methods for Evaluation of Soybean Hypocotyl Length," *Crop Science* 11:116-117.

Burton, G. W., G. M. Prine, and J. E. Jackson. 1957. "Studies of Drouth Tolerance and Water Use of Several Southern Grasses," *Agronomy Journal* 49:498-503.

Clements, F. E. 1921. *Aeration and Air-Content: The Role of Oxygen in Root Activity* Carnegie Institute of Washington Pub. 315.

Feather, J. T., C. O. Qualset, and H. E. Vogt. 1968. "Planting Depth Critical for Short-Statured Wheat Varieties," *California Agriculture* 22(9):12-14.

Fehr, W. R., J. S. Burris, and D. F. Gilman. 1973. "Soybean Emergence under Field Conditions," *Agronomy Journal* 65:740-742.

Foth, H. D. 1962. "Root and Top Growth of Corn," *Agronomy Journal* 54:49-52.

Grabe, D. F., and R. B. Metzer. 1969. "Temperature-Induced Inhibition of Soybean Hypocotyl Elongation and Seedling Emergence," *Crop Science* 9:331-333.

Hanks, R. J., and C. B. Tanner. 1952. "Water Consumption by Plants as Influenced by Soil Fertility," *Agronomy Journal* 44:98-100.

Harris, D. G., and C. H. M. Van Bavel. 1957a. "Nutrient Uptake and Chemical Composition of Tobacco Plants as Affected by the Composition of the Root Atmosphere," *Agronomy Journal* 49:176-181.

———. 1957b. "Root Respiration of Tobacco, Corn and Cotton Plants," *Agronomy Journal* 49:182-184.

Hatfield, J. L., and D. B. Egli. 1974. "Effect of Temperature on the Rate of Soybean Hypocotyl Elongation and Field Emergence," *Crop Science* 14:423-426.

Hunter, R. B., and L. W. Kannenberg. 1972. "Effects of Seed Size on Emergence, Grain Yield, and Plant Height in Corn," *Canadian Journal of Plant Science* 52:252–256.

Hurd, E. A. 1974. "Phenotype and Drought Tolerance in Wheat," *Agricultural Meteorology* 14:39–55.,

Irwin, R. W. 1969. *Farm Drainage.* Ontario Ministry of Agriculture and Food Publication No. 501.

Kaufmann, M. L., and A. A. Guitard. 1967. "The Effect of Seed Size on Early Plant Development in Barley," *Canadian Journal of Plant Science* 47:73–78.

Kaufmann, M. L., and A. D. McFadden. 1960. "The Competitive Interaction between Barley Plants Grown from Large and Small Seeds," *Canadian Journal of Plant Science* 40:623–629.

———. 1963. "The Influence of Seed Size on Results of Barley Yield Trials," *Canadian Journal of Plant Science* 43:51–58.

Kelley, O. J. 1954. "Requirement and Availability of Soil Water," *Advances in Agronomy* 6:67–94.

Kramer, P. J. 1969. *Plant and Soil Water Relationships: A Modern Synthesis.* New York: McGraw-Hill Book Company.

Letey, J., L. H. Stolzy, and G. B. Blank. 1962. "Effect of Duration of Timing of Low Soil Oxygen Content on Shoot and Root Growth," *Agronomy Journal* 54:34–40.

Liptay, A., and D. Davidson. 1971. "Coleoptile Growth: Variation in Elongation Patterns of Individual Coleoptiles," *Annals of Botany* 35:991–1002.

Lupton, F. G. H., and R. H. Oliver. 1974. "Root and Shoot Growth of Semi-Dwarf and Taller Winter Wheats," *Annals of Applied Biology* 77:129–144.

Maizlish, N. A., D. D. Fritton, and W. H. Kendall. 1980. "Root Morphology and Early Development of Maize at Varying Levels of Nitrogen," *Agronomy Journal* 72:25–31.

Mitchell, R. L., and W. J. Russell. 1971. "Root Development and Rooting Patterns of Soybean Sorghum (*Glycine max* (L.) Merrill): Evaluated under Field Conditions," *Agronomy Journal* 63:313–316.

Pearson, R. W. 1966. "Soil Environment and Root Development," in *Plant Environment and Efficient Water Use,* pp. 95–126, W. H. Pierre et al., eds. Madison, Wis.: American Society of Agronomy.

Phillips, R. E., and D. Kirkham. 1962. "Soil Compaction in the Field and Corn Growth," *Agronomy Journal* 54:29–34.

Prihar, S. S., and D. M. Van Doren. 1967. "Mode of Response of Weed-Free Corn to Post-Planting Cultivation," *Agronomy Journal* 59:513–516.

Russell, M. B. 1952. "Soil Aeration and Plant Growth," in *Soil Physical Conditions and Plant Growth,* pp. 265–273. B. T. Shaw, ed. Agronomy Monograph, Vol. 2. New York: Academic Press, Inc.

Russell, W. J., W. R. Fehr, and R. L. Mitchell. 1971. "Effects of Row Cultivation on Growth and Yield of Soybeans," *Agronomy Journal* 63:772–774.

Shantz, H. L., and L. N. Piemeisel. 1927. "The Water Requirement of Plants at Akron, Colorado," *Journal of Agricultural Research* 34:1093–1190.

Sheard, R. W. 1980. "Nitrogen in the Band for Forage Establishment," *Agronomy Journal* 72:89–97.

Spencer, J. T. 1941. "The Effect of Root Pruning and the Prevention of Fruiting on the Growth of Roots and Stalks of Maize," *Agronomy Journal* 33:481-489.

Sprague, V. G., and L. F. Graber. 1938. "The Utilization of Water by Alfalfa (*Medicago sativa*) and by Bluegrass (*Poa pratensis*) in Relation to Managerial Treatments," *Journal of the American Society of Agronomy* 30:986-997.

Stoffella, P. J., et al. 1979. "Root Characteristics of Black Beans II. Morphological Differences among Genotypes," *Crop Science* 19:826-830.

Stranak, A. 1968. "Soil Compaction and Direct Drilling of Cereals," *Outlook on Agriculture* 5(6):241-246.

Stucky, D. J. 1976. "Effect of Planting Depth, Temperature, and Cultivars on Emergence and Yield of Double-Cropped Soybeans," *Agronomy Journal* 68:291-294.

Taylor, H. M., and B. Klepper. 1971. "Water Uptake by Cotton Roots during a Drying Cycle," *Australian Journal of Biological Science* 21:853-859.

——. 1973. "Rooting Density and Water Extraction Patterns for Corn (*Zea mays* L.)," *Agronomy Journal* 65:965-968.

——. 1978. "The Role of Rooting Characteristics in the Supply of Water to Plants," *Advances in Agronomy* 30:99-128.

Troughton, A. 1962. "The Roots of Temperate Cereals (Wheat, Barley, Oats and Rye)." Commonwealth Agricultural Bureau. Mimeographed Pub. No. 2. (91 pp.)

Van Doren, D. M., and G. B. Triplett. 1965. "Corn Cultivation: Is It needed? *Ohio Report* 50:46-47.

Weaver, J. E. 1926. *Root Development of Field Crops.* New York and London: McGraw-Hill Book Company.

Wessling, J., and W. R. Van Wijk. 1937. "Soil Physical Conditions in Relation to Drain Depth," in *Drainage of Agricultural Lands,* 7:461-472, J. N. Luthin, ed. New York: Academic Press, Inc.

Winter, S. R. 1980. "Suitability of Sugar Beets for Limited Irrigation in a Semi-Arid Climate," *Agronomy Journal* 72:118-123.

Glossary

Acid soil. A soil with a pH reading of less than 7.0 on a scale from 0 to 14.

Adaptability. In plants, a modification in the structure or function to fit a changed environment.

Aerobic. Requiring oxygen to function, as opposed to *anaerobic*.

Agriculture. The science of using crops and animals to transform sunlight energy into products that can be stored and used by humans elsewhere and at a later date.

Agro-ecosystem. An integrated approach to farming.

Agrologist. One who understands and practices the science of agriculture.

Agronomy. A science combining crop production and soil management. The word is derived from two Greek words: *agros* (field) and *nomos* (to manage).

Alkaline soil. A soil with a pH reading above 7.3 on a scale from 0 to 14.

Anaerobic. Not requiring oxygen to function.

Anatomy. The study of structure.

Annuals. Plants that complete all developmental and reproductive stages in one season or year.

Anthesis. The period when anthers are extruded from the glumes.

Apical dominance. The inhibition in plants of lateral buds by high levels of auxins produced in the lead shoot or apical meristem.

Apomixis. A form of asexual reproduction. Seeds are formed in plants without sexual fertilization.

Arid climate. A region with an annual rainfall of less than 25 cm (10 in). Aridic soils are dry more than half the time.

Ash. The residue remaining after complete combustion of organic matter.

Assimilation. In plants, the conversion of photosynthetic products into substances used by the plant.

Auricles. Ear or finger-like clasping appendages located at the base of the leaf blade and at the top of the leaf sheath.

Awn. A bristlelike extension of the glumes (lemma) of cereal and grass plants.

Axillary buds. Vegetative buds arising from leaf axils.

Backcross.   Combining the progeny of a cross with one of the parents.

Bearded.   In plants, having awns.

Biennials.   Plants that live for two years and reach reproductive development in the second year.

Biological significance.   Differences between any two treatments that have a high (usually 1 to 5 out of 100) probability of being caused by the treatment and a low probability of being caused by chance variation.

Biological variation.   The difference between any two measurements resulting from factors other than the treatment (error, chance) within defined limits.

Biological yield.   The total dry-matter weight of above-ground parts including economic or grain yield and nongrain plant parts.

Biomass.   The total dry-matter production of a crop, the net result from photosynthesis, respiration, and mineral uptake.

Blade.   In plants, the leaf portion above the sheath.

Bloat.   In ruminant animals, a condition of excess stomach gas, often caused by succulent legumes, that can cause death.

Bog soils.   Imperfectly drained soils developed from peat.

Boll.   The oblong fruit of the cotton or flax plant.

Bolt.   In plants, the undesirable formation of reproductive organs on plants grown for their vegetative organs such as sugar beets, turnips, carrots, or parsnips.

Boot stage.   The point in cereal plant development when the developing inflorescence is encased in the leaf sheath (boot).

Breeder lines.   150 to 300 individually selected heads from a plot of a potential new cereal cultivar, grown in individual head rows, closely checked for uniformity, and bulked to produce the initial seed of a new cultivar.

Broadcast.   A method of seeding by distributing seed on the soil surface.

Cambrium.   A cellular layer of tissue separating the xylem and phloem in the stems of dicots.

Canola.   A name applied to rapeseed or any derivative of the crop.

Carbohydrate.   Sugar, starch, and cellulose components.

Carbon: nitrogen ratio.   The ratio of the percentage of carbon to that of nitrogen in organic materials.

Cardinal temperatures.   Plant development is governed by three cardinal temperatures: the minimum is the temperature below which plant functions are not detectable; the maximum is the temperature above which it is not detectable; and the optimum is the temperature at which the function proceeds at maximum speed.

Cassava.   A tuberous rooted, starchy, tropical perennial crop, also known as *manioc, mandioca,* or *yuca.* The plant is shrubby, possibly over 4 m (13 ft) in height, and propagated from stem cuttings. Cassava roots are the source of tapioca starch.

Cell.   In plants, a unit of structure.

Cellulose.   A long chain structure of carbon, hydrogen, and oxygen that serves as the building material of plant cells.

Cereal.   A member of the grass family grown for its grain, including rice, barley, wheat, oats, rye.

Chaff. The glumes or bracts covering the kernels or grains of plants removed by threshing.

Check. A standard reference cultivar for comparison in tests.

Check row planting. Plants spaced in hills equally in all directions.

Chlorophyll. The green pigmentation in the chloroplasts of the plant cell that is necessary for photosynthesis.

Chromosomes. Precisely arranged chemical threads carrying the units of heredity and found in the nucleus of cells.

Clay. A mineral soil comprised of small layered particles, 0.002 to 0.005 mm in size.

Climate. The total long-term weather conditions of an area.

Cobalt. A minor element required by rhizobia bacteria in plant nodules for the formation of vitamin B_{12}, which in turn is essential to the formation of haemoglobin needed for nitrogen fixation.

Combine. A machine that incorporates the operations of cutting, threshing, and separating grain, straw, and chaff.

Companion crop. A crop grown in association with forage seedlings to act as a cover crop to suppress weeds.

Compensation point. In plants, the intensity of a photosynthetically related input required to equal the loss of carbon compounds through respiration. At the compensation point, the amount of CO_2 absorbed is equal to the amount given off.

Competition. Events associated with a retardation in plant growth arising from an association with other plants.

Controlled storage. Units to reduce respiration to extend storage life and quality of stored plant products.

Corm. In the timothy plant, an enlargement at the base of the stem to store photosynthate.

Corn heat units. A corn hybrid rating system based on the relationship between temperature and corn development using day and night temperatures. The daytime relationship is based on 10°C as the base and 30°C as the optimum. The nighttime relationship uses 4.4°C as the base temperature. Corn development values have been established for daily maximum and minimum temperatures, are summed over a defined growing season, and a heat unit rating is established in comparison with known check hybrids.

Cotyledons. The two halves of a pea or bean seed that form the first leaves of a plant. Cotyledons are storage organs.

Coumarin. The bitter flavor substance of sweet clover. Coumarin may occur in lesser amounts in other plants. Coumarin may be converted to toxic substances when spoilage of sweet clover hay or silage occurs.

Cover crop. A crop grown to protect the soil from erosion or nutrient leaching.

Critical fall harvest period. Predetermined calendar dates for regions on which perennial forages should not be cut for three weeks before or after, to provide a six week period of photosynthesis for the build-up of root reserves.

Crop physiology. The science of plant functions and the phenomena of plants studied in a community of plants under field conditions.

Crop production. A science aimed at maximizing photosynthesis to increase economic crop yield.

Cross-inoculation groups. Symbiotic bacteria are specific for many legumes such as soybeans, but in some cases bacteria will cross-inoculate with several species, e.g., alfalfa and sweet clover.

Crown. In plants, the top of a root where buds and new shoots arise.

Crown buds. Differentiated cells on the top of roots capable of initiating new shoot growth.

Culm. The stem or straw of grasses with joints or nodes at intervals.

Cultivar (cv.). An inclusive term for lines, varieties, hybrids, or selections of crops. Each cultivar is distinct from other cultivars of the same species.

Cultural energy. Energy from human and animal labor, and from fossil fuels used by machines in their manufacture and operation in all aspects of food production, transportation, processing, and distribution.

Cultural energy efficiency. The ratio of useful output of energy to energy inputs.

Cytology. The study of individual cells.

Denitrification. The release of inert nitrogen through the breaking down action of plant tissues and nitrates to nitrites, ammonia, and N_2 gas.

Detassel. Removal of the pollen-producing organ (tassel) of corn.

Determinate species. Plants having definite limits, i.e., a definite distinction between vegetative and reproductive stages.

Dicots. Plants having two cotyledons at the first node. The word dicot is derived from dicotyledon meaning having two cotyledons.

Differentiation. The formation of specialized tissues during growth and development.

Diploid. Having two sets of chromosomes, one from the female and one from the male parent.

Direct-seeded forages. Seeded without a companion crop.

Double cropping. The practice of producing two successive crops from the same piece of land in one year.

Dryeration. The combination of high temperature drying and aeration.

Durum wheat. A hard spring wheat high in protein that is favored for the production of semolina flour used to make pasta products such as spaghetti and macaroni.

Ecofallow. A system combining crop rotation minimum tillage and weed control to conserve soil and moisture and to control disease.

Ecology. The study of reciprocal relations among plants and animals and their environment.

Economic yield. The grain tubers, fiber, oil, or plant component of value, in contrast to noneconomic yield of stover, straw, leaves, or other residue.

Ecosystem. The interaction of natural forces in a harmonious situation.

Egg. In plants, the reproductive cell of the female.

Empirical. Practices based on observation and experience without a scientific understanding; e.g., ancient agriculturists knew that manure increased crop yields and accordingly recommended manure application, but it took centuries to learn that it was nitrogen, phosphorus, and potash and other microelements that were associated with yield increases.

Energy.   The potential of light, heat, or chemical energy to do work.

Ensilage (silage).   Chopped plant material preserved by fermentation.

Enzyme.   A substance that catalyzes and often initiates a biochemical reaction.

Erosion.   In soil, the loss by water, wind, or other action.

Etiolation.   The elongation of plant cells caused by a high concentration of auxins (plant hormones) usually as a result of low light intensity.

Eutrophication.   A process whereby phosphorus-induced algae growth in still water depletes oxygen from the water as the algae decay. The reduced oxygen supply may suffocate fish or other living forms.

Evapotranspiration.   The combined loss of water from an area from evaporation and transpiration.

Fallow.   A field uncropped, usually for one growing season to conserve moisture.

Fermentation (silage).   A process whereby bacteria act on the energy content of the chopped plant material to produce an acid that preserves or pickles the crop material.

Fertilization (plant).   The union of the nucleus from the male pollen grain with the embryo or egg of the female ovary.

Fertilization (soil).   The addition of plant nutrients to the soil to promote photosynthesis and growth.

Fibrous root.   A root with many fine branches as opposed to a single or multiple taproot.

Filial generation. (F_1, F_2, etc.).   The generation of the progeny following a cross.

Flag leaf.   The final, uppermost leaf to develop. Usually refers to cereal plants.

Forage (herbage).   Plant biomass that serves as animal feed.

Fungicide.   A chemical used to control plant disease (fungi).

Gamete.   A sex cell containing half the chromosome number as other body cells. The male gamete and the female gamete combine in fertilization to produce the normal chromosome number.

Gasohol.   A blend of gasoline and alcohol.

Genes.   Unit of inheritance having a specific chemical composition and having a specific location on a chromosome.

Genetics.   The science that deals with the inheritance of traits.

Genotype.   The heritable or genetic composition of an individual (see *phenotype*).

Geotropism.   Growth or movement of a plant part with respect to gravity forces. Vegetative plant parts are usually negatively geotropic and grow upward; roots are geotropic and grow downward.

Glumes.   Bracts (chaff) covering a kernel. Wheat glumes are easily removed at threshing; in barley they often adhere to the kernel.

Grain.   The kernels of a crop.

Grazing.   Livestock harvesting plant material.

Groat.   The endosperm of an oat kernel with the lemma and palea (hull) removed.

Growing point.   The lead meristematic tissue from which subsequent growth arises.

Growth.   An increase in dry weight usually as a result of an increase in cell size or cell number.

Haemoglobin.   The "blood" in a legume root nodule resulting in a pink-red color.

Harvest index.   The ratio of grain yield to biological yield by weight.

Haylage.   A silage product made from forages and preserved at 40 to 60% moisture.

Herbage.   The fresh or preserved plant parts of a crop.

Herbicide.   A substance used to kill or destroy plants, sometimes called *weedicides*. Selective herbicides have a relatively specific action in contrast to the relatively general action of nonselective herbicides.

Heterozygous.   An individual having unlike genes at one or more points (loci) on its chromosomes.

Homozygous.   An individual having similar genes at all points (loci) on its chromosomes.

Horizontal silo.   An open flat bunker with sloped walls for the production and storage of silage.

Hormones.   Naturally occurring chemicals produced in plant cells to regulate plant functions.

Humus.   A form of soil organic matter that is somewhat resistant to further rapid decomposition. Humus has a carbon:nitrogen ratio of 10:1.

Hybrid.   The first generation progeny of a cross between two different strains of the same species. A hybrid may combine characteristics derived from the two parent stocks and may be more desirable than either parent.

Hybrid vigor or **heterosis.**   The extra vigor or yield often obtained from progeny following a cross.

Ideotype.   In plants, a biological model that is expected to perform in a predictable manner within a defined environment (Donald 1968).*

Indeterminate.   Plants having no defined limits of vegetative and reproductive development.

Inflorescence.   The flowering parts of a plant such as a spike of wheat, a head of barley, a panicle of oats.

Inoculation.   The application of a rhizobium strain to seed or soil of a legume crop to promote nitrogen fixation.

Insecticide.   A chemical used to kill insects.

Intercropping.   The production of two or more crops simultaneously on the same field.

Internodes.   The region between two nodes of a plant stem. In grasses, the leaf sheath encases much of the internode.

Interseeding crops.   Establishing a second crop between the rows of an existing crop.

In vitro.   Conducted in a test tube as contrasted to *in vivo,* which means conducted in a living organism. E.g., in vitro digestibility of plant material.

Isolines.   Lines identical in all aspects but one.

Kilogram.   One thousand grams.

Kilojoules.   A unit of energy equal to 1000 joules.

*C. M. Donald, "The Breeding of Crop Ideotypes," *Euphytica* 17(1968):385–403.

Lateritic soils. Mineral-rich (iron and aluminum) soil that hardens irreversibly into rocklike material when dried. The name comes from the Latin word for "brick."

Leaching. The action of water removing soil materials in solution.

Leaf Area Index (LAI). The ratio of leaf area to land area. A LAI of 4 means that on one hectare there are four hectares of leaf area, usually one side of the leaf lamina only.

Legume. A plant capable of forming a symbiotic relationship with Rhizobium bacteria to fix atmospheric nitrogen.

Lemma. The outer structure of a single floret of grasses often bearing an awn. The lemma and palea cover the grain of oats and barley. The lemma is larger than the palea and, in oats and barley, adheres to the grain on the opposite side of the crease. In wheat and rye, the lemma and palea are usually removed at threshing.

Light saturation. When photosynthesis is not increased by additional light intensity, a leaf has reached light saturation.

Lignin. A differentiation product in cell walls that makes them strong and firm.

Ligule. A thin membranous extension of the leaf sheath positioned against the culm of grasses where the leaf blade forms. This clear-cut division between leaf blade and sheath is not present in a liguleless line, and the leaf blade adopts an upright position.

Loam. A soil comprised of a mixture of clay, silt, sand, or gravel.

Lodging. The permanent displacement of the stems of crops from their upright position.

Lodicules. Small organs in grasses and cereals located between the ovary and the surrounding glumes. The lodicules swell at the time of fertilization to force open the glumes.

Magnetic response. A reaction to the earth's polarity whereby roots become aligned to the N-S magnetic force.

Mediterranean climate. Typified by mild wet winters and hot dry summers with 200–300 mm of rainfall.

Meristem regions. Areas in the plant where cells are rapidly dividing. Cells in meristem regions are not differentiated.

Minimum tillage. The minimum soil manipulation necessary for maximum crop production under existing soil and climatic conditions.

Mitochondria. Cellular bodies contained in cell protoplasm that serve as sites for the breakdown of food energy (singular form is mitochondrion).

Mixed grain. Usually refers to oat and barley mixtures in a 50:50 ratio or a 65:35 ratio of oats to barley. Spring wheat may be added to a mixture of oats and barley.

Molybdenum. A minor element required by rhizobia for nitrogen fixation in plants.

Monocots. Plants having a single cotyledon or leaf at the first node of the lead shoot or stem. The word monocot is derived from monocotyledon meaning having one cotyledon.

Monoculture. The production of a single species on an ongoing basis.

Morphology. Referring to plant shape or structure.

Muck. A black soil developed from fairly well decomposed organic material formed under poor drainage conditions.

Multilines. A blend or mixture of lines of one species that are genetically very similar (e.g., a multiline of wheat).

Multiple cropping. Growing two or more crops on the same field in a year.

Mutant. A selection resulting from a heritable variation (mutation).

Natural ecosystem. A geographic area virtually untouched by man.

Neutral soil. On a scale from 0 to 14, a neutral soil has a reading of 7.0 and is neither acid nor alkaline.

Nitrogen fixation. In plants, the conversion of inert atmospheric nitrogen N_2 into a form useful to plants.

Nodes. Solid regions or joints in stems. In grasses, the leaf sheath is attached to the node. Lateral buds may develop at a node.

Nodules. In plants, refers to nitrogen-fixing nodules on a legume root consisting of plant cells crammed full of nitrogen-fixing bacteria.

Nubbin. A small, poorly developed ear of corn.

Optimum leaf area. The leaf area at which dry-matter production is maximized, i.e., the leaf area that produces the highest net assimilation rate.

Organic farming. An attempt to minimize the use of chemicals and inorganic inputs in crop production.

Organic matter. Plant residues (roots, leaves, stems) in soil.

Palea. The small inner structure of a single floret in grasses. In barley and oats, the palea adheres to the grain in the crease area. In wheat and rye, the palea is removed by threshing.

Panicle. In grasses, the spikelets may be attached by fine branches or subdivisions of the stem. In this case the inflorescence is called a panicle.

Parasitic leaves. When the loss of energy by respiration exceeds the contribution from photosynthesis, a leaf is considered to be parasitic.

Pathology. The study of disease.

Penultimate leaf. The next to last leaf to develop on a plant such as corn and cereals.

Petiole. The stalk on which the leaf blade is attached to the stem. In ladino clover, the petiole arises from the stolon. In rhubarb, the petiole is the harvested portion.

Perennial. A plant that lives for more than one year. A perennial plant may reach reproductive development in the first season and in each subsequent season.

Pesticide. Includes chemicals such as rodenticides, insecticides, herbicides, and fungicides used to control pests in plants and animals.

Phenotype. The physical appearance of a plant resulting from the action of genotype and environment.

Phloem. Conducting tissue within a plant to transport the products of photosynthesis to other plant parts.

Photosynthesis. The basis of all agriculture; the process that transforms CO_2 into food.

Phototropism. A growth, enzyme-regulated plant response of bending toward a light source.

Physiological maturity. The point in economic yield development at which no further increase in dry weight takes place.

Phytoplankton. Simple plant forms that are part of the plankton group, are capable of photosynthesis, and are the major source of sustenance for all animal life in the seas.

Pistil. The female reproductive organ in plants consisting of a stigma, style and ovary.

Plankton. The small plants (phytoplankton) and animals (zooplankton) that drift and float in the oceans.

Plant breeding. Organized attempts to produce progressively better adapted populations.

Plant physiology. The science of functions and phenomena of plants as studied individually (see *crop physiology*).

Plastic responses. The ability of plants to respond (within limits) to competitive stress such as seeding rate.

PLUARG. Pollution from *L*and *U*se *A*ctivities *R*eference *G*roup, an organization of the International Joint Commission, established under the Canadian-U.S. Great Lakes Water Quality Agreement of 1972.

Pollen. Plant germ cells produced in the anthers containing the genetic material of heredity in the nucleus.

Pollination. The transfer of pollen from the anther to the stigma.

Productivity score. A measure of crop performance obtained by summing grain yield, biological yield, and harvest index values.

Prolific corn. Plants that are consistently capable of producing two or more ears per plants.

Proteins. Substances comprised of amino acids and present in all living systems.

Protoplasm. A viscous, gel-like living substance within a cell that includes the nucleus, chloroplasts in green cells, and mitchondria.

Relay cropping. A term to describe the seeding of one crop into another standing crop, e.g., winter wheat into standing soybeans, or soybeans into standing wheat. Starting one crop in another.

Respiration. The release of energy involving the use of oxygen and giving off carbon dioxide.

Rhizobium. A microorganism capable of entering the root hair of a legume to form a nodule and fix nitrogen from the atmosphere.

Rhizomes. Underground stems of plants which grow horizontally beneath the soil surface and which are capable of developing new roots and shoots at internodes.

Rotation. Varying the sequence of crops grown on the same ground.

Ruminant animals. Multistomached livestock capable of utilizing cellulose and inorganic nitrogen forms to produce protein.

Silage (ensilage). A form of feed preserved by the acid-producing action of fermentation.

Sink. In plants, the storage capacity of the economic yield component.

Soil capability.   The capacity of a soil or tract of land to be used for sustained and profitable production of food.

Soil salinity.   Soils with a high level of soluble salts, but not highly alkaline.

Source.   In plants, the photosynthetic capacity to provide assimilate.

Spike.   In grasses, if the spikelets are directly attached to the stem or rachis the entire inflorescence is called a spike.

Spikelet.   The inflorescence of grasses consists of a series of spikelets. The spikelet therefore is the basic unit of the spike. Each spikelet may have one or more florets encased by a pair of outer glumes.

Stamens.   The pollen-producing organs of plants.

Stolons.   Horizontal, above ground stems of plants.

Stomata.   An opening in the epidermis of a leaf contained between two guard cells which regulate the size of the opening. Stomata allow for free exchange of carbon dioxide and oxygen within the leaf and the outside air and are the openings through which water vapor passes in transpiration. (Singular form is stoma).

Stooling.   A term to indicate branch or tiller development in plants.

Sucker.   A shoot or tiller arising from the base or axils of plants.

Sward.   A canopy of crop leaves.

Symbiotic relationship.   When two different organisms live in close association and depend on each other for their mutual benefit.

Taxonomy.   The classification and naming of organisms.

Teleology.   The doctrine that final causes influence events. As applied to plants, the view that plant development is due to the purpose served by them is considered unscientific and incorrect: e.g., roots do not grow toward a moisture source, leaves do not orientate themselves to maximize photosynthesis, and plants do not develop large horizontally disposed leaves so that they can compete in evolutionary processes.

Tillage-planting system.   All operations needed to produce a crop, including preplanting tillage, planting, and weed control by cultural or chemical methods or both.

Topsoil.   The upper productive layer of soil.

Transpiration.   Water loss from the plant that accounts for 99% of the water used by plants.

Turgid.   Swollen with water.

Unstable soils.   Soils subject to rapid loss of organic matter and erosion.

Upright silo.   A vertical cylinder made of concrete or steel with a glass lining, for fermentation of chopped plant material.

Urea.   A solid form of nitrogen fertilizer containing 45% nitrogen and produced by reacting ammonia with carbon dioxide under pressure at an elevated temperature.

Vacuole.   A region in a plant cell filled with cell sap and stored food products and by-products. The vacuole is bounded by a membrane.

Variation.   In plants, differences that exist among individuals of a species.

Vernalization.   In plants, a cold temperature-photoperiod requirement to develop reproductive organs.

Vertical integration. When a producer becomes a middleman by processing his own agricultural products for sale directly to the consumer.

Weed. A plant out of place.

Whorl. A circular arrangement of plant leaves.

Winter annual. A plant that develops a seedling stage in the fall, becomes vernalized over the winter, and completes vegetative and reproductive growth the following year.

Xylem. Conducting tissue in plants to transport water and nutrients absorbed by the root to other plant parts.

Zero-tillage. A system in which a crop is planted with minimum soil disturbance directly into an untilled stubble from the previous crop, using chemicals for weed control.

appendix A

Metric Information for Crop Producers

Measurement	Metric unit	Symbol	Common relationships
Weight	gram	g	1 g = 1000 mg
	kilogram	kg	1 kg = 1000 g
	tonne	t	1 t = 1000 kg
Length	millimetre	mm	
	centimetre	cm	1 cm = 10 mm
	metre	m	1 m = 100 cm
	kilometre	km	1 km = 1000 m
Area	hectare	ha	1 ha = 10,000 m^2
			100 ha = 1 km^2
			1 m^2 = 10,000 cm^2
Volume	litre (10 cm $\times$ 10 cm $\times$ 10 cm) hectolitre (replaces bushel weight)	L	1 L = 1000 ml
Temperature	Celsius	C	0°C = freezing point of water
			20°C = room temperature
			37°C = body temperature
			100°C = boiling point of water
Energy	joule	J	
Horsepower	watt	W	
	kilowatt	kW	

Metric Conversion Information

Metric Equivalents			
Length		*Area*	
inch = 2.54 cm	millimetre = 0.04 in.	square inch = 6.45 cm^2	cm^2 = 0.16 sq in.
foot = 0.30 m	centimetre = 0.39 in.	square foot = 0.09 m^2	m^2 = 1.20 sq yd
yard = 0.91 m	metre = 3.28 ft	square yard = 0.84 m^2	km^2 = 0.39 sq mile
mile = 1.61 km	kilometre = 0.62 mile	square mile = 2.59 km^2	ha = 2.47 acres
		acre = 0.40 ha	

Volume (dry)			
cubic inch	= 16.38 cm^3	cm^3	= 0.06 cu in.
cubic foot	= 0.03 m^3	m^3	= 31.39 cu ft
cubic yard	= 0.76 m^3	m^3	= 1.31 cu yd
bushel	= 36.37 litres		
board foot	= 0.0024 m^3		

Volume (liquid)			
fluid ounce (Imp)	= 28.41 mL	litre	= 35.2 fluid oz (Imp)
pint	= 0.57 litre	hectolitre	= 26.42 gal (U.S.)
gallon (Imp)	= 4.55 litres		= 22.00 gal (Imp)

Weight			
ounce	= 28.35 g	gram	= 0.035 oz
pound	= 453.6 g	kilogram	= 2.20 lb
ton	= 0.91 tonne	tonne	= 2205 lb

Proportion			
1 gal/acre	= 11.23 litres/ha	1 litre/ha	= 14.25 fluid oz/acre
1 lb/acre	= 1.12 kg/ha	1 kg/ha	= 14.5 oz/acre
1 lb/sq in.	= 6.90 kilopascals (KPa)	1 metric tonne/ha	= 0.45 ton/acre
1 ton/acre	= 2.24 metric tonne/ha	1 kilopascal	= 0.145 lb/sq in.

Weights and Measures (Metric)

1 litre = 1000 millitres
  = volume of a cube, 10 cm × 10 cm × 10 cm
1 litre of water weighs 1 kilogram
1 tonne = 1000 kilograms = 1,000,000 grams
1 kilogram = 1000 grams
1 hectare = 10,000 square metres
  = area of a square 100 metres × 100 metres

1 centimetre = 10 millimetres
1 metre = 100 centimetres = 1000 millimetres
  1 kilometre = 1000 metres
1 kilometre/hour = 16.6 metres/minute

Weights and Measures (Imperial)

1 imperial gallon = 4 quarts = 8 pints = 160 fluid ounces
  = approx 1.2 U.S. gallons
1 U.S. gallon = approximately 5/6 imperial gallon
1 imperial gallon of water weighs 10 pounds
1 imperial quart = 2 pints = 40 fluid ounces
1 imperial pint = 20 fluid ounces = 2-1/2 measuring cups
1 kitchen measuring cupful = 8 fluid ounces
3 teaspoonfuls = 1 tablespoonful = 1/2 fluid ounce

1 mile = 5,280 feet = 1,760 yards = 320 rods
1 rod = 16.5 feet = 5.5 yards
1 yard = 3 feet = 36 inches
  1 foot = 12 inches
1 acre = 43,560 square feet =
  4,840 square yards = 160 square rods
1 square yard = 9 square feet
1 square foot = 144 square inches
1 mile an hour = 88 feet a minute

Capacity of Upright Silos

See Table on Page 420.

Size	Metres (Metric)	Tons of Corn Silage at 70% Moisture	Tonnes (Metric)	Tons of Hay Silage at 50% Moisture	Tonnes (Metric)	Tons of H.M. Clean Cracked Corn at 30% Moisture	Tonnes (Metric)	Tons of H.M. Ground-Ear Corn at 28% Moisture	Tonnes (Metric)	Volume in cu. ft.	Volume in Litres
12 × 25	3 66 × 7 62	54	49	32	29	67	61	56	51	2827	80061
12 × 30	3 66 × 9 14	70	63	42	38	80	73	67	61	3390	96005
12 × 35	3 66 × 10 66	88	80	53	48	93	84	78	71	3959	112119
12 × 40	3 66 × 12 19	106	96	64	58	106	96	90	82	4520	128006
12 × 45	3 66 × 13 71	126	114	76	69	119	108	101	92	5090	144149
12 × 50	3 66 × 15 24	147	133	88	80	133	121	112	102	5650	160008
14 × 30	4 26 × 9 14	96	87	58	53	107	97	92	83	4618	130782
14 × 35	4 26 × 10 66	119	108	71	64	126	114	107	97	5388	152588
14 × 40	4 26 × 12 19	145	132	88	80	145	132	122	111	6160	174451
14 × 45	4 26 × 13 71	171	155	103	93	163	148	137	124	6927	196173
14 × 50	4 26 × 15 24	200	181	120	109	181	164	153	139	7700	218064
14 × 55	4 26 × 16 76	230	209	138	125	199	181	169	153	8464	239700
14 × 60	4 26 × 18 28	260	236	158	143	217	197	184	167	9236	261563
16 × 40	4 87 × 12 19	189	171	113	103	187	170	158	143	8042	227749
16 × 45	4 87 × 13 71	224	203	134	122	212	192	179	162	9048	256239
16 × 50	4 87 × 15 24	261	237	156	142	237	215	199	180	10050	284616
16 × 55	4 87 × 16 76	301	273	181	164	260	236	219	199	11058	313162
16 × 60	4 87 × 18 28	341	309	204	185	284	258	239	217	12060	341539
16 × 65	4 87 × 19 81	383	347	230	209	309	280	261	237	13070	370142
18 × 50	5 48 × 15 24	330	299	198	180	299	271	252	229	12700	359664
18 × 55	5 48 × 16 76	380	345	228	207	329	298	277	251	13996	396367
18 × 60	5 48 × 18 28	430	390	258	234	359	326	302	274	15240	431597
18 × 65	5 48 × 19 81	482	437	289	262	389	352	325	295	16541	468441
18 × 70	5 48 × 21 33	536	486	322	292	418	379	348	316	17780	503530
20 × 50	6 09 × 15 24	407	369	244	221	369	335	309	280	15708	444851
20 × 55	6 09 × 16 76	468	425	281	255	406	368	341	309	17279	489341
20 × 60	6 09 × 18 28	529	480	318	288	443	402	373	338	18840	533549
20 × 65	6 09 × 19 81	594	539	356	323	480	435	404	366	20420	578294
20 × 70	6 09 × 21 33	659	598	398	361	517	469	436	396	21980	622474
20 × 75	6 09 × 22 86	728	660	437	396	555	503	469	425	23562	667276
20 × 80	6 09 × 24 38	796	722	478	434	593	538	502	455	25120	711398
24 × 50	7 31 × 15 24	583	529	350	318	531	482	449	407	22620	640598
24 × 60	7 31 × 18 28	760	689	458	415	638	579	538	488	27120	768038
24 × 70	7 31 × 21 33	940	853	574	521	745	676	627	569	31640	896045
24 × 80	7 31 × 24 38	1145	1039	688	624	852	773	715	649	36160	1024051

Note: The above figures are estimates, not to be considered accurate—Use only as a guide.

Capacity of Horizontal Silos for Corn Silage at 70% Moisture

See Table on Page 422.

Capacity per foot (30 cm) of length in tons (ft) or tonnes (m)

Average Silo Width		Settled Silage Depth											
(Feet)	(Metre)	6 (ft)	1.8 (m)	8 (ft)	2.4 (m)	10 (ft)	3.0 (m)	12 (ft)	3.7 (m)	14 (ft)	4.3 (m)	16 (ft)	4.9 (m)
16	4.9	1.63	1.48	2.25	2.05	2.94	2.68	3.69	3.36	4.50	4.10	5.38	4.90
20	6.0	2.10	1.91	2.88	2.62	3.74	3.40	4.68	4.26	5.70	5.19	6.80	6.19
24	7.3	2.59	2.36	3.53	3.21	4.57	4.16	5.70	5.19	6.93	6.31	8.26	7.52
28	8.5	3.11	2.83	4.21	3.83	5.43	4.94	6.75	6.14	8.19	7.45	9.74	8.86
32	9.8	3.65	3.32	4.92	4.48	6.31	5.73	7.83	7.13	9.48	8.63	11.26	10.25
36	10.9	4.21	3.83	5.64	5.13	7.22	6.57	8.94	8.14	10.81	9.84	12.82	11.67
40	12.2	4.80	4.37	6.40	5.82	8.16	7.43	10.08	9.17	12.16	11.07	14.40	13.10
44	13.4	5.41	4.92	7.18	6.53	9.13	8.31	11.25	10.24	13.54	12.32	16.02	14.58
48	14.6	6.05	5.51	7.99	7.27	10.12	9.21	12.44	11.32	14.96	13.61	17.66	16.07

Subject and Name Index

AUTHOR INDEX

Bold face figures refer to pages
where references are given in
full.